POLYMER SCIENCE AND TECHNOLOGY
Volume 1

STRUCTURE AND PROPERTIES OF POLYMER FILMS

POLYMER SCIENCE AND TECHNOLOGY

A Symposia Series

Volume 1 • STRUCTURE AND PROPERTIES OF POLYMER FILMS
Edited by Robert W. Lenz and Richard S. Stein • 1972

POLYMER SCIENCE AND TECHNOLOGY
Volume 1

STRUCTURE AND PROPERTIES OF POLYMER FILMS

Based upon the Borden Award Symposium in Honor of Richard S. Stein, sponsored by the Division of Organic Coatings and Plastics Chemistry of the American Chemical Society, and held in Boston, Massachusetts, in April 1972

Edited by
Robert W. Lenz
Professor, Department of Chemical Engineering

and
Richard S. Stein
Commonwealth Professor, Department of Chemistry
Polymer Science and Engineering Program
University of Massachusetts
Amherst, Massachusetts

℗ PLENUM PRESS • NEW YORK – LONDON • 1973

Library of Congress Catalog Card Number 72-88422

ISBN-13: 978-1-4615-8953-2   e-ISBN-13: 978-1-4615-8951-8
DOI: 10.1007/ 978-1-4615-8951-8

© 1973 Plenum Press, New York
Softcover reprint of the hardcover 1st edition 1973
A Division of Plenum Publishing Corporation
227 West 17th Street, New York, N.Y. 10011

United Kingdom edition published by Plenum Press, London
A Division of Plenum Publishing Company, Ltd.
Davis House (4th Floor), 8 Scrubs Lane, Harlesden, London,
NW10 6SE, England

CONTRIBUTORS

George C. Adams, Film Department, E. I. du Pont de Nemours and
    Company, Incorporated, Wilmington, Delaware

Rodney D. Andrews, Department of Chemistry and Chemical
    Engineering, Stevens Institute of Technology, Hoboken,
    New Jersey

Tadahiro Asada, Department of Polymer Chemistry, Kyoto University,
    Kyoto, Japan

W. H. Chu, Research Laboratory, International Business Machines,
    San Jose, California

Edward S. Clark, Department of Chemical Engineering, University
    of Tennessee, Knoxville, Tennessee

Koji Hoashi, Department of Chemistry and Chemical Engineering,
    Stevens Institute of Technology, Hoboken, New Jersey

H. B. Hopfenberg, Chemical Engineering Department, North Carolina
    State University, Raleigh, North Carolina

Masaaki Itoga, Case Western Reserve University, Cleveland, Ohio

S. Kapur, Division of Macromolecular Science, Case Western Reserve
    University, Cleveland, Ohio

Nobuhiro Kawasaki, Department of Chemistry and Chemical
    Engineering, Stevens Institute of Technology, Hoboken,
    New Jersey

J. L. Koenig, Case Western Reserve University, Cleveland, Ohio

D. G. LeGrand, Research and Development, General Electric
    Corporation, Schenectady, New York

R. H. Marchessault, Département de Chimie, Université de Montréal, Montreal, Ontario, Canada

G. E. McGraw, Research Laboratories, Tennessee Eastman Company, Division of Eastman Kodak Company, Kingsport, Tennessee

Shigeharu Onogi, Department of Polymer Chemistry, Kyoto University, Kyoto, Japan

E. P. Otocka, Bell Telephone Laboratories, Murray Hill, New Jersey

A. Peterlin, Research Triangle Institute, Research Triangle Park, North Carolina

M. B. Rhodes, Chemistry Department, University of Massachusetts, Amherst, Massachusetts

C. E. Rogers, Division of Macromolecular Science, Case Western Reserve University, Cleveland, Ohio

Robert J. Samuels, Research Center, Hercules Incorporated, Wilmington, Delaware

J. R. Semancik, Division of Macromolecular Science, Case Western Reserve University, Cleveland, Ohio

T. L. Smith, Research Laboratory, International Business Machines, San Jose, California

V. Stannett, Chemical Engineering Department, North Carolina State University, Raleigh, North Carolina

R. S. Stein, Polymer Research Institute and Department of Chemistry, University of Massachusetts, Amherst, Massachusetts

Ban The Vu, Department of Chemical Engineering, Princeton University, Princeton, New Jersey

G. L. Wilkes, Department of Chemical Engineering, Princeton University, Princeton, New Jersey

J. L. Williams, Camille Dreyfus Laboratory, Research Triangle Park, North Carolina

PREFACE

    The study of the relationship between the structure,
morphology and properties of polymer films has significantly
progressed in recent years through the use of a number of
phyiscal techniques - some new and some old. These methods
include small and large angle x-ray diffraction, birefringence, light scattering, infrared dichroism, fluorescence
polarization, light and electron microscopy and interferrometry.
This collection of papers, most of which were presented at a
symposium at the Boston American Chemical Society Meeting in
April, 1972, represent a collection of recent studies using
many of these methods by some of the leading scientists in
their fields. It is evident that these various techniques
permit the study of various aspects of film structure such
as crystal structure and orientation, amorphous orientation,
the interrelation of crystalline and amorphous regions in
lamellar, fibrillar, and spherulitic superstructure and the
relationship of these structural variables to the mechanical
and optical properties of the films. Film structure is
sufficiently complex that a complete understanding of the
relationship between structure and properties will come from
the employment of a combination of several of these methods.

CONTENTS

OPTICAL STUDIES OF THE MORPHOLOGY OF POLYMER FILMS[†]

Richard S. Stein

Polymer Research Institute and Department of Chemistry

University of Massachusetts, Amherst, Massachusetts

It is a truism that the properties of a film are a consequence of its structure. This structure may be considered at different levels of size ranging from the molecular to the macroscopic. Various properties depend differently on these structural manifestations. The thermodynamic properties, density and electrical properties depend upon the molecular organization, whereas transparency, surface smoothness and some mechanical properties depend upon larger structures. The ductility of crystalline polymers, for example, depends to a great extent upon the size, perfection and organization of the crystals.

Molecular structure has been elucidated by the classical techniques such as x-ray diffraction, infrared and Raman spectroscopy, and NMR. The more gross fibrillar and spherulitic structures may be directly seen by light microscopy. Intermediate size structures can be seen by electron microscopy but the technique is limited to the inspection of surfaces or thin sections, and cannot be unambiguously applied to the examination of structures occurring within bulk polymer samples. The determination of changes occurring with deformation in short times is difficult. Information about structures of up to a few thousand Angstroms in size may be obtained by low-angle x-ray scattering, particularly if there is periodicity in density. However, interpretation is often difficult and its

[†] This work was supported in part by a grant from the National Science Foundation and in part by a contract from the Office of Naval Research. Much of the work was carried out by students of the author among whom particular contributors to this paper were Dr. T. Hashimoto, Mr. R. Prud'homme and Dr. Claude Picot.

1

extension to larger sizes is demanding on equipment.

A technique which is emphasized in this paper is the scattering of visible light by polymer films. The method provides information about structure having size of the order of the wavelength of visible light, of around 1000-100000Å. Like low-angle x-ray scattering, it provides information about fluctuations in density of scattering material, but in a size range much larger than can be readily studied by x-ray scattering. However, unlike x-ray scattering, it also provides information about fluctuations in anisotropy or refractive index and of orientation of anisotropic regions. X-ray scattering is insensitive to these anisotropy fluctuations since it involves inner electrons of atoms which are not influenced by chemical bonding which is the source of anisotropy. Light scattering, on the other hand, involves the outer valence electrons of atoms, so that information is provided about the orientation of the polymer molecules containing these atoms. While the light scattering size range overlaps that of optical microscopy, it complements it in two important aspects: (a) The light scattering provides a statistical evaluation of the sizes and shapes of scattering entities. Such an evaluation is often difficult from micrographs were small, overlapping and vaguely shaped structures are often seen. In fact, electronic image analysis is sometimes employed to analyze the micrograph in terms of statistical parameters. (b) The scattering technique may be employed to follow rapid changes that may occur, for example, along with the application of an oscillating strain to a sample, so as to permit a measurement of the time-dependence of the structural response to such a strain.

THE LIGHT SCATTERING TECHNIQUE (1,2,3)

Experimental Methods

The measurement of scattering can be carried out by either photographic or photometric means. The photographic technique is somewhat like that of photographic x-ray diffraction except that a laser is used as a radiation source as a substitute for an x-ray tube. A diagram of a typical photographic scattering apparatus is shown in Figure 1. Advantages of such apparatus is that it is relatively cheap (under $1,000) and complete scattering patterns can be recorded in short times.

Typical photographic light scattering patterns are shown in Figure 2 for a spherulitic low density polyethylene sample. Patterns are shown for the case where the polarizer and analyzer both have their polarization directions vertical (designated V_v) and where one is vertical and the other is horizontal (designated H_v).

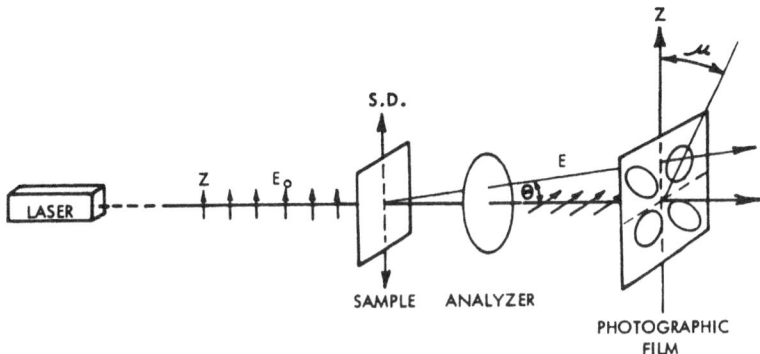

Figure 1. A typical photographic light scattering apparatus.

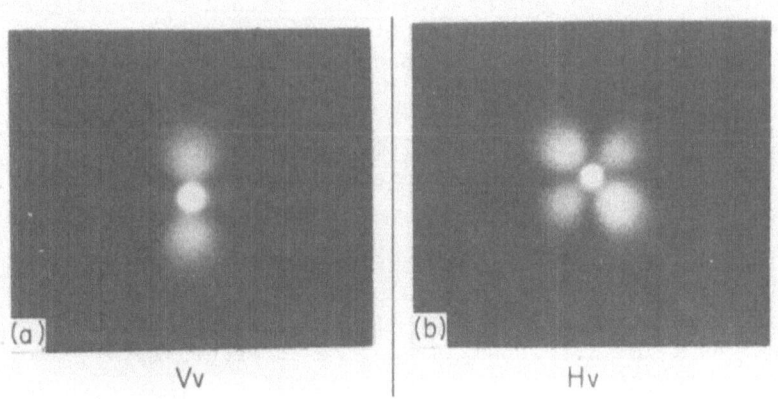

Figure 2. Typical V_V and H_V low angle light scattering pictures
for low density polyethylene.

(The measurement where the polarizer is horizontal and the analyzer
is vertical is designated V_h equivalent to H_v.) The intensity depends upon the scattering angle θ between the incident and scattered ray (Figure 1) and the azimuthal angle μ. The range of θ
that may be recorded in a picture depends upon the sample to film
distance d since $\tan\theta = (x/d)$, where x is the distance from the
center of the photographic film to the point where the intensity
is recorded. Since small objects scatter at large angles, the
sample to film distance must be made small for recording these
(down to a few millimeters) or else the photographic film should
be bent cylindrically about the sample as in powder x-ray cameras.
Since scattering power is small for small objects (varying with the
sixth power of the radium of the object), exposure times should be
large. For large objects where scattering occurs at small angles,
it is desirable to have sample to film distances large (several
meters in some cases). The laser beam must be well collimated and
it is desirable to include a pinhole collimater in the incident
beam to eliminate the non-coherent light coming from the discharge
tube. A beam stop should be introduced to prevent the incident
beam from hitting the film to avoid halation of the film resulting
from its high intensity.

It is necessary to correct the angle θ for its refraction at
the surface of the film using Snell's law, $n = \sin\theta/\sin\theta_i$, where n
is the refractive index of the film and θ_i is the true scattering
angle within the film. Also it is necessary to correct the wavelength of the light for refraction so that the wavelength λ used
in the scattering equations should be that within the scattering
medium and obtained from that in vacuum λ_o using $\lambda = \lambda_o/n$.

The thickness of the film should be chosen so as to provide
sufficient scattering to permit measurement with a reasonable exposure time. For highly scattering systems such as linear polyethylene, the film should be thin enough so that a large amount of
secondary scattering does not occur. This rescattering of the
scattered light obscures the detail of the scattering pattern (3).
A good rule-of-thumb is that the polymer film should transmit of
the order of 75% of the incident radiation. For linear polyethylene, a typical thickness is 50 μm.

Samples can be quite small (1 mm^2) and are ordinarily held
between glass microscope cover slips using an immersion fluid to
minimize surface contributions (3). It is possible to study fibers as well as films (3). Samples may be stretched (4,5), swollen (6), or heated (4,7). Studies can be made as a function of
time during crystallization (8) or following the rapid stretching
of samples (9-11). Scattering patterns can be recorded rapidly
using a high speed motion picture camera (9).

The photographic technique is well suited to the rapid and

qualitative evaluation of the entire scattering pattern. It represents a convenient and rapid means for evaluating the nature, size,
perfection and orientation of the polymer superstructure. It should
be possible, using image intensifying techniques, to see the scattering pattern in real time during mechanical or thermal treatment
of the film, so that the method has potential as a production control technique. It should be possible to rapidly record scattering
patterns from such presentations using a videotape recorder in a
manner similar to that recently reported for the recording of x-ray
diffraction patterns (12). Such tapes can then be played back at a
slower speed through a TV monitor or may be analyzed by computer.

Where scattering patterns exhibit maxima, the position of these
serves as a measure of particle size using theory to be discussed
later. Photographic films may be analyzed by microdensitometer in
order to obtain a quantitative measure of the variation of the scattered intensity with the angles θ and μ for comparison with theory.

When studying the scattering by oriented samples, the birefringence of the sample may affect the scattering (13-15). For
such cases, it is desirable to make measurements under conditions
such that the orientation direction making an angle Ω, Figure 3,
with the vertical direction lies along the polarizer or the analyzer ($\Omega = 0°$ or $90°$). For some systems it is desirable to in-
spect the pattern when $\Omega = 45°$. In such cases, it must be realized that the pattern may be severely distorted by the large birefringence effects occurring under these conditions.

For biaxially oriented films, the scattering pattern depends
upon the angle ϕ between the normal to the film and the incident
beam (Figure 3). (It is apparent that additional corrections must
be made for refraction of both the incident and the scattered beam
when varying ϕ.) While the variation of the scattering pattern
with   serves as an indication of biaxial orientation of superstructures, its interpretation for intermediate values of ϕ is
difficult. A useful approach is that of taking "XYZ scattering
patterns" as described by Wilkes where patterns are examined for
sections of samples cut perpendicular to the three principal axes
of the film. For obtaining such sections normal to the Y and Z
axes (Figure 3) lying in the plane of the film, it is necessary
that one start with a rather thick polymer film to obtain a sample
of sufficient area or else such a sample be formed by the difficult
procedure of laminating several thinner films under conditions such
as not to disturb the sample structure and also to not give too
much surface scattering.

More quantitative studies of scattering can best be made by
the photometric technique where the variation of scattered intensity with angle is measured directly using a photometer employing
a photomultiplier tube. Sensitivity is usually not so much of a

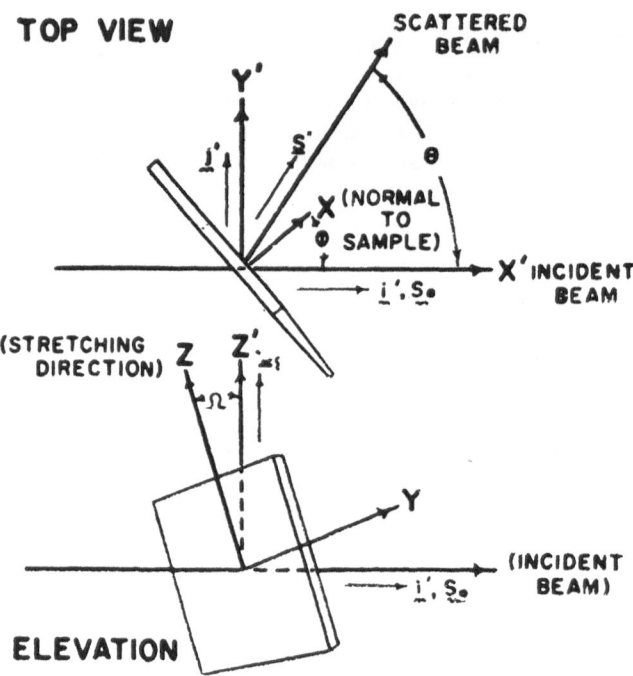

Figure 3.   The angles θ, ϕ and Ω specifying the scattering experiment for a biaxially oriented polymer film.

problem as with solution scattering apparatus since most polymeric solids scatter very much more than do solutions. For the same reason, contributions to scattering arising from adventitious impurities such as dust particles are not of so much concern if the film is reasonably clean. Of course, if films contain appreciable amounts of residual solid polymerization catalyst or filler particles, these will interfere with and may even preclude meaningful scattering measurements. If the scattering arises from crystalline structure, then a "blank" measurement including contributions from impurities and stray light may be made on films heated to above the crystalline melting point and subtracted from the measurements on the sample. If such a measurement is not possible, it is useful to make a blank measurement on a synthetic blank sample made by mounting some cover glasses similar to those used to hold the polymer film. If some silicone oil immersion fluid is used between the cover glasses, the scattering should approximate that coming from stray light when the sample is present and may be subtracted from the sample measurements.

The photometer for solid state studies should in general be capable of better angular resolution than that used for solution measurements and be capable of measuring intensities down to a small angle from the incident beam. This is particularly true when studying the scattering from larger structures (having characteristic sizes of a few thousand Angstrom units or more) where it is desirable to make measurements down to $\theta = 0.5°$ with a resolution of at least 0.1°. This requires a well collimated incident beam obtained from a CW laser or with a discharge lamp accompanied by a small pinhole and long focal length lens collimating system. The photometer must have similar collimation. It is desirable that there be adjustable or replaceable pinholes so that collimation may be sacrificed for greater sensitivity at the larger scattering angles.

The light source should be monochromatic and equipped with suitable filters for excluding undesirable wavelengths. Neutral filters should be provided for attenuating intensity if necessary.

The sample should be mounted on a goniometer so that it may be rotated about its normal through the angle Ω and tilted through the angle ϕ as indicated in Figure 3. It should be held between glass sheets using an immersion fluid as with the photographic method. It is desirable to have provision for holding the sample stretched and for controlling its temperature.

Polarizers should be mounted in the incident and scattered beams with provision for rotating the polarization direction through angles ψ_1 and ψ_2 from the vertical.

An apparatus having most of these features was originally designed by R. S. Stein, F. H. Norris and A. Plaza. An improved version designed by van Aartsen, et al. (17,18), is now commercially available (Nederlandshe Optiek-en Intrumentfabriek, Dr. C. E. Bleeker, N.V., Zeist, Holland).

It is necessary to correct intensities measured with such an apparatus for refraction, reflection and secondary scattering. Procedures for this have been described by Stein and Keane (19,20) which may now be readily applied to scattering data using computer techniques (21). Additional corrections necessary to account for non-ideality of polarization have been discussed by Keijzers, et al. (17,18).

A typical plot of the variation of scattered intensity with θ is given in Figure 4 for a few values of polarizer orientation (ψ_1) under conditions where the polarizer and analyzer are approximately crossed so that $\psi_2 = \psi_1 + 90°$ in which case the intensity is designated I_+. [Keijzers, et al. (17,18), point out that for theoretical reasons such measurements should not be made under conditions of exact crossing but rather $\tan\psi_2 = \cot\psi_1/\cos\theta$.] Another condition of interest for measurement is that designated by $I_{||}$ where the polarizer and analyzer have their polarization directions approximately parallel. By making such measurements at various values of ψ_1, a scattering contour diagram may be obtained, a typical one being given in Figure 5.

General Considerations

The light scattering from solids can be treated in terms of two limiting types of theories. The first of these referred to as the "model approach" involves calculating the scattering by summing the scattered amplitudes arising from all of the volume elements constituting some structural unit such as a sphere or a rod. The second is the statistical approach in which a structure is described in terms of correlation functions describing fluctuations in density and orientation. Each of these methods has its limitations, and real systems lie between the extremes conveniently described by these models.

The Model Approach

In general, the scattered intensity, I_s, is found by averaging the square of the amplitude E_s of the scattering which is found by summing the amplitude contributions from all of the volume elements of the scattering object, taking into account the phase differences arising from the different optical path lengths from the different volume elements. That is (22,23)

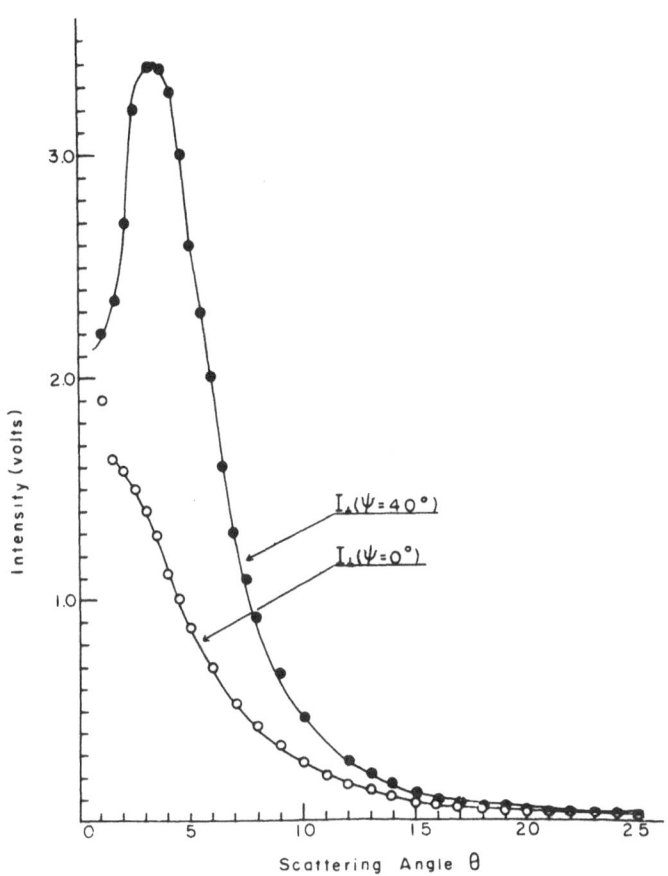

Figure 4. A typical variation of the I_+ scattered intensity with
           θ for a low density polyethylene film at $\psi = 0°$ and
           $\psi = 40°$.

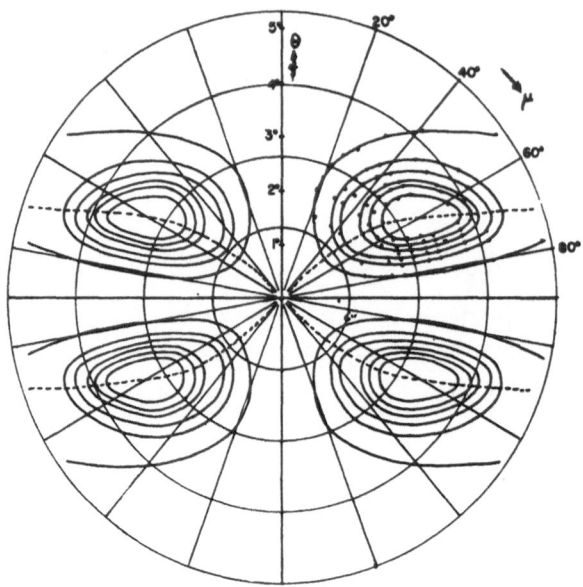

Figure 5.   The experimental H_v scattered intensity contour for a
1.5 fold uniaxially stretched film of low-density
polyethylene.

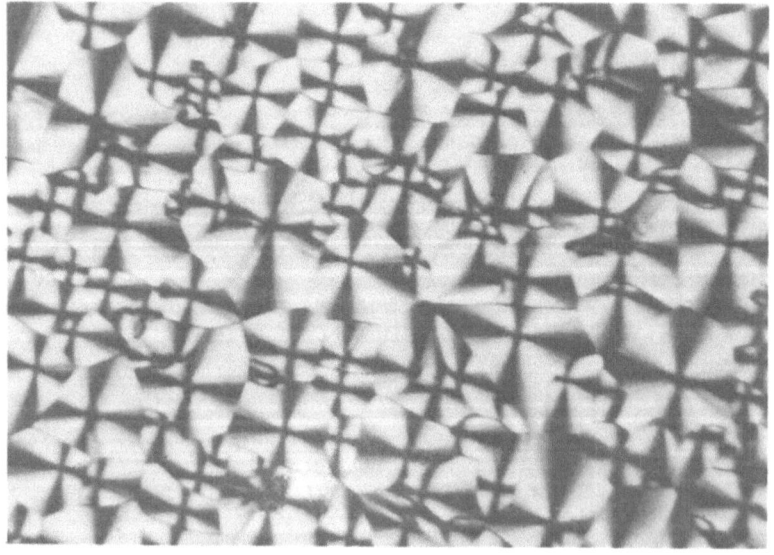

Figure 6.   A photomicrograph of polyethylene spherulites.

$$E_s = \sum_n (E_s)_n = \sum_n A_n \exp [ik (r_n \cdot s)] \tag{1}$$

where $(E_s)_n$ is the electric field strength of the scattered ray
from the nth volume element, A_n is the amplitude of this scattered
ray and k $(r_n \cdot s)$ is the phase factor in which k = $2\pi/\lambda$, r_n is a
vector from an arbitrary origin to the scattering element, and s
is the propagation vector given by s = s_0 - s_1 where s_0 and s_1 are
unit vectors in the direction of the incident and scattering rays,
respectively.

In practice, the sum in Equation 1 is usually approximated by
an integral over the bounds of the scattering particle. For a
uniform isotropic particle, A_n is a scaler constant proportional
to the polarizability, α_s, of the particle. This leads to the
familiar result for a uniform isotropic sphere of radius R, for
example (22-24)

$$E_s (R) = C_1 V (\alpha_s - \alpha_0) (3/U^3) (\sin U - U \cos U) \tag{2}$$

where C_1 is a physical constant, V is the volume of the sphere
[V = $(4/3)\pi R^3$], α_0 is the polarizability of the surroundings and
U = $(4\pi R/\lambda) \sin(\theta/2)$. It is noted that the scattered intensity is
cylindrically symmetrical about the incident beam and is independent of the azimuthal scattering angle μ defined in Figure 1.

A plot of Equation 2 as a function of θ leads to an intensity
exhibiting many maxima and minima corresponding to various orders
of interference. Such behavior has indeed been observed for light
and x-ray scattering from monodisperse dilute suspensions of
spheres. For real systems, there is usually a distribution of
sphere sizes so that the total intensity must be found by summing
over the contributions of the species present so that

$$I_s = K_1 \sum_n N(R_n) [E_s (R_n)]^2 \tag{3}$$

where K_1 is a constant and $N(R_n)$ is the relative number of spheres
of radius R_n (25). This averaging over the distribution usually
leads to a washing out of the distribution so that the intensity
from a distribution of spheres may often be approximated by a
Gaussian equation of the form

$$I_s = I_{so} \exp [-s^2/a^2] \tag{4}$$

where $s = 2\sin(\theta/2)$ and a is a parameter dependent upon the distribution.

The above treatment is based upon the so-called Rayleigh-Gans-Debye approximation which is valid when α_s and α_o are not too different so that there is not much distortion of the electric field at the boundary of the sphere. The more general case is described in the much more complex Mie theory (22,23).

These equations have found their application in polymer technology, for example, in the description of high impact polystyrene which is made by dispersing rubber latex particles in a polystyrene matrix. In favorable cases, it is possible to determine not only the average particle size of the spherical latex particles, but also the distribution of sizes.

In these cases of isotropic spherical particles, the Rayleigh-Gans-Debye theory leads to scattered radiation polarized in the same plane as the incident radiation so that the H_V or the I_+ component of scattering is zero.

Crystalline polymers often contain spherical aggregates of crystals called spherulites as shown in Figure 6. These are anisotropic as a result of the constituent crystals being oriented with their axes in some preferred direction with respect to the radius. This leads to different values of radial and tangential polarizability α_r and α_t. In this case, the amplitude A_n depends upon position within the scattering particle being given by

$$A_n = C_2 \, (M_n \cdot O) \tag{5}$$

where C_2 is a constant, M_n is the dipole moment induced by the field of the light wave in the nth volume element, and O is a unit vector lying in the plane containing the analyzer polarization direction and the induced dipole moment. M_n is given by (4)

$$M_n = |\alpha_n| \, E_O$$

$$= (\alpha_1 - \alpha_2)_n \, (E_O \cdot a_n) \, a_n + (\alpha_2)_n \, E_O \tag{6}$$

where $|\alpha_n|$ is the polarizability tensor of the nth volume element and E_O is the electric field vector of the incident ray. It is assumed that the polarizability is uniaxial with its principal axis lying in the direction of the unit vector a_n and with principal polarizability components $(\alpha_1)_n$ and $(\alpha_2)_n$ along and perpendicular to the principal axis. If a_n is along the radius, then $\alpha_1 = \alpha_r$ and $\alpha_2 = \alpha_t$. This leads to the approximate result that (4)

$$I_{V_v} = C_3 V^2 \{(3/U^3)(\alpha_t - \alpha_s)(2\sin U - U\cos U - SiU) + (\alpha_r - \alpha_s)$$

$$(SiU - \sin U) - (\alpha_t - \alpha_r)\cos^2(\theta/2)\cos^2\mu(4\sin U - U\cos U - 3SiU)\}^2$$

$$\tag{7}$$

$$I_{H_v} = C_3 V^2 \{(3/U^3)(\alpha_t - \alpha_r)\cos^2(\theta/2)\sin\mu\cos\mu$$

$$X\ (4\sin U - U\cos U - 3SiU)\}^2 \tag{8}$$

where V is the volume of the spherulite, $U = (4\pi R/\lambda)\sin(\theta/2)$, R is

the radius of the spherulite, $SiU = \int_0^U (\sin x/x)\ dx$, and μ is the azimuthal scattering angle defined in Figure 1.

It is noted that in this case, there is both a V_v and H_v component of scattering. The H_v component depends upon the anisotropy of the spherulite $(\alpha_r - \alpha_t)$ and varies with $\sin^2\mu\cos^2\mu$ leading to the previously-mentioned four-leaf clover appearance with scattering maxima occurring at odd multiples of $\mu = 45°$. The scattering passes through a maximum in the θ direction at a value of θ_n given by

$$U_{max} = 4\pi(R/\lambda)\sin(\theta_m/2) = 4.1 \tag{9}$$

Thus the position of the maximum serves as a convenient measure of the size of the spherulite. Its displacement toward smaller values of θ_m with increasing crystallization time has been used to follow spherulitic growth (8). Higher order scattering maxima in θ are also predicted but these are not usually seen except for systems having single spherulites or monodisperse distributions of spherulite size because of the "washing out" of this fine structure resulting from the distribution of spherulite size (25) or disorder (26).

The V_v pattern depends upon the polarizability of the surroundings α_o as well as those of the spherulite. The amplitude consists of two components, one of which is an angularly independent term dependent upon α_s and the other is a term dependent upon $\cos^2\mu$ having two-fold symmetry. It has been found that during the crystallization of polyethylene, the V_v scattering intensity passes through a maximum (4,7,27). This has been attributed to the terms dependent upon α_s passing through a maximum when the volume frac-

tion of spherulites is about 1/2, since the scattering amplitude of
a two-phase system is $\phi_1 (1 - \phi_1)$ where ϕ_1 is the volume fraction
of one of the phases. Thus one observes that the initial pattern
is independent of μ. At later stages of crystallization the pattern becomes dependent upon μ as $(\alpha_t - \alpha_r)$ becomes greater. The
H_v pattern which depends only upon $(\alpha_t - \alpha_r)$ only develops during
the latter part of the crystallization.

The effect of optic axis tilting upon the spherulite scattering pattern has been analyzed for the simple case of two-dimensional
spherulites. The normal type of H_v pattern with the lobes at $\mu =$
45° are seen when the optic axis lies at an angle β close to 0° or
90° to the radius, but occurs at 45° when the axis is tilted at 45°
to the radius (28).

It has been found that the theoretical H_v light scattering
patterns calculated using Equation 8 predict intensities that are
too low at small and large scattering angles. In part, this is due
to a distribution of spherulite sizes, but the calculated patterns
on the basis of reasonable distributions of sizes cannot account
for the observed results (25,29). It has been shown that excess
scattering at large angles can be explained in the case of isotropic spheres by assuming that there are internal fluctuations in
density (30). The effect of fluctuations in the orientation (26)
and anisotropy (31 of anisotropic scattering elements has been
shown to produce similar effects. It is pointed out that the low
scattering at small angles is a consequence of the symmetry of
spherulites (27) and that a perfect spherulite should lead to zero
H_v intensity at $\theta = 0°$. The finite intensity at small values of θ
is a consequence of fluctuations from this symmetry and probably
leads to the "tennis-racquet" types of patterns reported by Kawai
(29) et al., and also by Prud'homme and Stein for disordered
spherulites (32).

While the theory for the scattering from homogeneous or heterogeneous isolated spherulites accounts for most of the observed
features of scattering patterns quite well, it is realized that it
is an oversimplification since only isolated spherulites are considered. Actual systems can deviate because of the effect of
interspherulitic interference and truncation arising from impingement with neighboring spherulites. A theory for interspherulitic
interference valid for dilute suspensions of spherulites has been
worked out (33). The interference of scattering among small
groups of interfering spherulites has been experimentally demonstrated using groups of spherulitic starch granules or anisotropic
polystyrene spherulites and has been theoretically calculated (34).
These calculations have been extended to larger assemblies of
spherulites (35) and have been shown to account for the "speckled"
appearance that is experimentally found for such assemblies. It
is found, however, that the value of θ_m at which maximum scattering

appears is not appreciably affected by the internal structure so
that estimates of size will not be in error.

The effect of incomplete growth and impingement on the scattering through its effect on producing deviations from spherulitic
shape has been analyzed by Picot, et al. (36,37) and Kawai, et al.
(38). Such deviations in shape for starch granules have also been
described by Borsch, et al. (39,40). A general theory of deviations
from circular shape of two-dimensional spherulites has been developed by Kawai, et al. (38) giving, for example, equations of the
form

$$I_{H_v} = \frac{K_2 \, \delta^2 \, R_o^4}{4} \, \{A_1(\theta) - A_2(\theta) \, \cos 4\mu\} \qquad (10)$$

where R_o is some characteristic particle radius and δ is the anisotropy of the spherulite, $(\alpha_t - \alpha_r)$, which is assumed to be angularly
independent. $A_1(\theta)$ and $A_2(\theta)$ are coefficients which depend on the
external shape of the disc. For perfect discs, these coefficients
are equal, and the scattering equations reduce to those for perfect
two-dimensional spherulites. For imperfect spherulites, these coefficients depend upon moments of the shape. The cases of truncated
and sector spherulites treated by Picot, et al. (36,37) prove to be
special cases of the more general treatment.

Misra and Stein have recently shown (41) that the scattering
patterns observed during the early stages of the crystallization of
polyethylene terephthalate are of the sort predicted by the preceding theories. As crystallization proceeds, spherulites become more
complete and the patterns evolve toward those characteristic of
more perfect spherulites.

The change in scattering patterns occurring upon stretching
polymers has been treated by Clough, et al. (28) in two dimensions
on the basis of the assumption that the scattering change is the
result of the transformation of a circular spherulite into an
ellipsoid. Various models were introduced for describing the ways
in which principal polarizability directions change upon deformation. The model has been generalized to three dimensions by van
Aartsen and Stein (42).

A semi-empirical theory of the effect of spherulite deformation upon the light scattering patterns has been presented by
Samuels (5) which has the important virtue of being relatively
simple. Its predictions compare favorably with experimental results
for the deformation of polypropylene films and fibers.

The above theories involve the assumption of an affine deformation of spherulites. Detailed examination (43,44) of spherulite

deformation indicates that such deformation is not affine and greater deformation is often found in the equatorial than in the polar parts of the spherulite, leading to a decrease in thickness at the equator for two-dimensional spherulites and a decrease in density for three-dimensional spherulites. This latter tendency leads to density fluctuations which contribute to V_V scattering (44).

The theory has been used to account for the variation in intensity during a dynamic light scattering experiment in which a sample is subjected to an oscillating strain. The variation in scattered intensity is accounted for in terms of the oscillating deformation of the spherulite accompanied by the oscillating orientation of its optic axes (45,46).

Other models have been considered for scattering systems. A basic feature of the morphology of crystalline polymers is the initial formation of rod-like lamellae which then develop to form larger units (47). Many polymers, such as polytetrafluoroethylene (48) and polychlorotrifluoroethylene (49) scatter light in a manner that can be accounted for on the basis of a model of anisotropic rods. A two-dimensional theory of such scattering has been presented in which the effect of orientation is considered (50). The theory has been extended to three dimensions by Kawai, et al. (51) who has applied it to the scattering by collagen films. Such extended aggregates of rods really form the basis of scattering by spherulites as has been recently described (52).

The Statistical Approach

In the preceding considerations, scattering was calculated on the basis of scattering by objects having some definite shape. While such models provide a good approximation to the scattering by many real systems, it is necessary to consider interaction between particles and internal structure to account for detailed behavior. As the disorder of a system increases, such internal structure becomes more and more important, until it becomes more important than the particulate structure of the system. Under such conditions, a statistical description of such internal structure proves desirable.

A statistical theory of the scattering from a medium having fluctuation in density was proposed by Debye and Bueche (53) which led to the equation

$$I = K_3 <\eta^2>_{av} \int_{r=0}^{\infty} \gamma(r) \frac{\sin hr}{hr} r^2 dr \qquad (11)$$

where n_i is the difference between the refractive index at position i and the average refractive index. The term $<n^2>_{av}$ is the mean-squared value of this quantity. The function, $\gamma(r)$, is a correlation function defined by

$$\gamma(r) = \frac{<n_i \, n_j>_r}{<n^2>_{av}} \tag{12}$$

where $<>_r$ designates an average taken at constant separation of the volume elements i and j. This function is one at $r = 0$ but decays toward zero as r increases in a manner dependent upon the structure of the system.

This function can be determined from a Fourier inversion of the variation of scattering intensity with scattering angle. For many systems, an exponential function

$$\gamma(r) = \exp \, [-r/a] \tag{13}$$

appears to describe the scattering quite well. The correlation distance, a, is a measure of the size of the correlated region.

Fluctuation theory was generalized by Goldstein and Michalek (54) who proposed more general correlation functions involving correlations in density and anisotropy. Stein and Wilson (55) proposed a less general but more easily applied theory of "random orientation correlations" which leads to equations which may be approximately expressed as

$$I_{H_v} = K_4 \, <\delta^2> \, \int_{r=o}^{\infty} f(r) \, \frac{\sin \, hr}{hr} \, r^2 dr \tag{14}$$

and

$$I_{V_v} = \frac{K_4}{15} \, [<n^2>_{av} \int_{r=o}^{\infty} \gamma(r) \, \frac{\sin \, hr}{hr} \, r^2 dr$$

$$+ \frac{4}{45} <\delta^2>_{av} \int_{r=o}^{\infty} f(r) \, \frac{\sin \, hr}{hr} \, r^2 dr] \tag{15}$$

at small scattering angles. Here $<\partial>^2_{av}$ is the mean-squared anisotropy of the anisotropic scattering element. The orientation correlation function, $f(r)$, is

$$f(r) = \frac{3 <\cos^2 \theta_{ij}>_r - 1}{2} \tag{16}$$

where θ_{ij} is the angle between the optic axes of the ith and jth scattering elements separated by distance r.

The H_V scattering depends only upon the orientation correlation term, while the V_V scattering depends upon both density and orientation correlations. It is apparent that the density contribution can be obtained from $I_{V_V} - (4/3) I_{H_V}$. Both correlation functions can be obtained by suitable Fourier inversions, and exponential functions (with different correlation distances) often fit both terms.

For oriented systems, Stein and Hotta (56) introduced a vector correlation function f(r) which leads, for example, for the case of uniaxial orientation to the equation

$$I_{H_V} = K_5 \delta^2 [<\cos^2\varepsilon>_{av} - <\cos^4\varepsilon>_{av}]$$

$$X \int f(\underset{\sim}{r}) \cos [k (\underset{\sim}{r}\cdot\underset{\sim}{s})] d\underset{\sim}{r} \tag{17}$$

where

$$f(\underset{\sim}{r}) = \exp \left\{ -\left[\frac{x^2}{a^2} + \frac{y^2}{a^2} + \frac{z^2}{c^2}\right] \right\} \tag{18}$$

The coordinates x, y and z are the projections of r in the direction of the film plane normal, the transverse direction and the machine direction. The correlation distance in the machine direction is c while that normal to the machine direction is a. The angle ε is that between the optic axis and the machine direction. As the film becomes highly oriented $<\cos^2\varepsilon>_{av} = <\cos^4\varepsilon>_{av} = 1$ and the scattering intensity diminishes as all optic axes align parallel to the machine direction (57).

For oriented films, this equation predicts that the scattered intensity depends upon the angle Ω between the machine direction and the normal to the scattering plane in a manner that depends upon the difference between a and c.

These orientation correlation treatments are limited by the "random" assumption that the probability of correlation depends only upon the separation of the scattering elements and not upon the angle which the optic axis makes with the separation vector (55). For an unoriented system, this implies that the correlated

regions are spherical in shape and leads to the prediction that the
scattering is cylindrically symmetrical about the incident beam.
Experimentally, this is often found not to be the case.

One approach to the treatment of non-random correlations by
Keijzers, van Aartsen and Prins (17) is to consider the system as
a mixture of random and spherulitic type contributions. A somewhat
more general approach is to expand the correlation function in a
Fourier series in the angle β between the optic axis and the vector
between scattering elements in two dimensions (58) or in spherical
harmonics in three dimensions (59). An example of the type of result which is obtained is that from the two-dimensional theory for
small values of θ

$$I_+ = K <\delta^2>_{av} \int_{r=o}^{\infty} \{T_o(r) J_o(w)$$

$$- 1/2 [T_4(r) - S_4(r)] J_4(w) \cos4\psi\} r dr \qquad (19)$$

where $J_o(w)$ and $J_4(w)$ are Bessel functions of w where $w = (2\pi r/\lambda) \sin\theta$
and $T_o(r)$, $T_4(r)$ and $S_4(r)$ are the Fourier coefficients in the expansion of the correlation functions

$$<\cos2\theta_{ij}>_{r,\beta} = T_o(r) + T_2(r) \cos2\beta + T_4(r) \cos4\beta + . . . \qquad (20)$$

and

$$<\sin2\theta_{ij}>_{r,\beta} = S_2(r) \sin2\beta + S_4(r) \sin4\beta + . . . \qquad (21)$$

where $<>_{r,\beta}$ designates an average taken at a constant r and β over
all pairs of scattering elements i and j and θ_{ij} is the angle between the optic axis of these elements.

As discussed previously, the intensity I_+ is that obtained when
the polarizer and analyzer are crossed, but they are rotated together through the angle ψ. Each of the coefficients in the Fournier
expansion play the role of a correlation function. When $[T_4(r) -
S_4(r)]$ is zero, I_+ is independent of ψ and only the correlation
function $T_o(r)$ remains. In this case, the theory reduces to a random orientation type theory and $T_o(r)$ is the two dimensional equivalent of the orientation correlation function $f(r)$. The coefficient
$[T_4(r) - S_4(r)]$ is a measure of the non-randomness and leads to a
dependency of the scattered intensity on ψ having four-fold symmetry.

It is possible to obtain a similar expression for $I_{||}$ when the polarizer and analyzer are kept parallel. This expression involves the density correlation function and $[T_2(r) - S_2(r)]$ as well as $T_0(r)$ and $[T_4(r) - S_4(r)]$ and may possess two fold as well as four fold symmetry.

This two-dimensional theory has been extended to the description of oriented systems (60) where the Fourier coefficients depend upon the angular coordinates of the vector $\underset{\sim}{r}$ with respect to the machine direction.

The formulation of the three-dimensional theory is such that it cannot be readily extended to oriented systems. These types of theories are such that it is possible to formally describe states of order ranging from randon orientation correlations to highly organized structures such as spherulites. It was shown (58), for example, that it was possible to calculate the correlation functions for a two-dimensional spherulite and predict its scattering pattern from these functions.

While the scattering behavior of any system can be formally described in terms of these non-random type theories and, in principle, the higher order correlation coefficients can be determined from experimental data, it is often difficult to ascribe physical meaning to such terms. Consequently this author finds such theories not very useful in their description of highly organized systems and feels that their application is limited to the description of small deviations from randomness.

A Comparison of the Theories

We have discussed two approaches to the description of the scattering from a solid polymer. The first of these, the model approach, depends upon the association of a model such as a sphere or a rod as the structural unit responsible for the scattering. A difficulty with this approach, as has been pointed out, is that it neglects the hierarchy of structural units. A spherulite, for example, is not a homogeneous sphere but it is composed of rod-like lamellae interconnected with amorphous polymer. Additional scattering arises from this internal structure. Also a spherulite is not isolated and plays a part in a larger structure in which spherulites are packed together in space to form a coherent solid, resulting in irregularity in shape as well as inter-spherulitic interference.

The interrelationship between the location and orientation of the component units of a structure may be described in terms of statistical parameters. If the system is relatively random, the statistical functions are relatively simple, but as a system becomes

more ordered, more and more statistical functions are required and
the description soon becomes unwieldy.

Thus, neither approach is completely satisfactory. The statistical approach is most reasonable for a relatively disordered system, whereas the model approach is simplest for a highly ordered
system. Unfortunately, many systems have the intermediate degree
of order when neither approach is completely adequate, and one is
often forced to adapt a hybrid theory.

The difficulty is a consequence, of course, of the fact that a
partially ordered system is not simple. If one looks at an electron
micrograph of such a material, much detail may be seen, and it is
difficult to find a few numbers or even a few functions to completely describe it. It is easy to formulate a theory which requires
more parameters than can be measured experimentally which then becomes meaningless.

A Rational Approach

It becomes apparent that light scattering alone will not suffice to completely characterize the morphology of a polymer film.
While the experimental simplicity is attractive, a price is often
paid in the difficulty of theoretical interpretation. I feel that
the potential of the method can only be realized if it is used together with all other methods that are available for characterizing
a system.

If one sees spherulites under the microscope, for example, it
is senseless to try to characterize a system by the completely statistical approach which ignores the spherical symmetry. However,
it also will not suffice to neglect the internal structure which
may be often seen with an electron microscope. Thus, one may allow
the spherulites to be composed of sub-units such as rods which may
be located with some heterogeneity within the spherulite. The dimensions of such rods and their separation may best be studied by
x-ray scattering and diffraction and electron microscopy. While the
low-angle light scattering depends upon the size and anisotropy of
the spherulites, the scattering at larger angles depends upon the
heterogeneity of arrangement of the constituent crystals. The sta-
tistics describing such heterogeneity should be developed in such a
way as to describe the deviations from a perfectly spherulitic structure rather than deviations from a completely random collection of
spherulites as has been discussed by Stein and Chu (26). Such internal heterogeneity is characterized by correlation functions involving
correlation distances describing a "size" of the heterogeneity. As
such a distance becomes smaller, the fraction of the scattering occurring at higher angles increases. If the correlation distance is

small compared with the radius of the spherulite, then the effect of
the spherulitic structure will be lost and the scattering may best
be described in terms of a straightforward statistical approach.
Thus, a study of scattering from a spherulitic polymer at small
scattering angles provides information about the superstructure at
the spherulitic level, whereas studies at larger values of θ give
information about the internal structure on a dimensional scale corresponding to several crystals. Structure which is still smaller
is best studied by choosing a shorter wavelength as with low-angle
x-ray scattering.

It is evident that orientation changes measured by light scattering will be related to the spherulite orientation at the smaller
angles but to the internal structure orientation at the larger. It
is not surprising that the rates of orientational change as studied
by the dynamic light scattering technique are different if measured
at small or at large values of θ (45). The orientation as measured
by light scattering will be different but must be related to the
orientation as measured by birefringence or by x-ray diffraction.
The light scattering by a spherulite depends upon its change in
shape and its local birefringence with position. The total birefringence of the sample is the average over all directions within the
spherulite. The change in the spherulite birefringence results from
the orientation of both crystalline and amorphous polymer within the
spherulite when the polymer is deformed. This orientation of the
crystals also gives rise to the change in both low-angle x-ray scattering and wide-angle x-ray diffraction. Thus all these experiments
must be interrelated. One must find a model for a deforming polymer
which can account for all of these observations of changes seen by
the various techniques. Of course, such a model must also be consistent with what is seen with the light and electron microscope.

In summary, I feel that the light scattering technique is an
important supplement to other methods for characterizing the morphology of polymer films and should be a part of the kit of tools
available to the polymer physicist.

REFERENCES

1. R. S. Stein in PROCEEDINGS OF THE INTERDISCIPLINARY CONFERENCE
    ON ELECTROMAGNETIC SCATTERING, Ed. by M. Kerker, Pergamon Press,
    New York, 1963, pp. 430-458.
2. R. S. Stein, P. Erhardt, S. Clough and J. J. van Aartsen in
    ELECTROMAGNETIC SCATTERING, Ed. by R. L. Rowell and R. S.
    Stein, Gordon and Breach, New York, 1967, pp. 339-410.
3. R. S. Stein and M. B. Rhodes, ASTM Special Technical Publica-
    tion No. 348 (Symposium on Resinographic Methods), 59 (1963).

4. R. S. Stein and M. B. Rhodes, J. Appl. Phys. 31, 1873 (1960).
5. R. J. Samuels, J. Polymer Sci., C, No. 13, 37 (1966).
6. W. Yau and R. S. Stein, J. Polymer Sci. A2, 6, 1 (1968).
7. M. B. Rhodes and R. S. Stein, J. Polymer Sci. 45, 521 (1960).
8. C. Picot, G. Weill and H. Benoit, J. Polymer Sci. C, 16, 3973 (1968).
9. P. F. Erhardt and R. S. Stein, J. Polymer Sci. B, 3, 553 (1965).
10. P. F. Erhardt and R. S. Stein, Appl. Polymer Symposia, High Speed Testing, Vol IV: The Rheology of Solids 5, 113, Interscience, New York, 1967.
11. P. F. Erhardt and R. S. Stein, J. Appl. Phys. 39, 4898 (1968).
12. R. S. Stein and T. Oda, J. Polymer Sci. B, 9, 543 (1971).
13. R. S. Stein and S. N. Stidham, J. Polymer Sci. A2, 4, 89 (1966).
14. R. S. Stein and W. Chu, J. Polymer Sci. A2, 8, 489 (1970).
15. D. LeGrand, J. Polymer Sci. 8, 1937 (1970).
16. R. S. Stein, F. H. Norris and A. Plaza, J. Polymer Sci. 24, 455 (1957).
17. A. E. M. Keijzers, J. J. van Aartsen and W. Prins, J. Am. Chem. Soc. 90, 3167 (1968).
18. A. E. M. Keijzers, "Light Scattering by Crystalline Polystyrene and Polypropylene," Ph.D. Thesis, Delft, The Netherlands, 1967.
19. R. S. Stein and J. J. Keane, J. Polymer Sci. 17, 21 (1955).
20. J. J. Keane and R. S. Stein, J. Polymer Sci. 20, 327 (1956).
21. S. N. Stidham, ONR Technical Report 34a, Project: NR 356-378, Contract: Nonr 3357(01), University of Massachusetts, Amherst, Massachusetts, 1963.
22. M. Kerker, THE SCATTERING OF LIGHT AND OTHER ELECTROMAGNETIC RADIATION, Academic Press, New York, 1969.
23. H. van de Hulst, THE SCATTERING OF LIGHT BY SMALL PARTICLES, Wiley, New York, 1957.
24. A. Guinier, et al., SMALL ANGLE SCATTERING OF X-RAYS, John Wiley, New York, 1955.
25. R. S. Stein, S. N. Stidham and P. R. Wilson, ONR Technical Report No. 36, Project: 356-378, Contract: Nonr 3357(01), University of Massachusetts, Amherst, Massachusetts, 1961.
26. R. S. Stein and W. Chu, J. Polymer Sci. A2, 8, 1137 (1970).
27. S. Clough, R. S. Stein and M. R. Rhodes, J. Polymer Sci. C, No. 18, 1 (1967).
28. R. S. Stein, S. Clough and J. J. van Aartsen, J. Appl. Phys. 36, 3072 (1962).
29. M. Motegi, T. Oda, M. Moritani and H. Kawai, Polymer Journal (Japan) 1, 209 (1970).
30. S. N. Stidham and R. S. Stein, J. Appl. Phys. 34, 46 (1963).
31. T. Hashimoto and R. S. Stein, J. Polymer Sci. A2, 9, 1747 (1971).
32. R. E. Prud'homme and R. S. Stein, Macromolecules, in press.
33. R. S. Stein and C. Picot, J. Polymer Sci. A2, 8, 1955 (1970).
34. C. Picot, R. S. Stein, R. H. Marchessault, J. Borch and A. Sarko, Macromolecules 4, 467 (1971).
35. R. Prud'homme, D. Yoon and R. S. Stein, in preparation.

36. R. S. Stein, C. Picot, M. Motegi and H. Kawai, J. Polymer Sci. A2, $\underline{8}$, 2115 (1970).
37. C. Picot and R. S. Stein, J. Polymer Sci. $\underline{8}$, 2127 (1970).
38. T. Tatamatsu, N. Hayashi, S. Nomura and H. Kawai, Presented at the 20th Annual Meeting of the Society of Polymer Sci., Japan, May 27, 1971; in press.
39. J. Borch, "Light Scattering by Startch Granules," Ph.D. Thesis, State College of Forestry at Syracuse University, Syracuse, New York, 1969.
40. Z. Mencik, R. H. Marchessault and A. Sarko, J. Mol. Biol. $\underline{55}$, 193 (1971).
41. A. Misra and R. S. Stein, J. Polymer Sci., in press.
42. J. J. van Aartsen and R. S. Stein, J. Polymer Sci. A2, $\underline{9}$, 295 (1971).
43. I. L. Hay and A. Keller, Kolloid-Z. Polymere $\underline{204}$, 43 (1965).
44. R. Yang and R. S. Stein, J. Polymer Sci. A2, $\underline{5}$, 939 (1967).
45. T. Hashimoto, "Static and Dynamic Light Scattering Study of Crystalline Polymer Films," Ph.D. Thesis, University of Massachusetts, Amherst, Massachusetts, 1970.
46. R. S. Stein, Accounts of Chemical Research, $\underline{5}$, 121 (1972).
47. A. Keller and J. R. S. Waring, J. Polymer Sci. $\underline{17}$, 447 (1955).
48. M. Rhodes and R. S. Stein, J. Polymer Sci. $\underline{62}$, S87 (1962).
49. G. Adams and R. S. Stein, J. Polymer Sci. A2, $\underline{6}$, 31 (1968).
50. M. B. Rhodes and R. S. Stein, J. Polymer Sci. A2, $\underline{7}$, 1539 (1969).
51. M. Moritani, N. Hayashi, A. Utsuo and H. Kawai, Polymer Journal (Japan), $\underline{2}$, 74 (1971).
52. R. S. Stein and D. Yoon, in press.
53. P. Debye and A. M. Bueche, J. Appl. Phys. $\underline{20}$, 518 (1949).
54. M. Goldstein and E. R. Michalek, J. Appl. Phys. $\underline{26}$, 1450 (1955).
55. R. S. Stein and P. R. Wilson, J. Appl. Phys. $\underline{33}$, 1914 (1962).
56. R. S. Stein and T. Hotta, J. Appl. Phys. $\underline{35}$, 2237 (1964).
57. R. S. Stein, J. J. Keane, F. H. Norris, F. A. Bettelheim and P. R. Wilson, Ann. N. Y. Acad. Sci. $\underline{83}$, Art. 1, 37 (1959).
58. R. S. Stein, P. Erhardt, S. Clough and G. Adams, J. Appl. Phys. $\underline{37}$, 3980 (1966).
59. J. J. van Aartsen in POLYMER NETWORKS, STRUCTURAL AND MECHANICAL PROPERTIES, Ed. by A. J. Chompff and S. Newman, Plenum Press, New York, 1971, p. 307.
60. T. Hashimoto and R. S. Stein, J. Polymer Sci. A2, $\underline{8}$, 1503 (1970).

LIGHT SCATTERING BY ORIENTED NATIVE CELLULOSE SYSTEMS

R.H. Marchessault

Département de Chimie

Université de Montréal

INTRODUCTION

Many biological materials are expected to correspond to the rod-like model for purposes of characterization via solid state light scattering. Collagenous and cellulosic membranes are well-known examples (1,2,3,4). These materials have been extensively characterized by light and electron microscopical techniques but only recently have solid state light scattering studies been added to the techniques which can provide textural information. This non-destructive approach has been found particularly fruitful for thin cellulosic membranes such as are found as cuticles, or surface slimes or cell walls of certain plant or algal material. The full scope of how the light scattering approach will be used on these systems is not entirely clear. For example, it can provides evidence concerning the average length of the rods (2), it can be used to get the average longitudinal and transverse refractive indices of rod collections (5), it can provide information on the orientation distribution function and how it changes under the influence of external forces (2,6).

Some preliminary studies on cellulosic systems have shown that H_v scattering on thin membranes yield scattering envelopes which are in agreement with what would be expected for rod-like systems (2,3,7). However, sometimes the scattering behaviour is rather unexpected as the following examples will illustrate and have suggested some areas where new theoretical considerations shou l be undertaken.

THEORY

The scattering of light by anisotropic rods has been studied by Rhodes and Stein (6) who established the basic theoretical equations corresponding to the simple model of an infinitely thin anisotropic rod of length L. In the case of cellulosic systems, one may usually assume that the rods lie in a plane perpendicular to the incident beam (3). Each rod axis makes an angle α to the vertical. We have made direct use of the Rhodes-Stein (6) equations and geometry for calculating theoretical scattering corresponding to rod-like systems (5). Thus, the scattered intensity I_s is deduced from the calculated amplitude by squaring. The observed value of I_s depends on the intensity of the incident beam, I_o, and the distance, x, between the scattering medium and the point of observation has to be taken into account. For one rod lying at angle α to the vertical, with the coordinate system at the center of the rod, the ratio of scattered intensity to incident intensity is proportional to:

$$\frac{1}{x^2} \left(\frac{A_s}{A_o} \right)^2$$

For small scattering angles, x can be taken as constant for all points in a plane of observation normal to the incident beam and is essentially the "film to sample" distance in a flat film photographic recording arrangement.

A collection of N_o rods can be described by a distribution function: $N(\alpha)d\alpha$ which gives the number of rods lying between angles α and $\alpha + d\alpha$. The intensity is then calculated from:

$$\frac{I_s}{I_o} \propto \frac{1}{x^2} \int_0^\pi N(\alpha) \left(\frac{A_s}{A_o} \right)^2 d\alpha \qquad\qquad 1$$

for the collection of N_o rods. If the effect of orienting the collection of N_o rods is to be studied quantitatively, the distribution function must be such that

$$\int_0^\pi N(\alpha)d\alpha = N_o \qquad\qquad 2$$

in all cases. The following function:

$$N(\alpha) \quad \frac{N_o}{\pi} \left(\frac{1}{C \sin^2\alpha + (\cos^2\alpha/C)} \right) \qquad\qquad 3$$

can be shown to obey this condition; C = 1 corresponds to the random distribution, C < 1 to a preferred direction $\alpha = (\pi/2)$ and

$C = 1$ to a preferred direction $\alpha = 0$.

A preferred orientation symmetrical about direction α_s would be described by:

$$N(\alpha) = \frac{N_o}{\pi} \left\{ \frac{1}{C\sin^2(\alpha - \alpha_s) + [\cos^2(\alpha - \alpha_s)/C]} \right\} \qquad 4$$

It should be noted that, in such a case, the intrinsic birefringence of the sample imposes a correction to the expression for scattered intensity (8), the magnitude of which is a function of the birefringence itself.

For the case where the total number of rods N_o is divided into two fractions, A and B, which have, respectively, symmetrical orientations about α_s and $\alpha_s = \pi/2$), the general expression for the distribution function would be:

$$N(\alpha) = \frac{N_o}{\pi} \left\{ \frac{A}{C_1\sin^2(\alpha - \alpha_s) + [\cos^2(\alpha - \alpha_s)/C_1]} \right.$$

$$\left. + \frac{B}{C_2\sin^2(\alpha - \alpha_s) + [\cos^2(\alpha - \alpha_s)/C_2]} \right\} \qquad 5$$

where $A + B = 1$. This case is almost found in one of our examples. However, we obviously require yet another equation where the value of the angle between the two symmetrical distributions is any value.

RESULTS AND DISCUSSIONS

In a recent paper, Charrier and Marchessault (5) have made a thorough evaluation of influence of rod length, surrounding polarizability and orientation on the shape of the V_v and H_v scattering patterns. The theoretical predictions (5,6), both as regards the fundamental rod parameters and the predictions of the above-stated distribution function, were tested using a novel system vig. condenser paper which is a thin oriented system of cellulose fibrils. This type of paper is so highly refined that the individual fibers of cellulose are almost non-existent and only bundles of the fibrillar elements are the structural components. The thickness is ~5μ and the slight orientation while detectable makes it unnecessary to correct for the effect of birefringence on the scattering properties (8). Furthermore, the sample being microporous one may change the refractive index of the imbibition medium and see that the predicted effect of changing the refractive index of the surroundings is truly observed for V_v scattering (5) (Fig. 1).

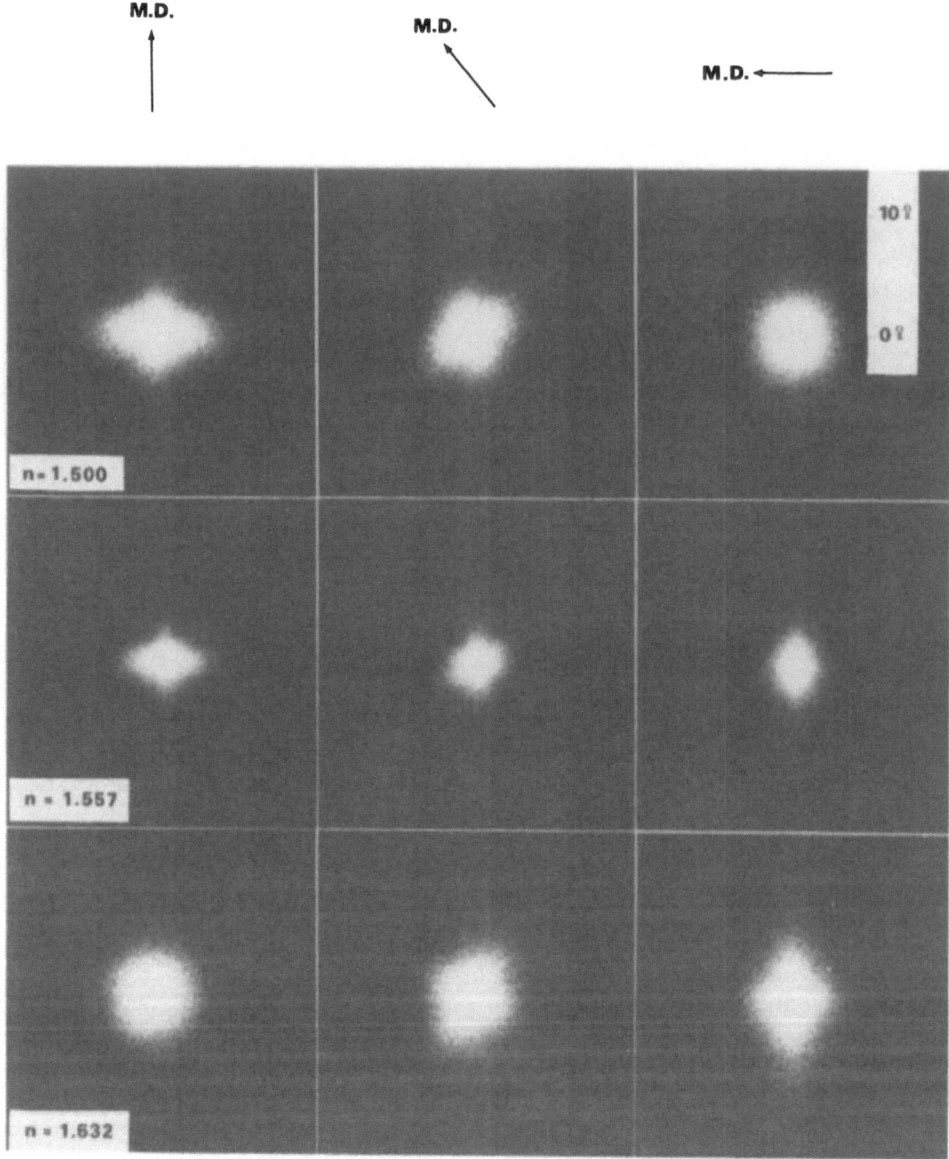

Fig. 1.   Experimental V_v scattering patterns for condenser paper;
          three orientations of the machine direction (M.D.) are
          shown for three different imbibition fluids (index of
          refraction n). Note the change in the scattering enve-
          lopes as a function of n for each particular orientation
          of M.D.

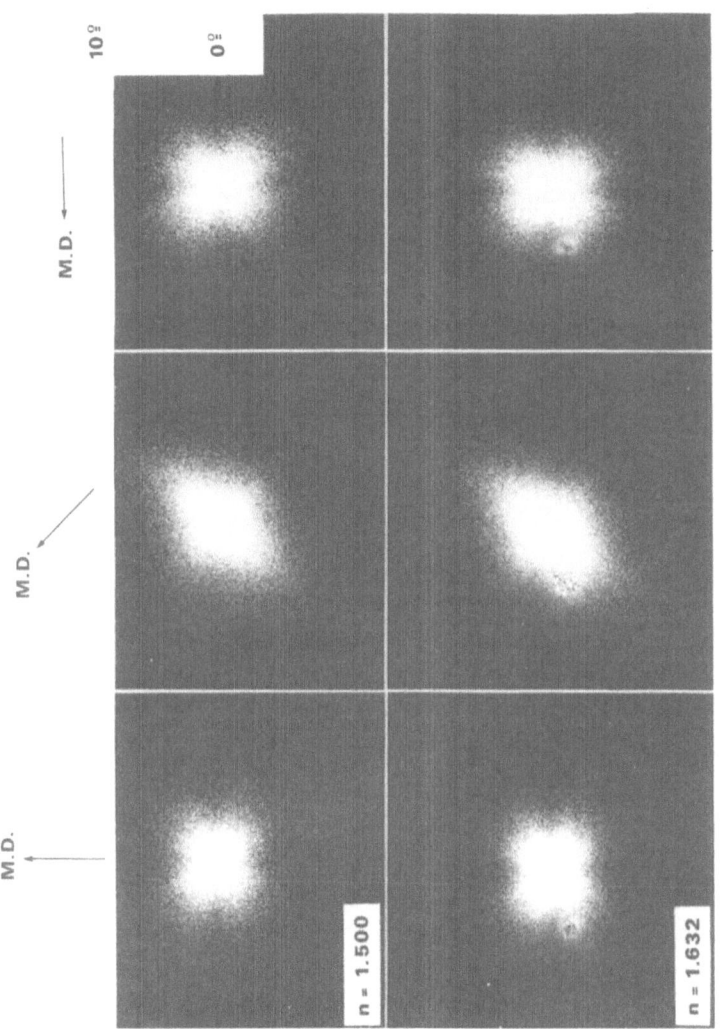

Fig. 2. Experimental H_V scattering patterns for condenser paper; three orientations of the machine direction (M.D.) are shown for two different imbibition fluids (index of refraction n). Note that the scattering envelope is invariant with n for a given M.D. orientation.

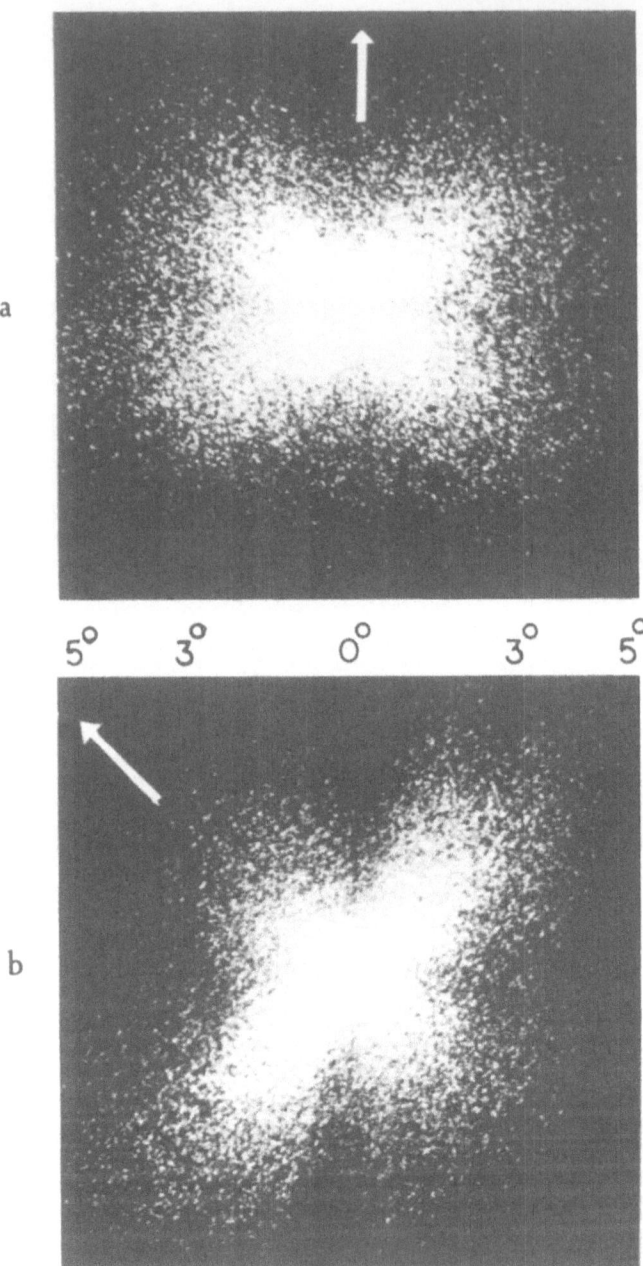

Fig. 3. Photographic H$_V$ scattering patterns of condenser paper.
Polarization of the laser beam is vertical. Machine direction is indicated by white arrows.

On the other hand, Fig. 2 shows that refractive index has no effect on H_v scattering as would be expected if this scattering mode is sensitive only to fluctuations in anisotropy (5).

In using condenser paper as a model for a rod-like system, it is important to remember that this material is made on a machine which usually induces orientation into the system. This is best shown by enlarged views of two of the sections from Fig. 2 for an imbibition medium with n = 1.50 (cf. Fig. 3). Clearly when the machine direction is vertical, the usual X-shaped pattern, with orthogonally related scattering arms as expected for a random rod pattern, is lacking. Rather, the scattering streaks approach the horizontal axis (the X-shape is flattened). This is a sign of prefered orientation in the machine direction. As predicted (3,5), when the machine direction is at 45°, the scattering streaks are orthogonal but of unequal intensity. The information in Fig. 3 has been used to show (3) that the sample has an average rod length of 80,000Å and an orientation parameter C ≃ 1.6.

The information derived from the model experiments with condenser paper were invaluable in interpreting the patterns derived from animal cellulose membranes (Tunicin). The scattering pattern shown in Fig. 4a is typical of one which was obtained periodically as one probed various areas of membrane with the 1 mm. diameter laser beam in the usual photographic light scattering arrangement (2). The two orthogonal scattering arms at the 45° azimuthal position seemed to indicate large, well-oriented regions. However, this orientation had to be of a local nature as the membrane, although anisotropic in the polarizing microscope presented no overall birefringence. Surface replicas examined in the electron microscope (Fig. 4b) testified as to the accuracy of this interpretation which led to a proposed model (Fig. 5) concerning the organization of the cellulose microfibrils in such membranes. More recent electron microscope studies on thin section of tunicates have confirmed this model (9). It seems that a hydrated system of oriented ribbon-like aggregates of cellulose microfibrils interdigitate to form a cuticle. This structure bears a certain similarity to leather and represents a novel use of cellulose in nature.

Aggregation of cellulose microfibrils into well-oriented or randomly oriented structures have been the more commonly observed systems to date. Thus the electron micrograph shown in Fig. 6 shows the classical construction of cellulose microfibrils which are found in the membranous slime which develops at the air-water interface when apple cider goes sour ("ropy" cider). Indeed the rope-like nature of the structural elements in the membrane is quite obvious. Such membranes lead to scattering envelopes (2) which are associated with a random collection of anisotropic rods (Fig. 7a). When the membranes have not been deformed, any area

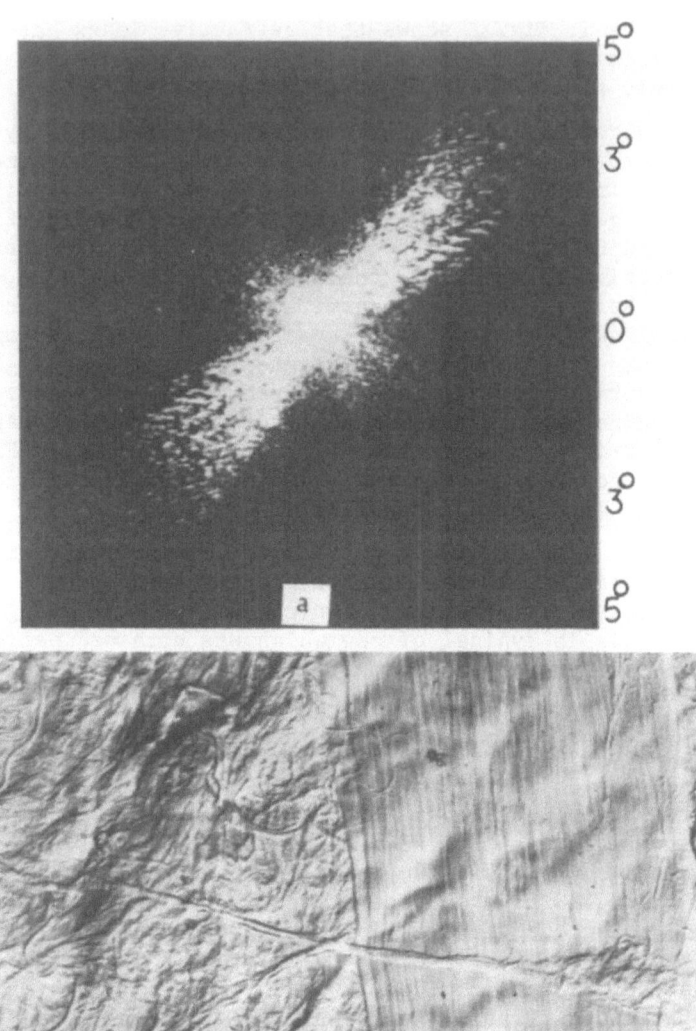

Fig. 4. Observations on <u>Tunicin</u> cellulose:(a) H_v experimental light
        scattering from a region which corresponds to a predominant
        fibrillar orientation at 45°, the polarization plane is
        vertical; (b) chromium shadowed electron micrograph of sur-
        face replica of <u>Tunicin</u> membrane, a zone of highly corre-
        lated orientation of cellulose microfibrils is clearly
        evident.

Fig. 5. Proposed model of <u>Tunicin</u> textural organization, the
dotted areas represent end on views of microfibrillar
aggregates.

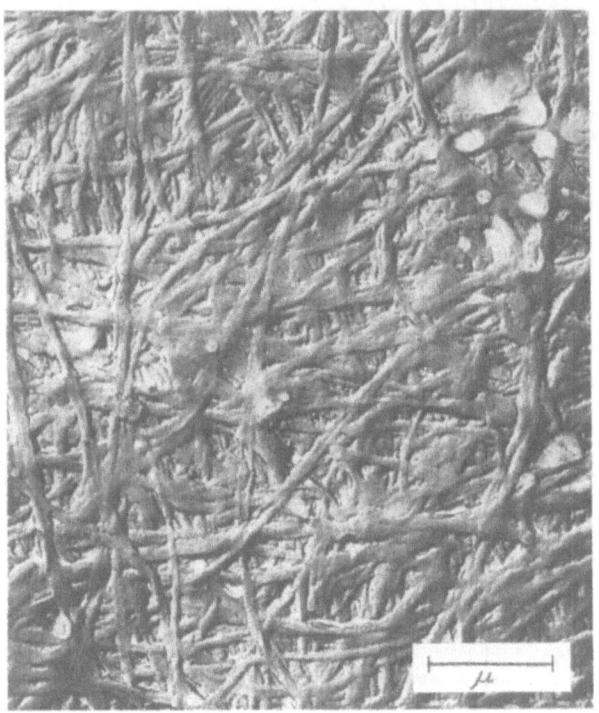

Fig. 6.   Chromium shadowed surface replica of bacterial cellulose
          membrane, the ropy microfibrillar aggregates are more
          or less randomly oriented in a plane.

will give rise to the pattern shown in Fig. 7a. By contrast, the
class of algae known as <u>Valonia ventricosa</u> gives rise to a unique
type of pattern which is in keeping with its well-known laminated
structure (10). This membrane acts as the exterior reenforcement
for the true cell-membrane, in this salt water species, and is
called upon to resist the swelling pressures which develop due to
differences in osmotic pressure inside and outside the cell. Electron microscopists have shown the beautiful cross-ply organization
of the cellulose microfibrils in successive layers (10). The pattern shown in Fig. 7b corresponds to two oriented fibrillar layers
which, classically, make an angle of about 80° with each other.
The non-orthogonal orientation of the two layers is reflected directly in the mutual orientation of the two scattering strecks
each one being associated with a given layer. The narrowness of
the streaks is to be associated with the great length of the rods
and the diffraction grating effect of the near perfect alignment
of microfibrils in a given layer.

CONCLUSIONS

The theory usually applied in studies of this kind is based
on the assumption that there is no correlation in the orientation
of neighboring rods. This may be a reasonable assumption for some
systems but in general for paper and many other cellulosics, especially of biological origin, the opposite is true. The effect of
non-random orientation in the case of spherulitic films has been
clearly demonstrated. Among other thinkg, it leads to a visible
speckling in the scattering pattern, as is visible in Fig. 3 but
is much less in evidence in Fig. 6. From electron micrographs of
condenser paper such as studied in this paper, there is clearly
such a high level of correlated orientation that light scattering
can only serve as a qualitative or semi-quantitative tool. This
is nevertheless quite useful in cases where one cannot clearly
detect a macroscopic orientation (machine direction) as in the
case of condenser paper or when a clean bidirectional or multidi-
rectional orientation is present. When simple distribution functions of the type shown in the above equations apply, it is also
possible to characterize the extent of orientation. This has been
reported for condenser paper (3) and there is evidence of a growing
interest in the use of light scattering for control and research
applications in the field of papermaking (12,13). Two recent studies have focussed on the scattering to be expected from single
fibers of cellulose (15,16).

The challenge of reconstituting the full texture of materials
with rod-like components and possessing correlated orientation is
an interesting one. The pattern (Fig. 3) contains all the information required except for the phase relation between domains of
correlated orientation. However, there are so many ways of recording the patterns that it should be possible to derive even the

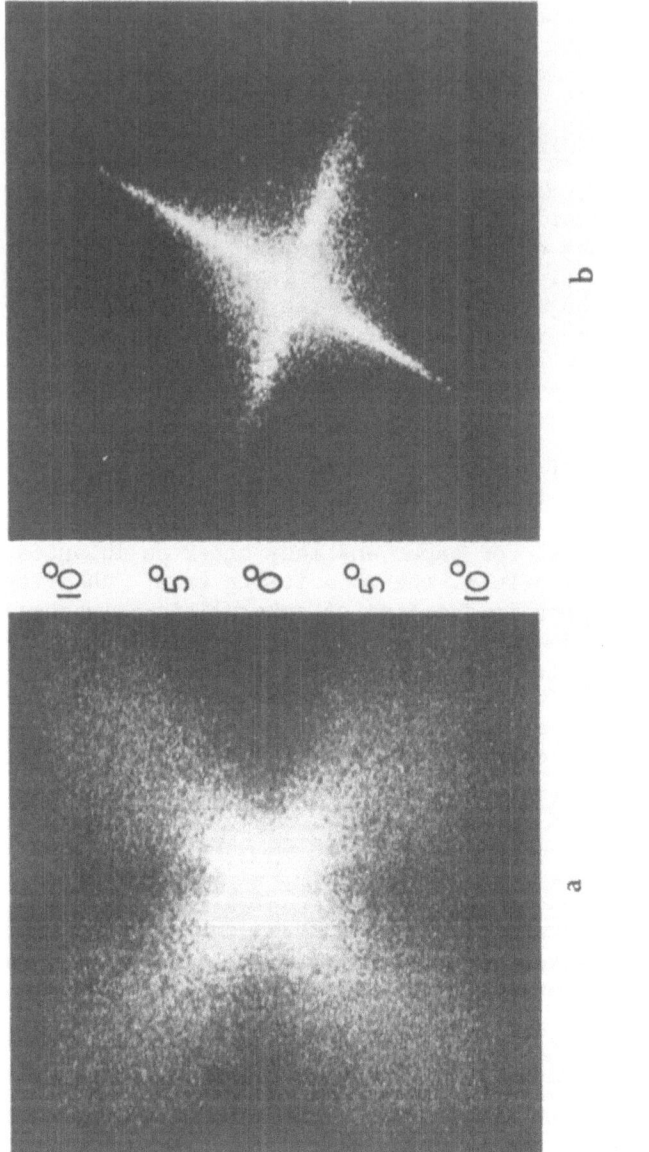

Fig. 7. Experimental H_V scattering for: (a) bacterial cellulose membrane; (b)
Valonia ventricosa membrane, the latter has a crossply structure with
a unique fibril orientation in each ply.

phase relationship between regions.

BIBLIOGRAPHY

1. M. Moritari, N. Hayashi, A. Utsuo and H. Kawai, Polymer Journal, 2, 74 (1971).

2. J. Borch and R.H. Marchessault, J. Polymer Sci. Part C-28, 153 (1969).

3. P. Lim, A. Sarko and R.H. Marchessault, TAPPI, 53, 2314 (1970).

4. G. Wilkes, American Chemical Society, Div. of Organic Coatings and Plastic Chemistry, Preprints 32, 35 (1972).

5. J.M. Charrier and R.H. Marchessault, Fibre Sci. and Technology, 5, 157 (1972).

6. M.B. Rhodes and R.S. Stein, J. Polymer Sci., Part A-2, 7, 1539 (1969).

7. J. Borch, P.R. Sundararajan and R.H. Marchessault, J. Polymer Sci., A-2, 9, 313 (1971).

8. W. Chu and R.S. Stein, J. Polymer Sci., A-2, 8, 489 (1970).

9. A.B. Wardrop, Latrobe University, Australia, private communication.

10. A. Frey-Wyssling and K. Muhlethaler in Ultrastructural Plant Cytology, Elsevier, Amsterdam-London-New York (1965).

11. C. Picot, R.S. Stein, R.H. Marchessault, J. Borch and A. Sarko, Macromolecules 4, 467 (1971).

12. L. Rudstrom and U. Sjölin, Svensk Papperstdn., 73, 117 (1970).

13. D.J. Williams, Appita 24, 196 (1970).

14. P.R. Sundararajan and R.H. Marchessault, J. Mol. Biol. 63, 305 (1972).

15. R. Muggli, A. Sarko and R. Marton, "Proceedings of Seventh Cellulose Conf.", J. Polymer Sci., Part C, in press.

SUPERSTRUCTURE IN FILMS OF BIO AND BIORELATED POLYMERS AS NOTED BY

SMALL ANGLE LIGHT SCATTERING.

Garth L. Wilkes[†] and Ban The Vu[*]

Department of Chemical Engineering, Princeton

University, Princeton, New Jersey 08540

INTRODUCTION

Recently there has been considerable interest generated in understanding the material properties of many bio or bio-related polymers. The origin of this interest is broad and extends from their material application (enzyme supports, membranes and bio-materials) to that of understanding the mesophase or liquid crystalline behavior common to many of these systems. Although considerable information exists concerning the solution properties of various polypeptides, there have been relatively few investigations aimed at correlating properties with morphology and structure of solids made from these. A particular exception to this is collagen, a number of studies having been directed at understanding its mechanical behavior. Even in this case, however, there is still need of further structure-property correlations.

Within a number of our own studies involving various bio or bio-related macromolecules, we have utilized the small angle photographic light scattering technique to learn about the superstructure induced within solution cast films of these polymers. This paper will attempt to correlate the results from all of these studies and will also include some more recent work on the application of the scattering technique to a biological tissue. In some cases, other common techniques such as x-ray diffraction and microscopy have also been used to aid in the interpretation of our scattering results.

---

†To whom correspondence should be sent.
*A portion of this work was submitted as a BSE Thesis.

In a general sense, each of the synthetic polypeptides (SP), fibrous proteins (e.g., collagen) and polynucleotides (e.g., DNA) can be basically considered as a rod-like molecule if in its helical form. The reader should recall that the synthetic polypeptides tend to form a single alpha helix, DNA forms a double helix while the tropocollagen molecule consists of a triple helix structure. Because of their rod-like character, these systems, particularly the synthetic polypeptides, tend to display mesophase behavior and for this reason interest has also grown in understanding their liquid crystal character. Our own interest however centers more on how the morphology and texture change as a mesophase solution loses solvent and approaches a solid form, and particularly, what the final texture is within the film.

These peptide materials differ in general from the majority of monomeric mesophase systems in that the SP systems illustrate lyotropic mesomorphism rather than thermotropic. That is, the transitional behavior depends on solvent and the related solution parameters of pH, concentration, and solvent-solute interaction. The kinetics and structural behavior of these mesophases are also generally influenced by the presence of electromagnetic field gradients, by the surface-volume ratio, and by the "type" of surface in contact with the mesophase system. The influence of electromagnetic fields on mesomorphism might, in fact, be denoted as electrotropic.

A number of investigations of SP systems and their lyotropic or electrotropic behavior have been reported. Robinson (1) has been foremost in this area and has found considerable microscopy evidence for the occurrence of cholesteric behavior of high molecular weight PBLG in helix forming solvents such as chloroform. Samulski (2) has extended these studies to consider the morphology of PBLG[*]films formed by casting from different solvents. He has presented some evidence based on x-ray diffraction measurements for the existence of a nematic mesophase in the solid state as well as in highly swollen or plasticized PBLG films. He also considered the effects of magnetic fields (during casting) on the final film morphology. Others (3) have also studied the electrotropic effects on PBLG with respect to its fluid behavior. Characterization parameters used have been viscosity, dielectric anisotropy, birefringence, etc.

With our desire to learn about the morphology and superstructure within solid films of these polymers, we have utilized the low angle photographic light scattering technique. This technique has been extensively employed to study the morphology of common synthetic polymers, particularly the semicrystalline polyolefins prone to spherulitic superstructure. Because of our emphasis on the scattering technique, it seems appropriate to review a few of

[*]poly-γ-benzyl-L-glutamate

the basic principles of scattering and its applicability in studying superstructure.

ORIGINS OF SCATTERING

Light scattering arises primarily from four possible sources:
1) density fluctuations; 2) fluctuations in directional orientation of the optic axes in local anisotropic regions; 3) fluctuations in the local anisotropy but with no correlation between local regions and 4) optical rotation. The first can arise in optically isotropic systems whereas the others require the presence
of optical anisotropy. Since many mesophase systems give rise to
optical rotation, the latter source can be of particular importance.

It is well known that in an isotropic medium the polarization
of the scattered ray at small scattering angles is essentially
identical to that of the incident beam if the Rayleigh-Gans treatment can be applied.* It therefore follows that no scattered radiation is transmitted through a second polarizer (analyzer) if it
has its electric vector rotated at 90 degrees to the electric vector of the polarized incident beam. The presence of optical anisotropy, however, can give rise to the transmission of scattered
light when the polarizer and analyzer are crossed. Interpretation
of the observed or measured scattering leads to information regarding superstructure that arises due to preferential molecular
packing, aggregation or crystallization. By rotation of the analyzer one can attempt to separate the isotropic and anisotropic
contributions. Quantitative interpretations can be undertaken
either by the use of statistical approaches (4,5) or by the comparison of the scattering patterns calculated from definite geometrical models (6,7). For our purposes here, the use of theory
based on definite models will be sufficient although not completely satisfactory.

SCATTERING FROM ANISOTROPIC RODS

Rhodes and Stein (8) have calculated the H_V[†] and V_V[†] scattering patterns expected based on the scattering by independent infin-
itesimally thin anisotropic rods in various planar configurations
i.e., oriented or random. The equations for the scattered V_V and

---

*Applicability of this approach is in order when the refractive
index difference between the scattering object and the surroundings is small and, when the dimensions of the scattering object
are of the order of the wavelength of the incident radiation. This
treatment is the basis for the scattering theory discussed in this
paper.

†See footnote next page.

H_V intensities per unit of scattering volume have been given as

$$I_{V_V} = \rho_o^2 N_o L^2 \int_{\beta=0}^{\pi} (\varepsilon^{-2}\cos^2\beta \; \varepsilon^2\sin^2\beta)^{-1/2} (\delta\cos^2\beta' + \alpha_t)^2 \cdot$$

$$\left(\frac{\sin(\frac{kaL}{2})}{(\frac{kaL}{2})} \right)^2 d\beta \qquad\qquad (1)$$

$$I_{H_V} = \rho_o^2 N_o^2 L^2 \int_{\beta=0}^{\pi} (\varepsilon^{-2}\cos^2\beta + \varepsilon^2\sin^2\beta)^{-1/2} \delta^2\beta' \cos^2\beta' \cdot$$

$$\left(\frac{\sin(\frac{kaL}{2})}{(\frac{kaL}{2})} \right)^2 d\beta \qquad\qquad (2)$$

where N_o, L and ρ_o refer to the number, N_o, of rods per unit of volume each being of length L and having a scattering power per unit length of ρ_o. The optical anisotropy of the rod, δ, is defined to be

$$\delta \equiv \alpha_\ell' - \alpha_t' \qquad\qquad (3)$$

where

$$\alpha_\ell' = \alpha_\ell - \alpha_s \qquad\qquad (4)$$

$$\alpha_t' = \alpha_t - \alpha_s \qquad\qquad (5)$$

The subscripts ℓ, t and s respectively refer to polarizability in the longitudinal and tangential direction of the rod and, of the surroundings. The parameter k is $2\pi/\lambda$ where λ is the wavelength of radiation in the medium. The first term in parentheses in each equation has the form of an elliptical orientation distribution function where ε refers to an orientation parameter. If ε is unity, the distribution of rods is circularly symmetric. It is assumed in the theory that all the rods lie in a plane which is normal to the incident beam. This assumption as applied to cast SP films seems reasonable based upon the x-ray and swelling data of Samulski (2) as measured on films of PBLG and our own x-ray data for PBDG. We have also noted similar planar orientation in cast reconstituted collagen films. The angle β is the angle that the rod axis makes with the vertical axis while β' is equal to $\beta +$ ω where ω is the angle that the *optic* axis of the rod makes with the *rod* axis. The final parameter a is a function of the radial

---

†H_V refers to a horizontal (H) and vertical (v) alignment of the electric vectors of the analyzer and polarizer respectively. If a stretch axis is present, it would be aligned along the v direction. Similarly a V_V orientation refers to both electric vectors being aligned in a vertical direction. The stretch axis would also be parallel to these vectors. In an H_h position, the stretch axis would be perpendicular to both analyzer and polarizer vectors.

and azimuthal scattering angles as well as being dependent upon β. Further comments on these equations and their limitations will follow shortly. It is in order to mention that a more general theory of scattering from rods has been presented by Van Aartsen (9).

Typical H_v and V_v scattering contour plots as calculated from use of eq. (1) and (2) are shown in Fig. 1 where three different values of the orientation parameter, ε , have been used. The various values of the other mathematical parameters are given in the figure caption. It should be stated here that patterns la and ld can be rotated 45° if the optic axis parameter (ω) is allowed to lie along the rod axis rather than at 45°. Patterns have been observed from films of synthetic polymers which display both types of scattering behavior (8,10). Further applicability of this model has been illustrated by Samuels (10) in his study of hydroxypropylcellulose in which he clearly showed that by using an appropriate orientation distribution function for the rods, the scattering behavior from oriented films agreed excellently with the theory.

SCATTERING FROM ANISOTROPIC SPHERES

Polymers, in general, tend to form spherulites during the onset of crystallization. This leads typically to the formation of a spherical, anisotropic superstructure with a characteristic radius ranging from a fraction of a micron to hundreds of microns. The exact size depends on various molecular parameters and on the crystallization kinetics. The size, anisotropy and nature of the spherulite, can be studied utilizing well established light scattering theory (6,11). Examples of theoretical H_v and V_v spherulite scattering patterns are given in Fig. 2 as calculated from eq. (6) and (7) given below.

$$I_{V_v} = AV^2{}_o(\frac{3}{U^3})^2[(\alpha_t-\alpha_s)\ (2\sin U-U\cos U-SiU)$$

$$+ (\alpha_r-\alpha_s)\ (SiU-\sin U)-(\alpha_t-\alpha_r)(\cos^2(\frac{\theta}{2})\cos\mu \ \cdot$$

$$(4\sin U-U\cos U-3SiU)]^2 \tag{6}$$

$$I_{H_v} = AV^2{}_o(\frac{3}{U^3})^2[(\alpha_t-\alpha_r)\cos^2(\frac{\theta}{2})\sin\mu\cos\mu \ \cdot$$

$$(4\sin U-U\cos U-3SiU)]^2 \tag{7}$$

where

$$SiU = \int_0^U \frac{\sin x}{x}\ dx \tag{8}$$

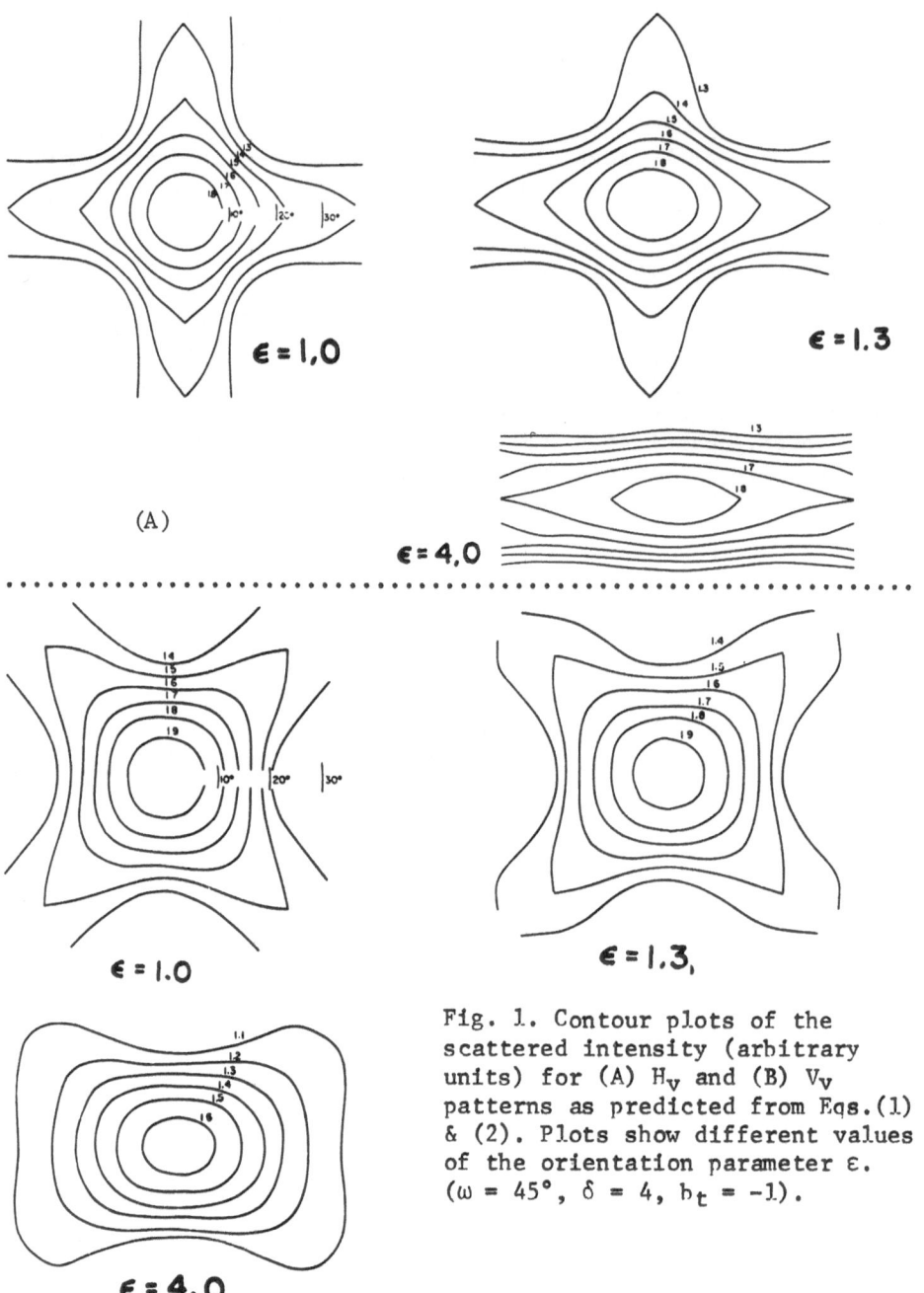

Fig. 1. Contour plots of the
scattered intensity (arbitrary
units) for (A) H_v and (B) V_v
patterns as predicted from Eqs.(1)
& (2). Plots show different values
of the orientation parameter ε.
($\omega = 45°$, $\delta = 4$, $b_t = -1$).

and

$$U = \frac{4\pi R}{\lambda} \sin\left(\frac{\theta}{2}\right) \tag{9}$$

In the above α_t and α_r represent the respective tangential and radial polarizabilities of the anisotropic sphere of radius R imbedded in an isotropic matrix having a uniform polarizability of α_s. Parameters A, V_0, μ and θ are respectively an instrumental constant, the volume of the sphere, and the azimuthal and radial scattering angles. The wavelength in the medium is λ.

It is noted that in both eqs. (2) and (7) the H_v intensity depends only on an anisotropy term $(\alpha_t-\alpha_r)$ or δ whereas the V_v intensity is composed of both isotropic and anisotropic contributions. In the spherulite model the surrounding matrix has been treated as having a uniform polarizability of α_s. This assumption could plausibly be in error. Rather more realistic is that there is a distribution of α_s's caused by local density fluctuations. These density fluctuations which also give rise to V_v scattering are not accounted for by the model. If there is a significant contribution to the total V_v intensity by such a distribution, then the V_v pattern predicted by the *model* may be buried in the intensity caused by the unaccounted for matrix density fluctuations. This means that an H_v pattern may be nearly in agreement with the model calculations as given here but at the same time, the V_v pattern may not display the appropriate symmetry. In the rod model a similar situation exists in that a uniform matrix polarizability, α_s, is assumed. Again, this does not consider any effects of other density fluctuations on the total V_v scattering. Hence, as in the spherulitic case, one may experimentally observe a "good" H_v rod pattern but not a "good" V_v pattern. These considerations should be kept in mind by the reader for they are useful in interpreting the results presented. It might be added that the rod model or the anisotropic sphere model, as presented here, do not consider interparticle effects. Stein *et al* (12) have shown that interparticle interference effects do not drastically alter the scattering behavior for the sphere model. Although we are not aware of similar calculations for the rod system we anticipate that the effects would be of a secondary nature.

MATERIALS AND PREPARATION OF FILMS

All synthetic polypeptides, with the exception of poly-γ-methyl-L-glutamate (PMG), were obtained from Pilot Chemicals (Boston, Mass.). The PMG was obtained from Johnson and Johnson and came in film form. The films of reconstituted collagen were supplied through the courtesy of Biomedix Laboratories, Inc., Princeton, N. J.

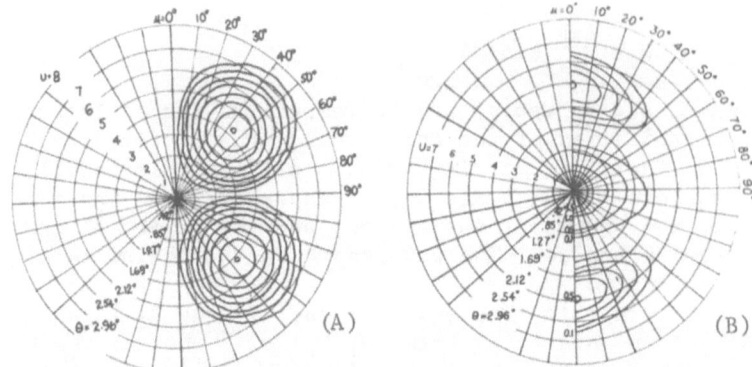

Fig. 2. Contour plots of the scattered intensity (arbitrary units) for one half of the H_V and V_V patterns as predicted from an undeformed anisotropic sphere (Eqs. 6 & 7). a) H_V; b) V_V [From Stein *et al* (6)].

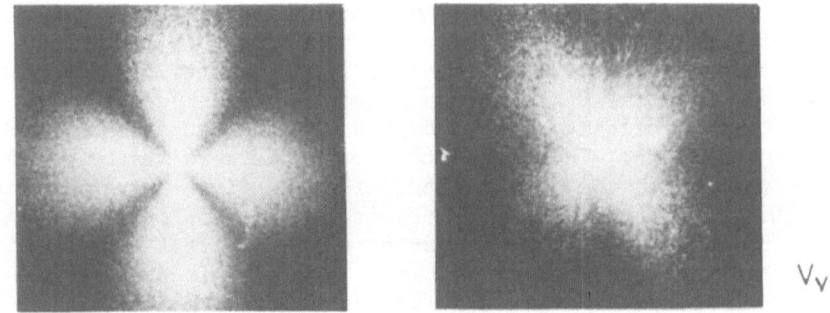

Fig. 3. H_V and V_V scattering patterns for chloroform cast films of PBLG on glass. Sample to film distance (STF) was 9 inches.

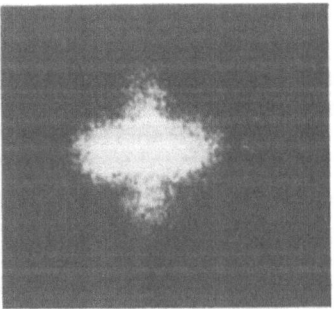

Fig. 4. H_V scattering pattern obtained from a "pool" of PBLG solution placed on a glass substrate.

The PMG film was dissolved in chloroform and films were cast on a glass substrate. A similar procedure was used for the PBLG, PBDG, PBDLG and poly-γ-ethyl-L-glutamate (PEG) material. The polylysine-HBr (PL) was cast from a 20 weight percent aqueous solution on glass. Poly-γ-L-glutamic acid (PGA) films were cast from pure dimethyl formamide (DMF) and pure dioxane. The procedures for preparing the collagen films have been reported elsewhere (13, 14). Purified calf thymus DNA (Worthington) was cast in film form from a distilled unionized aqueous media.

RESULTS AND DISCUSSION

Undeformed Synthetic Polypeptides

Figure 3 shows the H_v and V_v patterns taken on PBLG films cast from chloroform solutions that were not older than one month. Fresh and one year old solutions gave identical results. The patterns suggest rod-like scattering (compare with Figures 1a and d) rather than that of spherulitic character. The V_v pattern is not as "perfect" as Fig. 1d but this may be due to the superposition of background scattering from matrix density fluctuations. Identical patterns were found for PBDG. In the case of the copolymer PBLDG, the H_v pattern was clearly larger but obviously weaker in intensity than those from either the pure L or D form for the same sample to film distance. This suggested that the rod-like aggregates were somewhat smaller than those in the pure L or D systems.

The H_v pattern shown in Fig. 4 was taken from a "pool" of the PBLG solution placed on a glass slide. That the H_v 0°-90° rod-like pattern is still apparent is significant and along with the smaller pattern (relative to Fig. 3) implies there is a larger superstructure responsible for the scattering* or that there is simply a lower concentration of the scattering elements. Most likely, the second speculation is the more reasonable since it is really only the top surface layer of the pool that is responsible for the scattering. After a pool of the clear solution was placed on the glass substrate, a whitish cast appeared on the upper surface in

---

*The reader recalls that scattering angle and scattering element "size" are reciprocally related. Based on the sample film distances used for Figs. 3 and 4 it is clear that smaller scattering angles occur in Fig. 4. Caution must be taken however when trying to compare "sizes" of scattering elements where rod scattering is involved. This is due to the fact that monotonic dropoff in scattered intensity exists for all rod sizes and clearly the size of the pattern is influenced by exposure times. A case where average size of scattering element can be determined from the scattering pattern is that of the spherulitic structure.

a very short time and only when this layer was present could the
cross H_v pattern be obtained. The scattering intensity has, in
fact, been enhanced when a draft of air was passed over the glass
substrate. This implies that the difference in the H_v patterns
between Fig. 4 and Fig. 3 is due to a concentration effect of
scattering elements.

Figure 5 shows the typical H_v and V_v patterns for the PMG
films cast from DMF. Figure 6 shows the H_v and V_v patterns for
PEG cast from chloroform and PGA from DMF. In all the above cases
a 0-90° H_v cross is again observed while the ±45 V_v cross also
suggests that the scattering arises from anisotropic rod-like
elements having a principal optic axis at ∿45° to the "rod" axis.
Figure 7 shows only the H_v pattern for the PL system. The ±45°
V_v pattern, although observed, was so highly buried in background
scatter that a suitable photograph of this pattern could not be
obtained.

In order to follow the development of superstructure as a
film was formed, a drop of PGA-dioxane solution was placed between two glass slides and allowed to dry slowly. The outer pe-
riphery of the sample (see Fig. 8) approached film form more rapidly than the inner region which lost solvent at a slower rate.
This resulted in two regions as noted by optical microscopy. The
H_v scattering pattern was taken from each region and the results
are shown in Fig. 9. It is seen that the inner or "wetter" region
displays a ±45 cross while the "dryer" region shows the presence
of the 0°-90° cross with a monotonic decrease in intensity characteristic of rod-like scattering. The inner pattern appears to
display a monotonic decrease in intensity as well but it is not
completely clear that this is the case thereby leaving some possibility open for its interpretation based on the spherulitic model.
We do not wish to discuss the details of this transformation (region 1 → region 2) at this time but present these data to show
that for this SP-solvent system there is indeed a transformation
in morphology and the final texture is again that of the 0°-90°
cross noted as in all the proceeding SP H_v patterns.

In light of the fact that *all* the SP systems 1) display the
0°-90° H_v cross typical of rod scattering and 2) are known to display mesophase behavior it is tempting to correlate the two phenomena.

Rod-like or thread-like elements are typical of the nematic
and smectic mesophases. Each of these mesophases requires a
principal optic axis, yet the light scattering patterns imply a
random orientation of anisotropic rods. Since the laser beam
"sees" a multitude of these oriented regions each of which may
have its optic axis oriented along a different direction, the

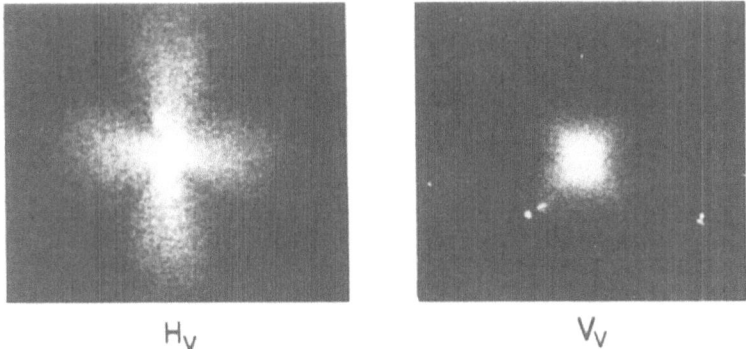

Fig. 5. H_v and V_v scattering patterns obtained from a chloroform cast film of PMG.

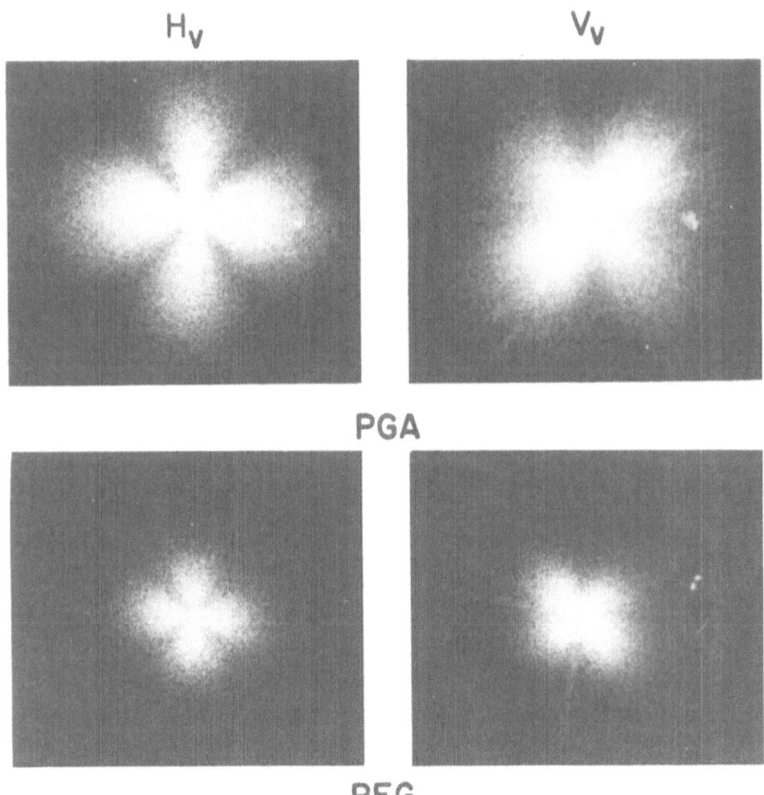

Fig. 6. H_v and V_v scattering patterns obtained from films of PGA cast from DMF and PEG cast from chloroform.

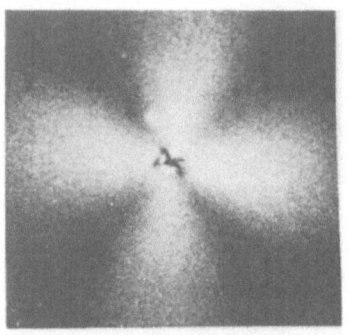

Fig. 7. H_v scattering pattern obtained from the polylysine preparation cast on glass.

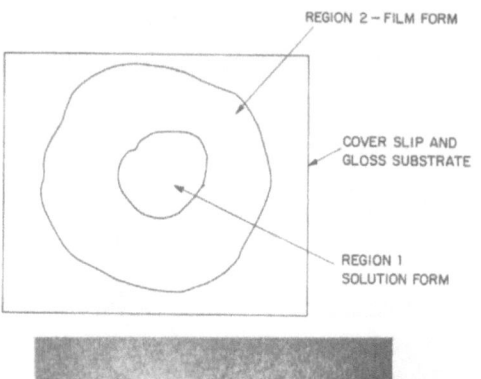

Fig. 8. Illustration showing Regions 1 and 2 of the
PGA-DMF preparation.

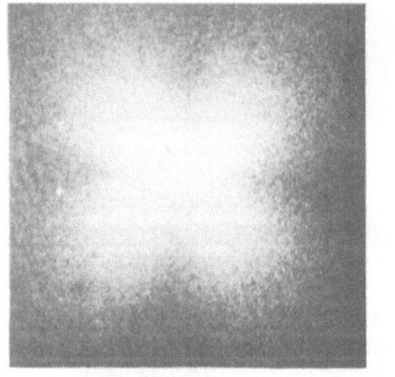
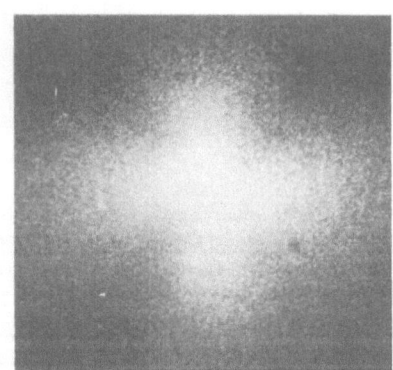

Fig. 9. H_v scattering patterns obtained on the PGA-DMF preparation
cast between two glass slides. a) Region 1 - still contains solvent; b) Region 2 - "dry".

apparent discrepancy is explainable. That is, we view the texture as being composed of randomly oriented anisotropic rods where each "rod" is itself composed of an oriented aggregate of peptide molecules. A model for this texture is shown schematically in Fig. 10a. Such a structure, however, cannot explain the existence of the ± 45° cross H_v pattern of Fig. 9. Assuming this pattern displays a monotonic decrease in intensity as a function of the radial scattering angles, it can be rationalized in terms of the model given in Fig. 10b where the "rods" or swarms are now composed of oriented bundles of peptides with their optic axes aligned parallel (or perpendicular)* to the "rod" axis†. These proposed structures are reasonable in view of the photomicrograph of PBLG films cast from chloroform by Elliot and Ambrose (15). Their photomicrographs, however, are of localized regions and hence it is not known if random orientation occurs over the cross section of the film. Our data offers further support for the presence of nematic structure in the solid film. The smectic phase has presently been ruled out based on x-ray diffraction studies by Samulski on similar films (2).

The formation of anisotropic systems both in solution and in film form is not unexpected. Anisotropic phase development in the liquid can arise through the onset of solution crystallization or, by ordered packing as discussed in the theories developed by Onsager (16), Isihara (17) and Flory (18). Flory's treatment is particularly straightforward and is based on equilibrium thermodynamics. His approach considers minimization of the free energy of a solution of rod-like solute particles. In such a solution, the development of an anisotropic phase results due to the decrease in molecular competition in the anisotropic phase for arrays of "lattice" sites suitable for placement of the solute particle in the lattice. In the case of highly anisotropic particles (high axial ratio) this decrease in competitive search for contiguous lattice locations lowers the system free energy more than does the increase caused by loss of configurational entropy. The consequences of this theory, although pleasing, should be applied with caution in our case since there is most likely considerable departure from equilibrium conditions i.e., rapid solvent evaporation during film preparation. This is particularly true where highly volatile solvents are used without a con-

---

*The H_v patterns cannot separate these two cases of parallel or perpendicularly oriented optic axes. To do so, it is necessary to make use of the V_v patterns. Since these were poor for the peptide films this separation of kinds of orientation could not be made based on these data.
†Both types of "rod" models do not eliminate the possibility of folding of the chains, rather, only the postulation is made concerning the orientation of the optic axes of the chains.

trolled vapor environment.

The occurrence of this anisotropic phase at the solution surface can result in a change in refractive index at or near the surface and a development of considerable density fluctuations due to
heterogeneities in molecular density. This can easily account for
the development of surface turbidity as observed in the "pool" of
solution on an exposed substrate as discussed earlier.

Flory's theory does not appear to place exact geometric requirements which the growing phase will follow. Apparent, however, is that the texture could be of a fibrillar or rod-like
character. The enthalpic effects and geometrical shape of the
asymmetric particles will likely determine the "shape" and packing
arrangements of the anisotropic phase. The "shape" of the anisotropic phase may in fact change as the concentration of the solute
increases due to volatilization of solvent. This speculation is
proposed based on the observations of Robinson (1) and Frankel *et
al* (19) for PBLG solutions where liquid spherulites were observed
initially but their collapse occurred as the solute concentration
approached 100 percent. The kinetic considerations are presently
lacking but it is suspected that the nucleation of these anisotropic phases (or "swarms") and their orientation behavior are
highly dependent on various external variables such as time, temperature, solvent-solute compatibility, etc.

Because of the occurrence of optical rotation that has been
observed in concentrated solutions of SP systems, and even within
films of proteins (20), it seems necessary to comment on the nature
of these effects on our scattering behavior. The occurrence of
optical activity can alter the scattering patterns in two ways:
a) the effect of an optical active medium will influence the polarization directions for both the incident and scattered rays and
b) the scattering will be enhanced due to the fluctuation in magnitudes and direction of optical rotation. One might describe
these optically rotating effects respectively as external and internal. Both of these may complicate the interpretation of the
scattering data.* The development of an optically active mesophase has been discussed by Robinson *et al* (21) in light of the
PBLG *solutions* where the solvent favors the formation of the alpha
helix (e.g., chloroform). The occurrence of optical activity also
depends on the type of mesophase. Specifically it seems to be
noted in the cholesteric phase. Its origin is discussed in some
detail by Robinson and will not be repeated here as it does not
appear to be directly pertinent to the *film* morphology.

---

*It is worthwhile to mention that to this author's awareness the
case of "internal" optical rotation effects has not been theoretically treated.

Picot and Stein (22), in discussing effects on the scattering arising from spherulitic origin, found that "external" optical rotation affects the H_v scattering patterns considerably while the V_v pattern is affected less extensively. The H_v pattern found in undeformed spherulitic material is transformed from a distinctly four-fold symmetric pattern (as Fig. 2a) to a diffuse circular pattern as the optical rotation increases. Their same treatment cannot be directly applied to the case of rod scattering, but it is anticipated that similar results would be found. The effects of optical rotation, however, are not a highly significant factor in our data since it appears that the cholesteric mesophase does not dominate the texture in any of the films. For example, identical H_v "rod" patterns were obtained on films of varying thickness; this would not be expected if considerable optical activity existed. The film thickness maximum is, however, limited by the optical transmission of the film itself. For minimization of secondary scattering and higher order scattering the transmission should be of the order of seventy percent or higher. From our experience this generally requires a relatively thin film for the peptides systems e.g., < 3 mils.

Undeformed Reconstituted Collagen

Having observed this common rod-like superstructure in all the SP systems it was of interest to see if similar superstructure could be found in peptides of natural origin. We chose to investigate collagen since we were carrying out some mechanical and dielectric studies on this system at the time. It should be realized that the preparation techniques for purification of collagen from tendon are numerous and vary widely in chemical treatment. Investigation was, therefore, carried out on two common forms of collagen film -- one that was prepared by acidic means, CA film (13) and a second by an alkaline method, CB film (14). Figure 11a shows the H_v pattern for the CA film while Fig. 11b shows the H_v pattern for the CB materials. Both films were identical in appearance and were optically clear. The data demonstrates conclusively that the similar "rod-like" patterns can be obtained for the CA film but not the CB film. To obtain the H_v pattern for the CA film, it was necessary to use a sample to film distance that was greater than for the synthetic peptides films. This suggests that the "rod" size is larger in the collagen material relative to the SP system. Recent scattering studies of Kawai *et al* (23) and Cheng and Chang (24) on collagen have also revealed identical rod scattering.

To support our scattering data, scanning electronmicroscopy was used to look at the film surfaces of both CA and CB films as well as films of PBLG cast from chloroform. Limited data could be

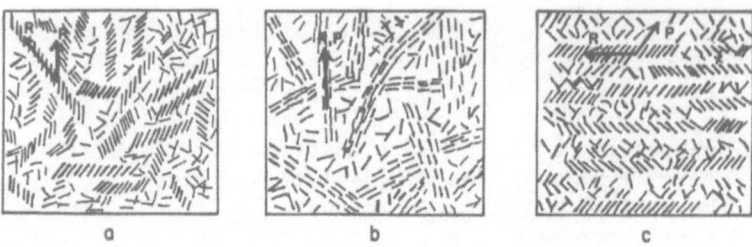

Fig. 10. Sketch of rod-like models that would explain the observed light scattering patterns obtained from the polypeptides studied. a) optic axis,P,is at 45° to the rod,R,axis (random distribution of rods);b) optic axis and rod axis are parallel (random distribution of rods); c) as (a) except rods are preferentially aligned perpendicular to the magnetic field which lies along the vertical.

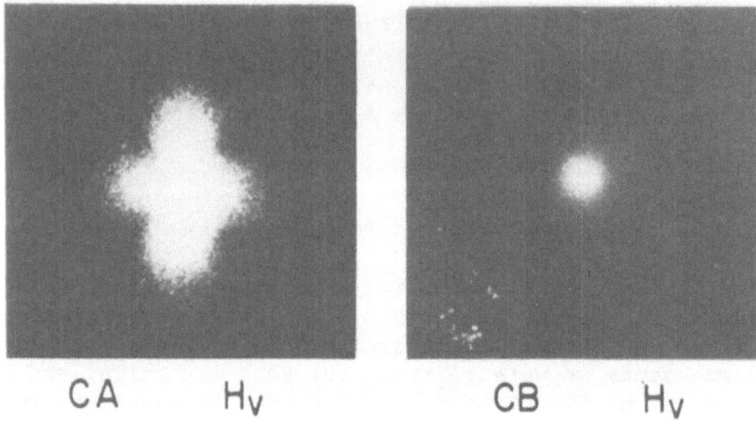

Fig. 11. H_v patterns from reconstituted collagen films. a) acidic preparation (CA); b) alkaline preparation (CB).

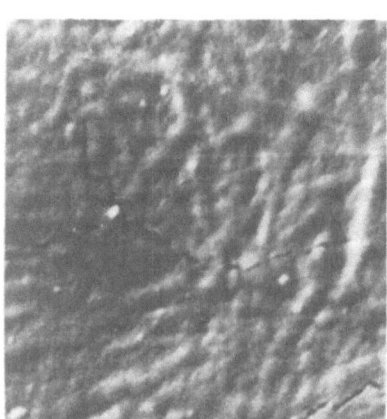

Fig. 12. Scanning electronmicrograph showing the surface of the CA collagen film (10,000 X). Cracks are due to damage by the electron beam.

obtained on the collagen due to its poor stability in the electron
beam. Figure 12 shows an electron photomicrograph of the surface
of the CA film. Although the film shows electron damage (the
presence of cracks) it is clear that a fibrillar texture is visible. This texture was completely lacking in the CB film, in
fact, the surface appeared to be homogeneous and smooth even at
higher magnifications. This difference in collagen superstructure
between the CA and CB films is somewhat surprising, as the difference appears to depend on the two types of chemical treatment
of the collagen preparation. We are presently investigating this
further. It is clear, however, that the scattering behavior of
the collagen is in agreement with the microscopy data. As with
the CB film, no fibrillar texture was apparent at the surface of
the PBLG film, but in light of the scattering pattern characteristics this was expected. That is, since the PBLG scattering
pattern occurred at larger scattering angles relative to CA it
indicated that the "rods" were of a smaller size. Since it was
difficult to observe the CA texture it is not surprising that it
was lacking at the surface of the PBLG films. There was some
evidence of fibrillar texture in an edge view of PBLG film.
Samulski (2) reported similar observations. We can reasonably
conclude that the preliminary microscopy studies are in agreement
with the light scattering data and its present interpretation
based on rod theory.

Calf Thymus DNA

In light of the fact that we had noted the $0°-90°$ H_v rod
pattern in several alpha helix systems as well as in films of the
triple helix macromolecule collagen, we were curious to see if
indeed this structure might occur in a film made from a double
helix rod-like polymer. Using a purified fraction of calf thymus
DNA a thin film was prepared by solution casting from an aqueous
medium. Figure 13a&b shows two H_v patterns taken from this film.
No V_v patterns could be obtained that displayed any azimuthal dependence of the scattered intensity. From Fig. 13a one again
sees the presence of a $0°-90°$ cross while Fig. 13b shows a pattern
typical of spherulitic scattering (recall Fig. 2a). Both of these
patterns were taken within the same film but in different locations; also, somewhat different sample to film distances were
used. Although the scattering suggests two different textures
(rod-like and spherulitic) no verification of this was noted with
the use of polarizing optical microscopy. To the author's awareness this double texture has not been seen within a system before.
This data on DNA is very recent and we are presently investigating
this further. The point we wish to make here is simply that the
same $0°-90°$ H_v rod pattern is indeed noted.

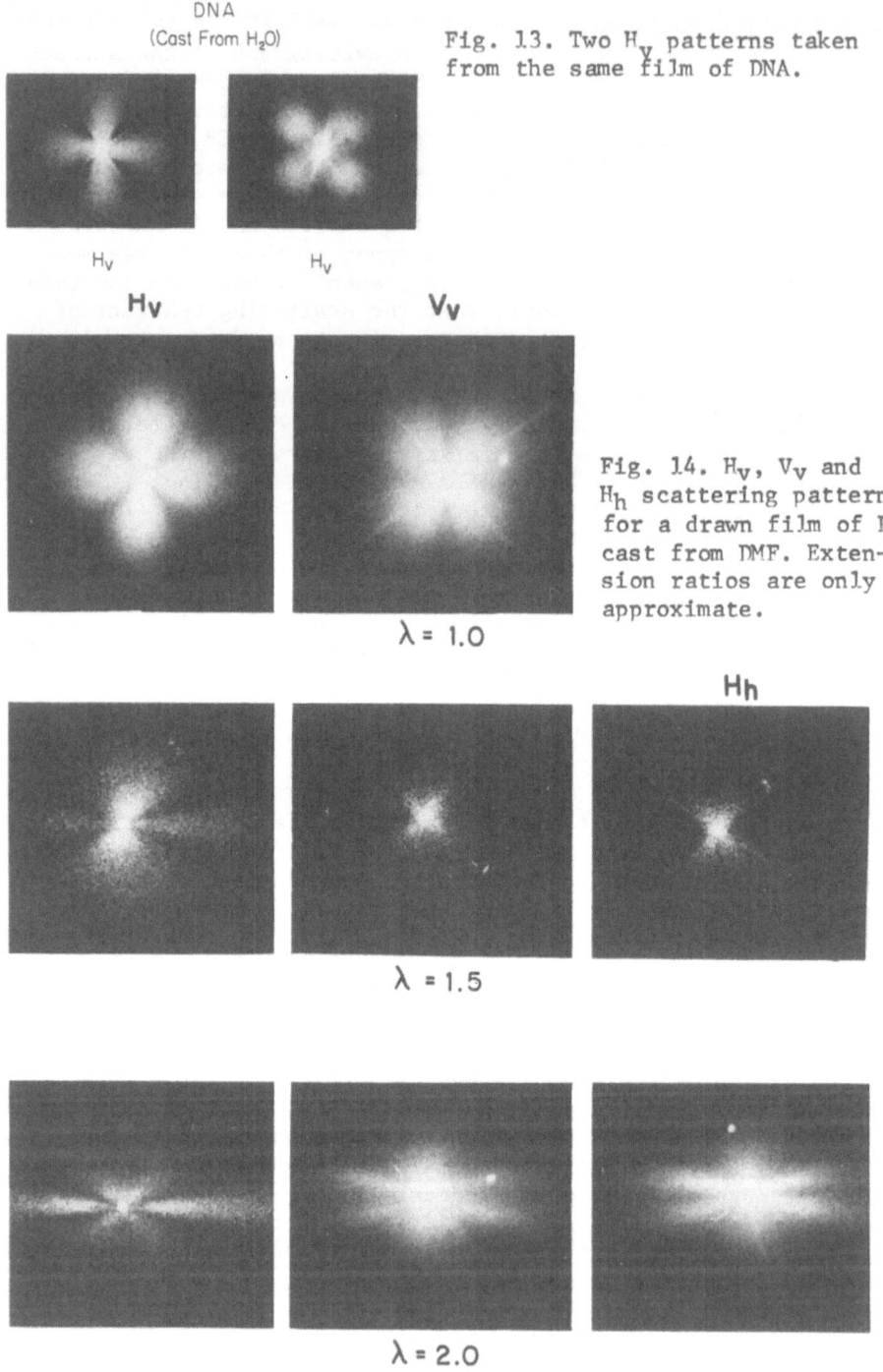

DNA
(Cast From H₂O)

Fig. 13. Two H_v patterns taken
from the same film of DNA.

H_V H_V

$\mathbf{H_V}$ $\mathbf{V_V}$

Fig. 14. H_V, V_V and
H_h scattering patterns
for a drawn film of PGA
cast from DMF. Exten-
sion ratios are only
approximate.

$\lambda = 1.0$

$\mathbf{H_h}$

$\lambda = 1.5$

$\lambda = 2.0$

Scattering from Deformed Systems

Although the model of Fig. 10a could be used to explain our
rod scattering data we became somewhat suspicious of this approach
due to the fact that *all* systems gave identical rod scattering
even though the structure (single, double and triple helix) as
well as chemical composition was changing tremendously. Specifically we were somewhat surprised that in *all* cases the scattering
suggested that the principal optic axis of the rod was approximately at 45° to the rod axis. Since the principal optic axis
of these molecules lies along the helix axis the model of Fig. 10a
does explain the data but is not necessarily physically pleasing
particularly for the collagen material. That is, it is known that
there is basically a parallel packing of the tropocollagen molecules within the small thread-like fibrils that make up the collagen film. There is no reason to suspect this arrangement changes
from that depicted by Fig. 10a upon film casting.

To further investigate this curiosity as well as to further
test the applicability of the rod scattering theory, a number of
experiments were carried out. The first was directed at uniaxially deformed films of PGA. This SP system was utilized since these
films displayed highest extensibility and high optical transparency.

It was hoped that upon stretching PGA films the scattering
patterns would reflect the orientation of the rods and x-ray diffraction could be used to note which way the alpha helix axis was
oriented. Figure 14 shows a series of H_v, V_v and H_h scattering
patterns for extension ratios of 1.0, 1.5 and 2.0. Clearly the
initial patterns change upon film stretching and are in
very good agreement with theoretically calculated patterns if one
uses the orientation distribution function for the rods first introduced by Rhodes (25) and used by Samuels in his study of hydroxypropylcellulose (10). This distribution function (eq. 10)

$$N'(\alpha) = N_0(\varepsilon^2\cos^2\alpha + \varepsilon^2\sin^2\alpha)^{1/2} \tag{10}$$

differs from the one given earlier (see eq. 2) in that the number
of rods at larger angles to the stretch axis is greater for a
given value of the orientation parameter ε. This simply results
in a less acute distribution and is probably more realistic.

Using Eq. 10, Samuels (10) calculated patterns where the angle
between the optic axis and the axis of the rod was becoming smaller
as ε increased -- see Fig. 15. We have noted that our scattering
patterns do agree with these -- compare Figures 14 (λ=2.0) and 15.

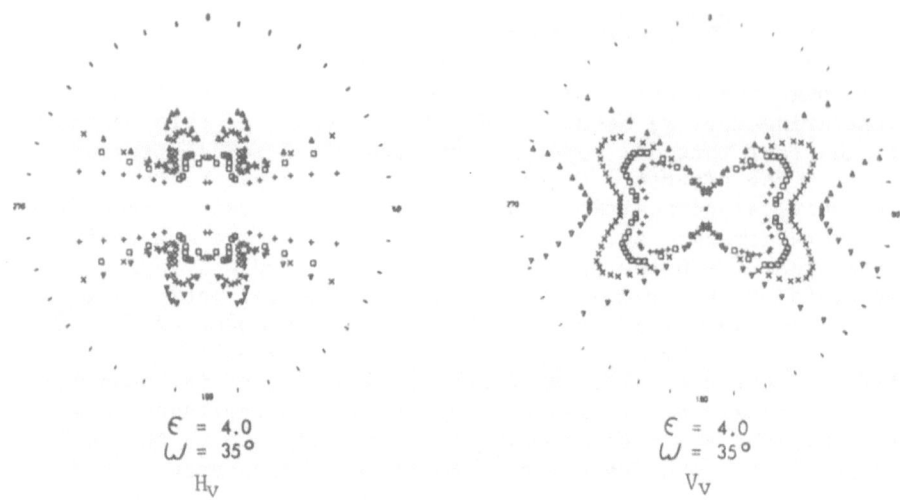

$\epsilon = 4.0$
$\omega = 35°$
H_v

$\epsilon = 4.0$
$\omega = 35°$
V_v

Fig. 15. Theoretical H_v and V_v patterns calculated by Samuels (10) using eq. 10 with $\omega = 35°$ and $\epsilon = 4.0$.

Fig. 16. Sketch of x-ray pattern from drawn ($\lambda \sim 2.0$) PGA -- stretch axis vertical. Beam was parallel to the film normal.

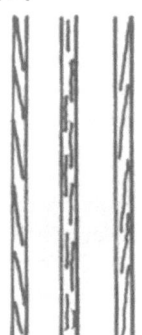

Fig. 17. Three types of oriented rods which combined in a single film would allow direct correlation of the H_v scattering and x-ray results. The small rods within a rod represent the alpha helix direction. ω is less than 45° due to "slipping" of the helices within the rod. Coiled coil models are also possible.

Fig. 18. Optical photomicrograph of a drawn PGA film ($\sim 300X$). Crossed polarizers were used (P_1, P_2). The draw axis is tilted at 45° to P_1 and P_2.

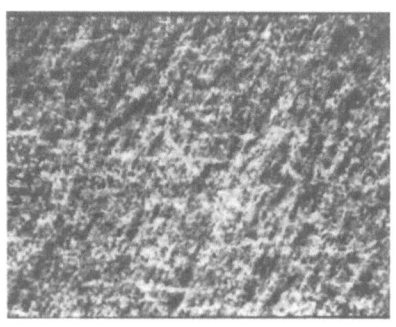

The x-ray diffraction from this same drawn PGA film gave a pattern as sketched in Fig. 16. Other drawn PGA films gave identical results. This pattern suggests that there is a preferential inclination of the alpha helices with respect to the stretch axis. This inclination should, in fact, be expected based on the light scattering data i.e., the optic axis (helix axis) was found to lie at some preferential angle to the super rod axis and therefore we might picture the texture as given in Fig. 17. Polarizing optical microscopy neither clearly supported nor disproved this model but it was noted that the film illustrated a somewhat fibrillar texture -- see Fig. 18.

Similar experiments were carried out on drawn films of PBDG. The scattering results were in general agreement with those from drawn PGA. The texture as noted by optical microscopy was also similar to that of drawn PGA. The texture as noted by optical microscopy was also similar to that of drawn PGA. The x-ray diffraction, however, did not yield the distinct "four-point" intermolecular spacing as did the PGA. This may be due to the fact that a comparable extension ratio could not be reached and hence the distribution in orientation was not as pronounced as with the PGA. It was obvious however that a preferential alignment of the alpha helix occurred along the draw axis as has been observed by others (26).

Another related experiment using the SP PBLG was stimulated by the earlier studies of Samulski who showed that oriented films could be made by solvent casting the material in the presence of a magnetic field (2). He showed that a hexagonal packing of the helices occurred having their axes oriented parallel to the magnetic field. We not only wished to obtain oriented films of PBLG for the reasons discussed earlier but we also were interested to see if, in fact, the rod superstructure was still induced when the material was prepared under these unique conditions. Using a Fieldial electromagnetic system we cast PBLG films from chloroform in the presence of magnetic fields as high as 25 kilo Gauss (KG). Using these films both x-ray diffraction and light scattering measurements were made. Figure 19a shows the x-ray diffraction pattern taken from the data of Samulski while Fig. 19 b&c show diffraction patterns taken from our own films. In our case we did not note the presence of layer lines as Samulski -- this difference may be due to the extreme thinness of our samples relative to those of Samulski. From these diffraction data we also note that the alpha helix of the PBLG molecule is oriented along the magnetic field direction but there is clearly a distribution of orientation as is suggested by the azimuthal spread in either the spacing caused by the pitch of the helix ($\sim$5.1 Å) or, by the intermolecular spacing ($\sim$11 Å). A typical H_v scattering pattern taken from our films is shown in Fig. 20. It is clearly apparent

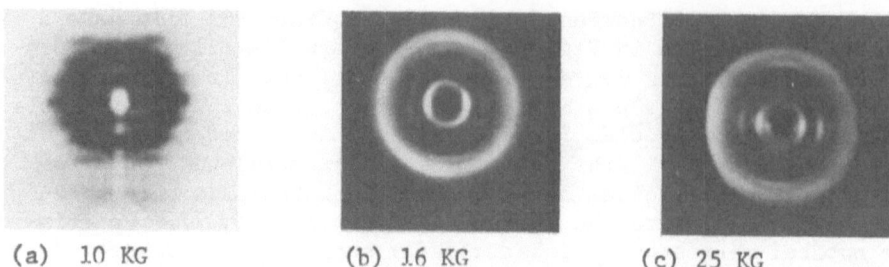

(a) 10 KG   (b) 16 KG   (c) 25 KG

Fig. 19. Wide angle x-ray diffraction patterns taken on chloroform cast PBLG films. Films were cast in a magnetic field with
the field parallel to the film plane. In the figure the field
direction is vertical. (Fig. 19a courtesy of E.T. Samulski.)

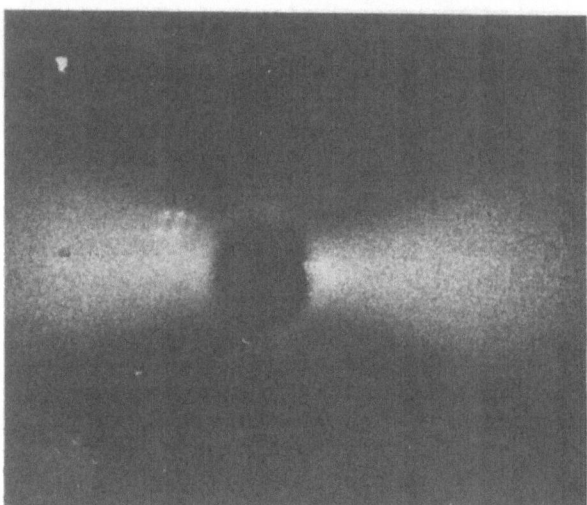

Fig. 20. H_v scattering
pattern taken on PBLG
film cast in a magnetic
field of 25 K Gauss.
The direction is
horizontal.

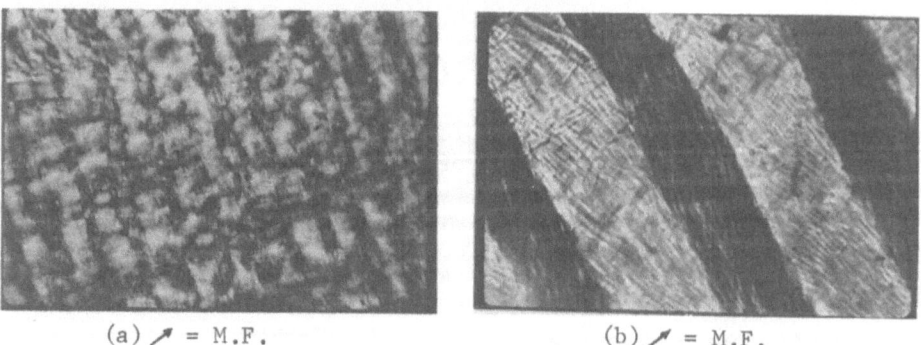

(a) ↗ = M.F.   (b) ↗ = M.F.

Fig. 21. Optical photomicrographs taken between crossed polarizers of the magnetically oriented PBLG film (16.5 KG):
(a)~120X, (b)~300X, M.F. = magnetic field direction.

that the magnetic orientation has affected the texture of the superstructure as well. One's first thought might be to compare the observed H_v pattern (Fig. 20) with a theoretical H_v contour plot where the orientation parameter, ε, is greater than unity -- recall eq. 2). Doing so (see Fig. la, $\varepsilon=4$) clearly reveals that the experimental pattern is rotated 90° from that given theoretically. This simply implies that the "rods" are oriented perpendicular to the magnetic field direction rather than being parallel. We, in fact, generated a theoretical scattering pattern in agreement with that found experimentally by using an ε value of 0.25 which nearly corresponds to a perpendicular arrangement. A possible model for this texture might be as shown earlier in Fig.10c where it is noted that the molecular axis lies along the direction of the magnetic field, the optic axis is at 45° to the rod axis and the rod axis is perpendicular to the field. This model is in contrast to what was expected based on the drawn PGA (PBDG) data. That is, we have suggested that the principal optic axis and helix axis coincide but the data for the magnetically oriented film implies that this is not the case.

Using polarizing optical microscopy we investigated the texture of the magnetically oriented film. From these studies we found an extremely interesting morphology which was more developed in some films than in others. Figure 21a&b shows two optical photomicrographs taken from a film cast at 16.5 kilo Gauss. These illustrate a very pronounced super rod texture that has some unique features: a) the rods are of the order of 40 microns in diameter and are hundreds of microns in length, in fact, the length extended from one meniscus of the film to the meniscus on the other side ($\sim$ 1 cm); b) the rods are very parallel although in some cases some deviation from this behavior could be noted;c) each rod seems to exhibit a texture similar to a twisted strand of hemp rope, that is, within a given rod, one can easily see small strands which tend to twist in a helical manner about the super rod axis. One, in fact, can get some indication of this twisting by noting the periodicity and colors along the rod when observed with polarized light. This periodicity extends along the rods for some distance and one can easily see identical colors (equal retardation) in alternating neighboring rods suggesting a correlation in structure over hundreds of microns; d) rotating the sample 45° between cross polarizers results in an exchange in retardation (color) between a rod and its nearest neighbors. This suggests that the optic axis of each adjacent rod differs from its neighbors by 45°.

We will not further discuss the details of this extremely unique and interesting morphology at this time since we plan to publish this separately.* We wish to emphasize, however, that
*It should be mentioned that some similar morphologies have been reported (26) but were generated under different conditions and were not as perfected as those shown here.

we have not as yet resolved the fact that the helix axis (from
x-ray) and the optic axis of the super rod (by SALS) do not seem
to coincide. This seems surprising and further efforts are underway to resolve this apparent discrepancy.

We have attempted to magnetically orient other SP systems
but our preliminary data for PMG and PGA have shown no molecular
alignment when cast under similar conditions. Although more experiments are necessary, it appears at this point that the presence of the aromatic ring in the PBLG molecule may be the cause
of the magnetic effect.

A parameter which we have also used to induce changes in the
superstructure of the peptides is that of temperature. We anticipated that at least in the case of collagen, possibly the thermal
dependence of the light scattering (H_v) could be utilized to follow denaturation or "melting" of collagen. This phenomenon is
generally associated with a characteristic shrinkage temperature.
By using a Mettler FP2 programmed Hot Stage system we could note
any apparent changes in the H_v scattering with heating. Because
the details of this work (on collagen) will be presented elsewhere (27) we only wish to comment on the results. Figure 22 a-c
shows the typical rod-like pattern from the cast film (CA type),
the theoretical H_v pattern for the rods (as discussed earlier)
and, an H_v pattern taken at the transition or shrinkage temperature. This latter pattern, Fig. 22c, shows complete loss of the
rod scattering as might well be anticipated at the time of denaturation.

It has been demonstrated by several workers (28,29) that the
shrinkage temperature of collagen is dependent upon its chemical
environment and can be depressed considerably with polar diluents
as ethylene glycol, water etc. By using two diluent systems and
other supporting techniques (DTA, TMA, polarizing microscopy) we
have demonstrated that the transition of the H_v scattering pattern is in very good agreement with the earlier shrinkage studies
of collagen. We have not noted a similar transition in PBLG, PMG,
or PGA films however, where only silicone oil was used as an
immersion medium (diluent). For example Fig. 23 shows a series of
H_v patterns for PBLG as a function of temperature. This figure
is somewhat misleading in that it appears that the patterns are
weaker in intensity at the higher temperatures. This was not
actually the case and is only due to exposure time difference. No
apparent change occurred upon heating to $\sim$ 300°C where the pattern
became less distinct with time due to discoloration by degradation.
Similar thermal stabilities were noted for PGA and PMG.

In attempting to illustrate the application of the scattering
technique to biorelated systems it seems appropriate to point out

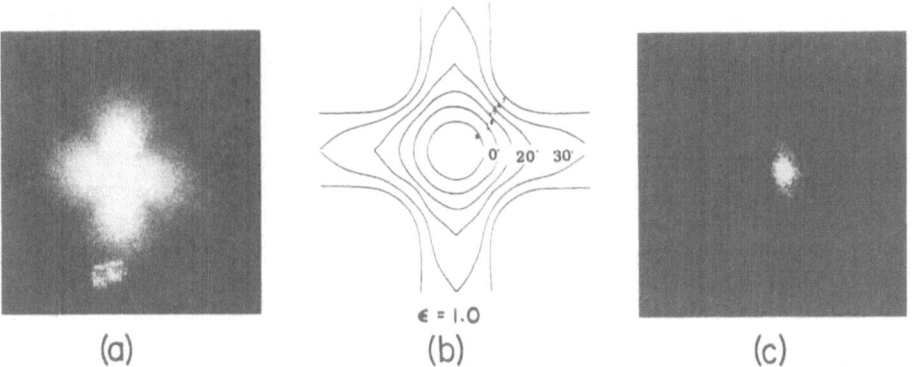

Fig. 22. H_v patterns for a) collagen below shrinkage temperature:b) theory ;c) collagen above shrinkage temperature.

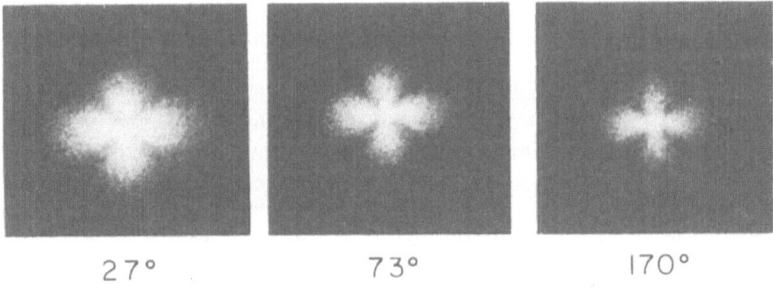

Fig. 23. H_v patterns taken as a function of temperature from a film of chloroform cast PBLG.

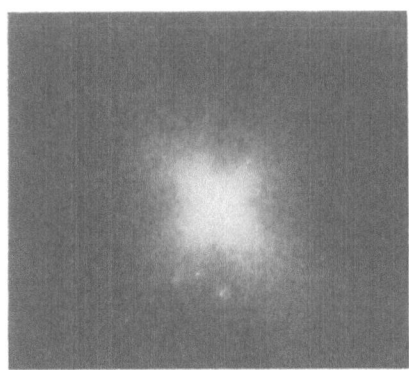

Fig. 24. H_v scattering patterns from neonatal rat stratum corneum: a) unextracted; b) extracted for 10 hours with a 3/1 chloroform/methanol mixture.

one instance where the sample is an actual biological tissue. In
our structure-property studies on the *stratum corneum*, the outermost layer of the epidermis, we have found the light scattering
technique to be useful in offering further evidence to support an
earlier model of Swanbeck (30) regarding the structure of the
protein-lipid complexes that are present within the normal stratum
corneum. These structures are basically rods and form a two-dimensional random packing in the plane of the epidermis. By extract-
ing the lipids one can accent the optical anisotropy of these rods
made of assumed aggregates of alpha keratin filaments. Figure
24 a&b shows the H_v patterns for the unextracted and lipid extracted *corneum*. Clearly the extracted film displays scattering
typical of anisotropic rod-like elements whereas the unextracted
does not. In this case however, the patterns indicate that the
optic axes of the rod is either along or parallel to the rod axis
(see footnote pg. 9).

Further details of this latter work are to be given elsewhere.
The point we wish to establish with this example is that scattering may be applied to such composite systems as tissue and may in
certain instances be a useful supporting technique in studying
structure and morphology.

In summary, we have illustrated that many bio-like or biological macromolecular solids do form distinct anisotropic superstructures and that, in general, these superstructures are of a
form that scatter light as may be described by scattering from
anisotropic rods. We have also pointed out, however, that although the rod scattering theory can be used to explain the observed patterns, there are still some questions left unanswered,
as for example, why the helix axis and principal optic axes of the
rod superstructure do not seem to coincide for the magnetically
oriented film. In light of questions as this which will motivate
further studies, we have illustrated that this scattering method
has much potential for aiding the studies relating to the area of
biophysics.

ACKNOWLEDGMENTS

One of the authors (GLW) wishes to thank Gordon and Breach
Publishers for allowing the use of excerpts and several figures
from a manuscript now in press (31)

The same author also wishes to acknowledge helpful dis-
cussions with Mr. C. Anderson and for the SEM assistance of Mr.
A. Coe of the Textile Research Institute.

REFERENCES

1. Robinson, C., Tetrahedron, 13, 219 (1961).
2. Samulski, E. T., Ph.D. Thesis, Dept. of Chem., Princeton Univ. 1969.
3. Powers, J. C., Jr., in *Liquid Crystals and Ordered Fluids*, Plenum Press, N. Y. 1970.
4. Debye, P. and A. Bueche, J. Appl. Phys., 20, 518 (1949).
5. Stein, R. S. and P. R. Wilson, J. Appl. Phys., 33, 1914 (1962).
6. Stein, R. S. and M. B. Rhodes, J. Appl. Phys., 31, 1873 (1960).
7. Stein, R. S., Erhardt, P., van Aartsen, Clough, S., and M. B. Rhodes, J. Poly. Sci., C, 13, 1 (1966).
8. Rhodes, M. B. and R. S. Stein, J. Poly. Sci., A-2, 7, 1539 (1969).
9. van Aartsen, J. J., Eur. Poly. J., 6, 1095 (1970).
10. Samuels, R. J., J. Poly. Sci., A-2, 7, 1197 (1969).
11. Clough, S. B., R. S. Stein and C. Picot, J. Poly. Sci., A-2, 6, 1147 (1971).
12. Stein, R. S. and C. Picot, J. Poly. Sci., A-2, 8, 1955 (1970).
13. Oneson, I., D. Fletcher, J. Olivo, J. Nichols and R. Kronenthal, J. Am. Leather, Chem. Assoc., 65, 440 (1970).
14. Courts, A., Biochem. J., 74, 238 (1960).
15. Elliott, A. and E. J. Ambrose, Dis. Fara. Cos., 52, 571 (1955).
16. Onsager, L., Ann. N. Y. Acad. Sci., 51, 627 (1949).
17. Isihara, A., J. Chem. Phys., 19, 1142 (1951).
18. Flory, P. J., Proc. Roy. Soc. (London), A234, 73 (1956).
19. Frenkel, S. Ya, V. G. Baranov and T. I. Volkov, J. Poly. Sci. C, 16, 1655 (1967).
20. Yannas, I. V. and N. H. Sung, Poly. Preprints. 13, 1, 123 (1972).
21. Robinson, C., J. C. Ward and R. B. Beevers, Disc. Faraday Soc., 25, 29 (1958).
22. Picot, C. and R. S. Stein, J. Poly. Sci., A-2, 8, 1491 (1970).
23. Moritani, M., N. Hoyashi, A. Utsao and H. Kawai, Poly. J., 1, 74 (1971).
24. Chien, J. C. W., and E. P. Chang, Univ. of Mass., private communication.
25. Rhodes, M. B., Ph.D. thesis, Univ. of Mass (1966).
26. Bamford, C. H., A. Elliott and W. E. Hanby, *Synthetic Polypeptides*, Academic Press, N. Y. 1956.
27. Nguyen, A., B. Vu and G. L. Wilkes, manuscript in preparation.
28. Witnauer, L. P. and J. G. Fee, J. Poly. Sci., 26, 141 (1957).
29. Flory, P. J. and R. R. Garrett, J. Am. Chem. Soc., 80, 4836 (1958).
30. Swanbeck, G., Acta Dermato-Venereologica, 39, 1 (1959).
31. Wilkes, G. L., Mol. and Liq. Crystals, in press.

RODLIKE SUPERSTRUCTURES IN AND MECHANICAL PROPERTIES OF BIAXIALLY ORIENTED POLYETHYLENE TEREPHTHALATE FILM

W. H. Chu and T. L. Smith

IBM Research Laboratory

San Jose, California 95114

INTRODUCTION

Semicrystalline polymers contain anisotropic crystal lamellae. Commonly, the orientations of lamellae are correlated within certain regions, giving superstructures, e.g., spherulites or rods. As a consequence of crystallization and superstructure formation, some of the amorphous material is under stress, thus being oriented. The response of such a complex material to surface tractions depends strongly on the orientation distribution and the physical character- istics of the lamellae; these factors affect the stress state, and hence the resulting deformation, on a microscale throughout the specimen.

To modify the physical properties of a polymer, it is commonly oriented by stretched under special thermal conditions. When the resulting material is heated somewhat above its glass temperature, frozen-in strains often are partially relieved, causing lamellae and amorphous material to disorient somewhat and perhaps the size and perfection of lamellae to increase.

In this paper, an exploratory study of a biaxially oriented polyethylene terephthalate (PET) film is discussed. The objectives were to determine: (1) the type and orientation of superstructures; (2) changes effected by annealing nonconstrained specimens at and below 180°C; and (3) qualitative relations between structure and mechanical properties.

CHARACTERISTICS OF PET FILMS

The production of a biaxially oriented PET film involves melt extrusion, a rapid quench from the extrusion temperature, a subsequent heating and stretching of the film, and finally, some heat-setting or stabilization process (1). If the extruded film is slowly cooled, spherulitic superstructures form (2,3) giving a rather brittle material unsuitable for subsequent orientation. On the other hand, rapid quenching gives an amorphous film that thereafter can be heated above the glass temperature and oriented by stretching, commonly in the machine direction (M.D.) first and then in the transverse direction (T.D.). During the stretch in the M.D., crystallites form; the c-axis and plane of each lamellae tend to be parallel, respectively, to the M.D. and the film surface (4-8). The subsequent stretch causes certain crystallites and amorphous chains to reorient toward the T.D., the amount being dependent on temperature, the magnitude of the stretch, and other process conditions. The amount of reorientation also depends on the position in the web, being greatest near the edges and least in the center; thus, an "unbalanced" film whose physical properties vary across the web is obtained.

During the heat-setting process, which involves annealing the film under tension above T_g, both the crystallinity and the uniplanarity of crystallites increases (7). Annealing without applying tension also increases the crystallinity but the uniplanarity decreases. Statton et al. (9) concluded that chain folding occurs during the annealing of oriented PET fibers. More chain folding occurs when a fiber is free to contract than when its length is held constant. Under either condition, the amount of chain folding, and thus the crystallinity, increases with the annealing temperature.

The low-angle light-scattering patterns obtained by Wallach (10) indicate that superstructures exist in uniaxially and biaxially oriented PET. From certain similarities between X-ray and light-scattering patterns, he concluded that the orientation of the super-structures (then assumed erroneously to be of spherulitic texture) is closely related to that of the crystallites.

MATERIALS AND TEST METHODS

Materials. From a 48-inch wide roll of 142 A Mylar (biaxially oriented PET film, 1.5 mils thick, manufactured by the du Pont Co.), rectangular specimens were cut at orientations of 0°, 45°, and 90° to the M.D.; specimens for determining tensile stress-strain curves were also cut at 30° and 60° to the M.D. Some specimens were from a location in the 48-inch web at which the extinction angle, β_m, was about 45° to the M.D.; other specimens were from a location where $\beta_m \simeq 75°$. The extinction angle, obtained by examining specimens under crossed polaroids, indicates the principal orientation direction. Figure 1 shows the variation of β_m and the birefringence across the web. Certain specimens were hung freely in an oven at 95°, 120°, and 180°C and annealed for three hours. Annealing caused β_m to shift by a few degrees at most.

Density and Thickness. Densities of untreated and annealed specimens were measured at 25°C using a water-zinc chloride density-gradient column, which had been calibrated with glass spheres of known densities. The thickness of specimens was measured with a precision thickness gage.

Low-Angle Light Scattering. A photographic light-scattering apparatus was used to obtain H_v scattering patterns on untreated and annealed (180°C) specimens in both unstretched and stretched states. The apparatus, similar to that of Rhodes and Stein (11), has a 1 mw helium-neon laser; scattering patterns were recorded on 4 x 5-inch Polaroid film (Type 52). Specimens were mounted between a 1 x 3-inch microscope slide and a 2.2 x 2.2-cm Corning cover glass. A screw-driven device was used to stretch specimens mounted between two cover glasses. In all instances, the long dimension (cut direction) and stretching direction of a specimen were parallel to the polarization plane of the incident light.

Dynamic Modulus. The storage and loss tensile moduli (E' and E") were measured with a Rheovibron DDV-II (direct reading dynamic viscoelastometer) at 35 Hz from about -90 to 160°C; the heating rate was about 1-2°C per minute.

Stress-Strain Curves. Rectangular strips 0.5 x 5.0-inch were tested with an Instron tester at a crosshead speed of 1.0 inch per minute. Crosshead displacement was converted into strain using an effective gage length evaluated from photographs taken of bench marks on specimens during certain of the tests.

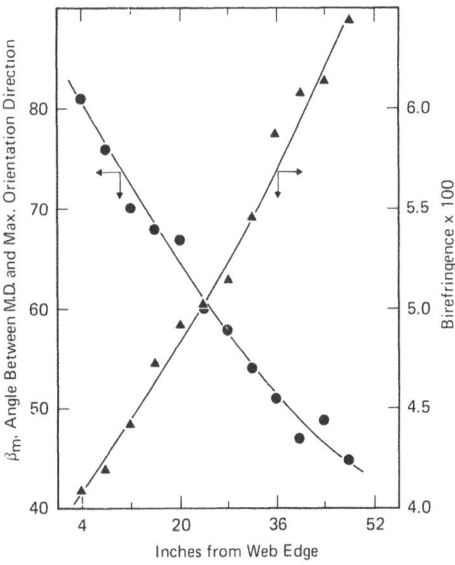

Fig. 1. Variation of optical properties across Mylar A film.

RESULTS AND DISCUSSION

Crystallization and Strain Release During Annealing

Table I shows that the density increases progressively with the annealing temperature. If the amorphous and crystalline phases have densities of 1.331 and 1.47 g/cm^3 (12), the density data correspond to crystallinities from about 46.7% for the untreated specimen to 53.5% for the specimen annealed at 180°C. The increased thickness effected by annealing results from a lateral contraction associated with the partial release of frozen-in tensile strain. Because each annealed specimen has nearly the same thickness, the amount of frozen-in strain released at each annealing temperature is essentially the same.

TABLE I

Thickness and Density of Untreated and Annealed Specimens

Annealing Temp., °C	Thickness mils	Density g/cm^3	Weight % Crystallinity(a)
untreated	1.47	1.3925	46.7
95	1.61	1.3941	47.9
120	1.60	1.3959	49.2
180	1.59	1.4019	53.5

(a) Calculated from density assuming that densities of amorphous and crystalline phases are 1.331 and 1.47 g/cm^3, respectively.

Light-Scattering Patterns

Figure 2 shows the scattering patterns obtained on specimens cut at 0°, 45°, and 90° to the M.D. Because β_m for these specimens was about 75° to the M.D., the polarization direction of the incident light did not coincide with β_m. Thus, the patterns are masked by birefringence effects (13-15), as evidenced by the upper row of photographs in Figure 2 (photographs d1 and e1 are for 180°C annealed specimens cut at 0° and 45° to the M.D.). This masking effect can be minimized by stretching a specimen until the light, on passing through the specimen, is retarded by an integral number of wavelengths. Even under these conditions, the optic axis does not necessarily coincide with the polarization direction of the incident light, and thus the observed scattering may represent a combination of H_v and V_v patterns. The intensity of scattered light (under

crossed polaroids) at a specified point is then given by
$I = I_{H_V} \cos^2\Theta + I_{V_V} \sin^2\Theta$, where Θ is the angle between the optic

axis and polarization direction (16). Such a mixed pattern, however, can still be interpreted qualitatively. Consequently, in studying stretched specimens, scattering patterns were recorded only when the applied strains ε_1, ε_2, and ε_3 (Table II) were such that the light retardation equaled an integral number of wavelengths, giving a pattern of minimum intensity.

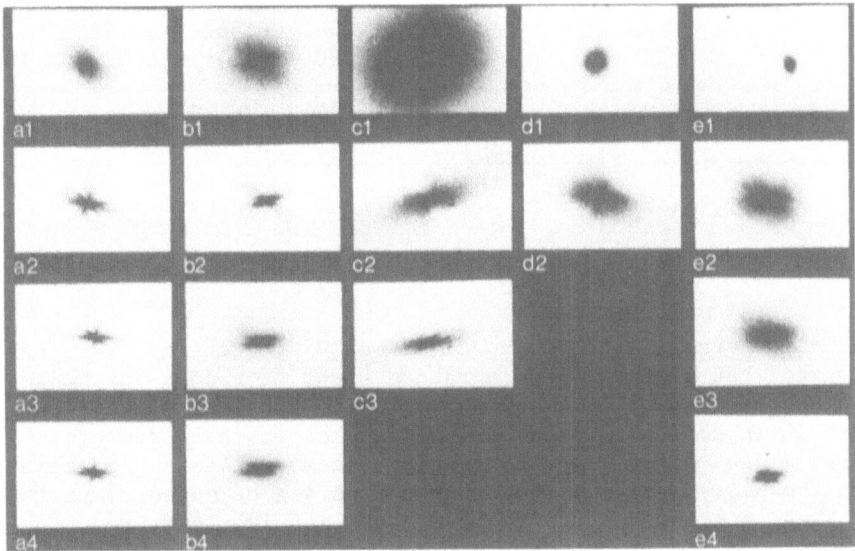

Fig. 2. Low-angle light-scattering patterns (H_v) obtained on biaxially oriented polyethylene terephthalate (Mylar A); the extinction angle was about 75° to the M.D. The angle between the polarization plane of the incident light and the M.D. was 0°, 45°, and 90° for patterns in columns a, b, and c, respectively, obtained on untreated specimens; the angle was 0° and 45° for patterns in columns d and e, respectively, obtained on 180°C annealed specimens. Patterns in rows 2, 3, and 4 are for specimens under strains ε_1, ε_2, and ε_3 (Table II), respectively.

Photographs a2, b2, and c2 in Figure 2 are patterns obtained on untreated specimens under strains ε_1, given in Table II. These multi-forked patterns are characteristic of rodlike superstructures. A basic difference between such patterns and the clover-leaf patterns characteristic of spherulites is that the scattering intensity along the axis of each fork decreases progressively from the center, whereas the maximum intensity in a pattern given by spherulites occurs away from the center (17).

TABLE II

Tensile Strains for Minimum Overall Scattering Intensity.
Extinction Angle of Unstressed Specimens is 75° to M.D.

Specimen Treatment	Angle Between Stretch Direction and M.D.	Strain at Extinction			K_1 (a)	K_2 (b)
		ε_1	ε_2	ε_3		
Untreated	0°	0.394	0.655	1.135	0.261	0.480
	45°	0.016	0.206	0.366	0.189	0.197
	90°	0.160	0.680	(b)	0.520	(b)
Annealed at 180°C	0°	0.565	0.830	(b)	0.265	(b)
	45°	0.06	0.273	0.454	0.213	0.181

(a) $K_1 = \varepsilon_2 - \varepsilon_3$; $K_2 = \varepsilon_3 - \varepsilon_2$

(b) Specimen broke

The pattern b2, for the specimen cut 45° to the M.D., is rather
similar to that expected from a random array of rods. The long
streak nearly perpendicular to the stretch (vertical) direction in
pattern c2 (specimen cut in the T.D.) indicates that rods are
oriented close to the T.D. Similarly, the streak in a2 (specimen
cut in the M.D.) indicates that other rods are oriented close to the
M.D. These conclusions result from a qualitative interpretation of
the patterns in light of the theory of scattering by rodlike superstructures developed primarily by Rhodes and Stein (17).

For specimens cut 0° and 90° to the M.D., an increase in the
strain to ε_2 changes the four-forked patterns (a3 and c3) only
slightly, indicating little change in the number and orientation of
rods, but the streaks are sharpened. Because light is scattered at
large angles by the smaller regions in the rods, a sharp streak
reflects increased order within rods. Under strain ε_3, the scattering pattern (a4) is essentially the same as under ε_3. (The specimen
cut 90° to the M.D. broke before ε_3 was attained.)

For the specimen cut 45° to the M.D., the pattern becomes flat-
ter when the strain is increased to ε_2 (compare b2 and b3); under
strain ε_3, the pattern (b4) is similar. This change indicates an
increase in the number of rods oriented in the stretch direction.
During stretching, some randomly oriented crystallites undoubtedly
reorient giving rodlike superstructures.

The fourth and fifth columns in Figure 2 are patterns from 180°C
annealed specimens cut at 0° and 45° to the M.D. These specimens
scatter more light and give patterns that are similar in shape,

although more diffuse, than those from untreated specimens. The
increased diffuseness indicates that some crystallites within the
superstructures disorient during annealing, a consequence of the
partial release of frozen-in strain. The other characteristics of
the patterns imply that annealing does not alter the orientation
distribution of the rods but it does increase the perfection of
crystallites, as indicated by the increased scattering intensity.

Values of $K_1 \equiv \varepsilon_2 - \varepsilon_1$ and $K_2 \equiv \varepsilon_3 - \varepsilon_2$ are included in Table II.
For specimens cut $45°$ to the M.D., the relatively small values of
K_1 and K_2 indicate a high degree of reorientation of crystallites
with a small increase in applied strain. Conversely, the large
values of K_2 for the specimen cut $0°$ to the M.D. and of K_1 for the
specimen cut $90°$ to the M.D. indicate that the amount of reorientation is rather small. These observations support the conclusions
drawn from the light-scattering patterns.

The possible orientation distribution of rodlike superstructures is shown schematically in Figure 3.

Storage and Loss Moduli

Figures 4 and 5 show E' and E'' measured at 35 Hz over an
extended temperature range on untreated and 120°C annealed specimens
for which β_m, and thus the direction of maximum orientation, was
about $45°$ to the M.D. Below about 100°C, the untreated and annealed
specimens exhibit rather similar behavior. The moduli E'' (except
possibly at intermediate temperatures) and E' are greater in the
direction of maximum orientation. These observations are consistent
with the findings of Armeniades and Baer (18) who studied uniaxially
oriented PET films.

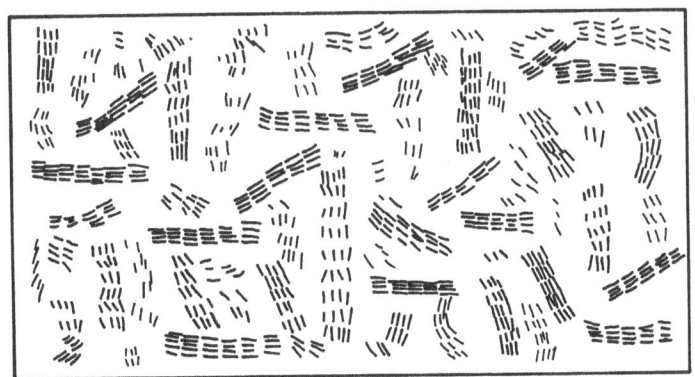

Fig. 3. Schematic diagram to illustrate the possible orientation
distribution of rodlike superstructures.

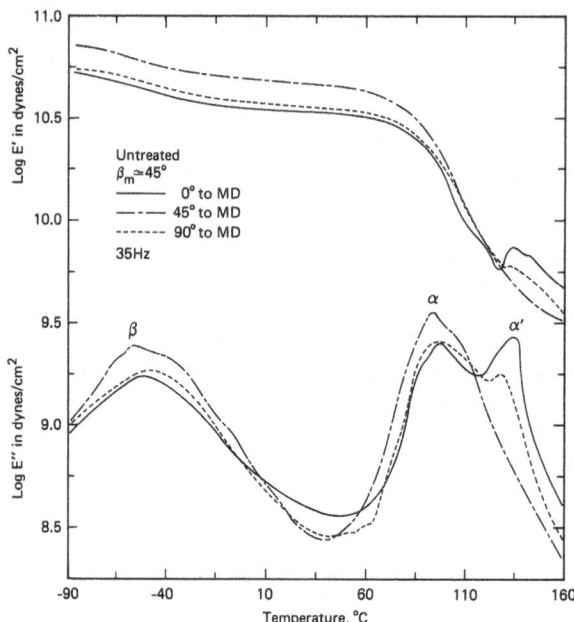

Fig. 4.   Storage and loss moduli measured at 35 Hz on untreated
specimens for which the extinction angle was about 45° to the M.D.
Direction of stress application was 0°, 45°, and 90° to the M.D.

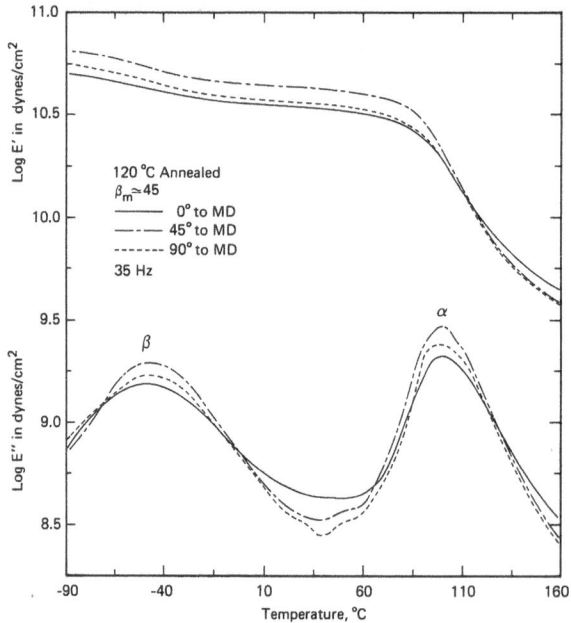

Fig. 5.   Storage and loss moduli for specimens (Fig. 4) annealed
at 120°C.

Above 100°C, data on the untreated specimens (Fig. 4) cut at 0° and 90° to the M.D. exhibit a peak, here designated α'. This peak results from an unknown type of structural change during the heating process. Possibly, however, the peak reflects a slow but continuous reorientation of crystallites into the direction of applied stress. Among those tested, the specimen cut at 45° to the M.D. contained the greater number of crystallites oriented in the stress direction; this specimen does not exhibit the α'-peak . In the specimen cut 90° to the M.D., somewhat fewer crystallites are oriented in the stress direction, and the α'-peak is small. Hence, a reorientation of crystallites is a possible explanation of the increase in both E' and E'' above the α-relaxation peak.

Stress-Strain Curves

Figure 6 shows stress-strain curves measured on untreated and annealed specimens, for which $\beta_m \simeq 45°$, cut at five orientations to the M.D. The yield stress, σ_y, is within a few percent the same for all specimens. For unoriented PET, Ward (19) has reported that σ_y, but not the yield strain, is essentially independent of the degree of crystallinity, at least for crystallinities up to about 40%. As expected, however, σ_y for uniaxially oriented PET depends on the stretch direction; for the particular sample studied, the yield stresses in the orientation and perpendicular directions were found to differe sixfold (20). In view of such behavior, σ_y for biaxially oriented PET might be expected to be a minimum when the stretch direction is 45° to the principal orientation direction. But the present, as well as other data (21) show that σ_y, in contrast to the modulus (Figs. 4, 5, and 7), is essentially independent of the stretch direction.

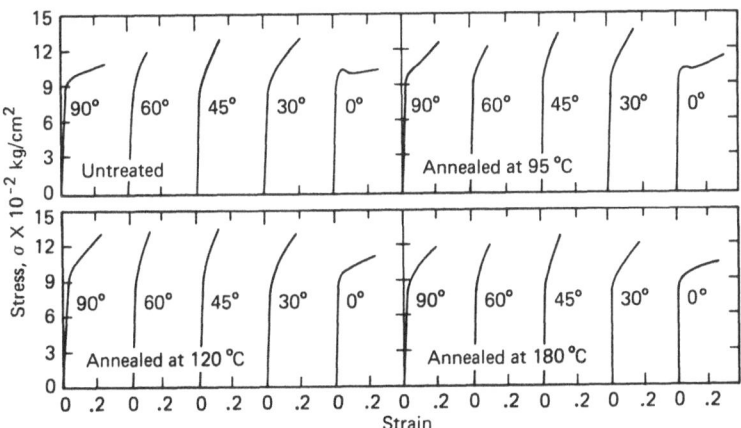

Fig. 6. Stress-strain curves for untreated specimens and specimens annealed at 95°, 120°, and 180°C. Direction of stress application was 90°, 60°, 45°, 30°, and 0° to the M.D. Extinction angle was about 45° to the M.D.

    In contrast to the yield stress, the stress response following
yield depends considerably on the angle between the principle orientation direction and the stretch direction, as shown in Figure 6.
Beyond the yield point, the curves for specimens cut parallel to the
principle orientation direction (45°) have the highest slope and
those for specimens cut at 0° and 90° to the M.D. have the lowest
slope; the other specimens show intermediate behavior. Actually,
the untreated and 95°C annealed specimens cut parallel to the M.D.
exhibit a drop in load subsequent to yielding. This behavior presumably occurs because a substantial number of randomly oriented
lamellae rotate toward the stretch direction forming superstructures,
as indicated by the light-scattering patterns obtained on the specimen cut at 30° to the principle orientation direction. (See discussion of specimen labeled 45° in Table II.) The load drop in the
0° but not in the 90° specimen (Fig. 6) indicates that the properties
are not balanced about the principle orientation direction. Such an
unbalance is also reflected in the modulus and ultimate properties
(Figs. 3, 4, and 6).

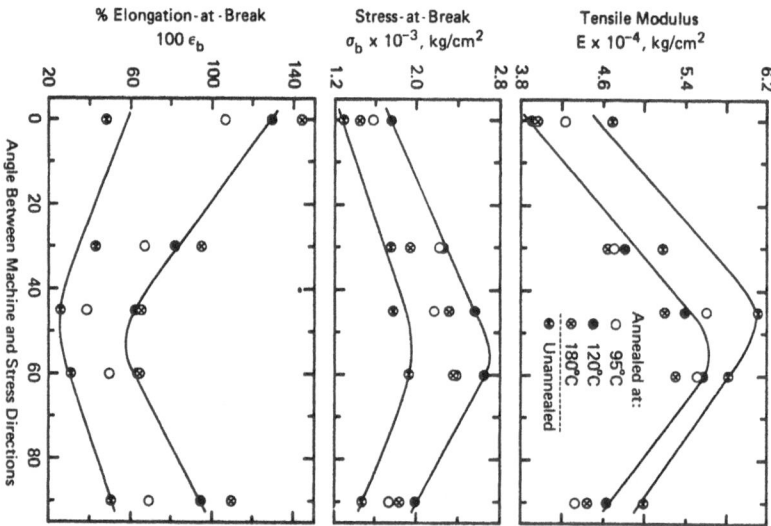

Fig. 7. Tensile modulus, stress-at-break, and elongation-at-break
obtained from stress-strain curves shown in Fig. 6.

Annealing increases the crystallinity and allows certain taught chains in amorphous regions to relax. The latter gives a reduction in the modulus; the higher crystallinity possibly accounts for the increased stress-at-break and elongation-at-break, shown by data in Figure 7. (The modulus data in this figure are from the initial slopes of the stress-strain curves in Figure 6.) With increased crystallinity, rotation of lamellae is more restricted, possibly accounting for the increased slope of the stress-strain curves beyond the yield point of annealed specimens, especially for spec- imens cut in the M.D. and T.D.

SUMMARY AND CONCLUSIONS

Unconstrained specimens of 1.5-mil Mylar film were annealed for three hours at 95°, 120°, and 180°c and thereafter their densities and thicknesses were determined. The degree of crystallinity, as indicated by density, increased progressively with annealing temper- ature. However, the anisotropic shrinkage during annealing at each temperature was essentially the same.

The principle orientation direction and the orientation distri- bution of lamellae vary across the 48-inch web. The extinction angle data in Figure 1 indicate that the principle orientation direction is close to the T.D. near the outer edge (provided in preparing the film, it is in fact stretched first in the M.D. and then in the T.D.); in moving across the web, β_m decreases progres- sively, i.e., the angle between the principle orientation direction and the M.D. diminishes. The birefringence, which is a measure of the average overall orientation, is smallest near the outer edge of the web, and it increases with decreasing β_m owing to a narrowing of the orientation distribution function.

Low-angle light-scattering patterns were measured on untreated and annealed specimens for which the extinction angle, β_m, was about 75° to the M.D.; thus, the principle orientation direction was close to the T.D. Patterns obtained on unstretched and stretched specimens indicate that: (1) rodlike superstructures exist and tend to be oriented in the M.D. and T.D.; (2) when a specimen is stretched at 45° to the M.D., oriented rodlike superstructures are formed, pre- sumably because disoriented crystallites orient in the stretch direction; (3) when specimens are stretched either parallel or per- pendicular to the M.D., the number and orientation of the rods are not changed but the order within the superstructures is increased; and (4) annealing increases the perfection within crystallites but causes some disorientation of crystallites, owing to the partial release of frozen-in strain. Because the extinction angle varies across the 48-inch web of Mylar A, the above statements about the orientation angles of superstructures apply only to specimens for which $\beta_m \simeq 75°$. At any location in the web, however, the super-structures are nearly--although not precisely--perpendicular to β_m.

The storage and loss moduli were measured at 35 Hz on untreated and 120°C annealed specimens cut at 0°, 45°, and 90° to the M.D.; for these specimens, $\beta_m \simeq 45°$. Except at elevated temperatures, specimens cut at 0° and 45° to the M.D. had the smallest and largest E', respectively; annealing reduced E', especially for the specimen cut 45° to the M.D. The temperature of maximum loss (E") for either the β- or α-relaxation process is essentially the same for all spec-imens. For the untreated specimens cut 0° and 90° to the M.D., a relaxation-type process, here designated α', was observed about 40°C above the α-peak. The α'-peak results from some type of structural change, tentatively assumed to be a slow reorientation of certain crystallites into the direction of applied stress.

Tensile stress-strain curves were measured on untreated and annealed specimens (for which $\beta_m \simeq 45°$) cut at five angles to the M.D. The yield stress σ_y is substantially the same for all spec-imens. In contrast, the modulus, the stress-at-break, the elonga-tion-at-break, and the shape of the stress-strain curve beyond the yield point vary considerably with the stretch direction. Further-more, annealing reduces the modulus, increases both the stress-at-break and elongation-at-break as well as the slope of a stress-strain curve beyond the yield point.

REFERENCES

1. C. J. Heffelfinger and K. L. Knox, Polyester Films, in "The Science and Technology of Polymer Films," Vol. II, O. J. Sweeting, ed., Wiley-Interscience, New York, 1971, p. 587.

2. N. Yoshioka and H. Sato, Chemistry of High Polymers (Japan) 26, 644 (1969).

3. V. G. Baranov, A. V. Kenarov, and T. I. Volkov, J. Polymer Sci. C, 30, 271 (1970).

4. M. A. Hughes and R. P. Sheldon, J. Appl. Polymer Sci. 8, 1541 (1964).

5. R. Bonart, Kolloid-Z. 213, 1 (1966).

6. W. J. Dulmage and A. L. Geddes, J. Polymer Sci. 31, 499 (1958).

7. C. J. Heffelfinger and P. G. Schmidt, J. Appl. Polymer Sci. 9, 2661 (1965).

8.   A. Hasegawa, Rept. Progr. Polymer Phys. Japan 12, 207 (1969).

9.   W. O. Statton, J. L. Koenig, and M. Hannon, J. Appl. Phys. 41, 4290 (1970).

10. M. L. Wallach, J. Polymer Sci. C, 13, 69 (1966).

11. M. B. Rhodes and R. S. Stein, J. Appl. Phys. 31, 1873 (1960).

12. W. H. Cobb, Jr. and R. L. Burton, J. Polymer Sci. 10, 275 (1953).

13. R. S. Stein, P. F. Erhardt, and W. Chu, J. Polymer Sci. A2, 7, 271 (1969).

14. W. Chu and R. S. Stein, J. Polymer Sci. A2, 8, 489 (1970).

15. M. Motegi, M. Moritani, and H. Kawai, J. Polymer Sci. A2, 8, 499 (1970).

16. D. G. Le Grand, private communication; also, J. Polymer Sci. A2, 8, 1937 (1970).

17. M. B. Rhodes and R. S. Stein, J. Polymer Sci. A2, 7, 1539 (1969).

18. C. D. Armeniades and E. Baer, presented at Pacific Conference on Chemistry and Applied Spectroscopy, San Francisco, Oct. 6-9, 1970.

19. I. M. Ward, J. Macromol. Sci.--Phys. B1, 667 (1967).

20. (a) N. Brown, R. A. Duckett, and I. M. Ward, Phil. Mag. 18, 483 (1968); (b) I. M. Ward, J. Polymer Sci. C, 32, 195 (1971).

21. O. Ishai, T. Weller, and J. Singer, J. of Materials 3, 337 (1968).

OPTICAL ANISOTROPY OF THE STATISTICAL SEGMENT IN BLOCK COPOLYMERS

D. G. LeGrand

General Electric Corporate Research and Development

Schenectady, New York 12301

SYNOPSIS

The optical anisotropy of the statistical segment in block copolymers of poly(dimethyl siloxane) bisphenol-A polycarbonate has
been determined for several different compositions and molecular
structures as a function of temperature and in the presence of
several different swelling fluids. A thermal transition in the
stress-optical coefficient (SOC) has been observed, which depends
on composition and structure of copolymer. In the swollen state
the SOC depends on the nature of the swelling fluid. Some liquids,
such as silicone fluids and cyclohexane, reduce the magnitude of
the statistical segment anisotropy while other liquids, such as
CCl_4, increase it. An interpretation of these results in terms of
local internal field variations, optical and geometric anisotropy
of the swelling fluid, and the molecular dynamics of the block copolymer molecules is presented.

INTRODUCTION

The optical properties of oriented heterogeneous materials such
as polycrystalline polymers, block copolymers, swollen rubbers, and
sheared polymer solutions can be used as a direct indication of the
orientation and morphology of such materials, if the effect of local internal field fluctuations, which occur as a result of the
heterogeneous nature of such materials, can be determined. In such
systems, when the size of the heterogeneous regions or the distance
over which the regions are correlated are of the order of one-tenth
the wavelength of light or greater, the local internal field fluctuations, which occur at boundaries of the heterogeneous regions may
give rise to a shape or form birefringence.

While it is true that Wiener has derived approximate equations for the birefringence of systems containing asymmetric particles, adequate experimental verification of this theory is lacking and we have previously pointed out that Wiener's theory fails to take into account sample morphology.[1,2] A rigorous theory of the birefringence of block copolymers requires the calculation of the polarizability tensor of a macromolecule from the properties of the segments which make up the polymer chain and a calculation of the effective field arising from the interaction of a segment with its surroundings and the applied field.

For a block copolymer, consisting of two different types of segments a and b, the following four different regions can exist within a sample: a's surrounded by a's, b's surrounded by b's, a's surrounded by b's, and b's surrounded by a's. While the calculation of the optical properties of such a system might seem to present a formidable theoretical problem, we felt that the application of the recent theoretical work of Bullough would be appropriate because it included not only the effect of local fluctuations in the optical properties but also pair-wise correlation effects.[3]

By generalizing Bullough's work the refractive index of an oriented block copolymer along the σ-direction is obtained

$$\frac{\eta_\sigma^2-1}{\eta_\sigma^2+2} = \frac{4\pi}{3} \quad (N_A \, \bar{\eta}_a^\sigma + N_B \, \bar{\eta}_b^\sigma) + N_A^2 \, (\bar{\eta}_a^\sigma)^2 \, J_{aa}^\sigma$$

$$+ N_B^2 \, (\bar{\eta}_b^\sigma)^2 \, J_{bb}^\sigma + N_A \, N_B \, \bar{\eta}_a^\sigma \, \bar{\eta}_b^\sigma \, (J_{ab}^\sigma + J_{ba}^\sigma) \tag{1}$$

where η_σ is the refractive index,
N_i is the fraction of segments of type i,
$\bar{\eta}_i^\sigma$ is the average polarizability of the i segment along the σ-direction, and
J_{ij}^σ terms are functions which give a description of the field in terms of an interaction tensor and the correlation between two segments in terms of their position and orientation.

It is important to realize that the use of this equation in the present context is similar to the use of the Lorentz-Lorenz equation for calculating the refractivity of chemical compounds from their constituent elements.[4] It is necessary to point out that equation (1) must be extended to include more terms if the number of chemical bonds of type a-b become too large.

Utilizing equation (1) the birefringence is then given as

$$\Delta = \varphi_A \Delta_A + \varphi_B \Delta_B + \varphi_A^2 \, [\, (\eta_A^1)^2 \, J_{aa}^1 - (\eta_A^2)^2 \, J_{aa}^2 \,]$$

$$+ \, \varphi_B^2 \, [\eta_B^1)^2 \, J_{bb}^1 - (\eta_b^2)^2 \, J_{bb}^2]$$

$$+ \, \varphi_A \varphi_B \, [\eta_a^1 \, \eta_b^1 \, (J_{AB}^1 + J_{BA}^1) - \eta_a^2 \, \eta_b^2 \, (J_{AB}^2 + J_{BA}^2)] \qquad (2)$$

While this equation gives an accurate description of the birefringence in terms of the sample's constituents and morphology, it
does not have a direct relationship to the deformation in the
sample. Further, we have not been able to evaluate the J_{ij}^σ terms.

On the other hand, the Kuhn-Grün[5] theory for the birefringence
of rubbery polymers has recently been extended to copolymers by
Stein and Shindo.[6,7] The stress optical coefficient (SOC) for rubbery copolymers is given by

$$C = \frac{\Delta}{\sigma} = \frac{2\pi}{45kT} \, \frac{(\overline{\eta}^2 + 2)^2}{\overline{\eta}} \, \overline{(b_1-b_2)}_s \qquad (3)$$

where k is Boltzman's constant, T is the absolute temperature, σ is
the stress, Δ is the birefringence, $\overline{\eta}$ is the average refractive index and $\overline{(b_1-b_2)}_s$ is an average segment anisotropy of polarizability,
defined by

$$\overline{(b_1-b_2)}_s = \frac{\Sigma \, N_i \, (b_1-b_2)_{si} \, L_i^2}{\Sigma \, N_i \, L_i^2} \qquad (4)$$

where
$\overline{(b_1-b_2)}_{si}$ is the anisotropy of the polarizability of the ith segment, N_i is the total number of segments of the ith type, and L_i
is the statistical segment length of the ith segment.

They stated that these equations are applicable only if the
polymer is homogeneous. Further, if domains, associated regions,
or microphase separation occurred, even though both regions were
rubbery, the theory would not be generally applicable because of
the failure of an affine transformation assumption.[6,7] This latter
statement would appear to require that the Van der Waals Forces or
other intermolecular forces are sufficiently strong to withstand
the applied mechanical stresses. If the segments within a domain
are randomly oriented and the domain shape is invariant with respect to the applied stress, then the domains will act like

multifunctional crosslinks and/or filler particles. For such materials, other than small effects, which might arise from the optical boundary between the domains and their surroundings, the birefringence from domains would be negligible. Studies of the optical properties of block copolymers of styrene-butadiene and
styrene-isoprene indicate agreement with these hypotheses.[8,9]
Since previous studies of the stress-strain birefringence behavior
of alternating block copolymers of polydimethyl siloxane and
bisphenol-A polycarbonate did not agree with these concepts, we
postulated a different model for our materials, which is presented
in Figure 1.[10] In our model we assumed that some of the polycarbonate blocks were associated to form domains which act as physical

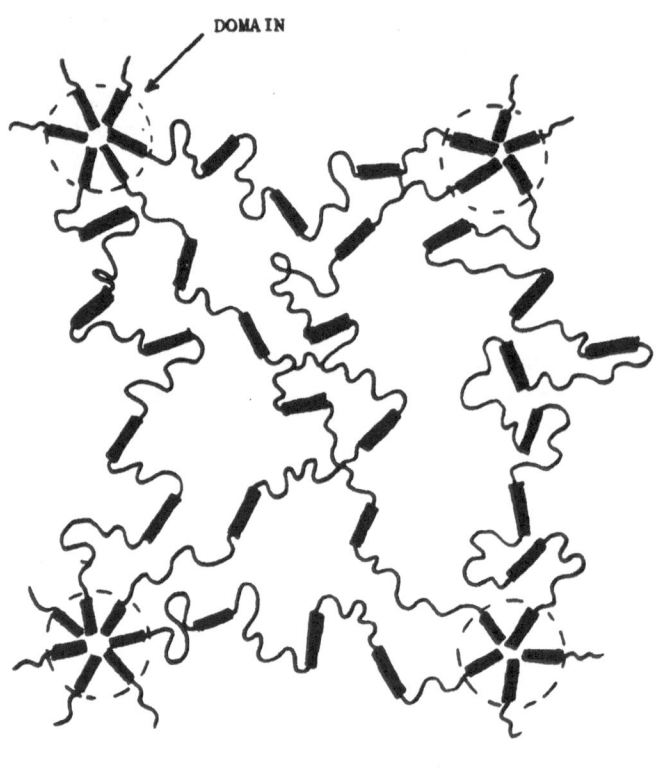

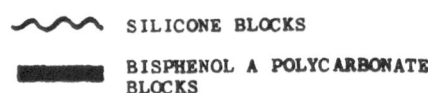

SILICONE BLOCKS

BISPHENOL A POLYCARBONATE
BLOCKS

Fig. 1 - A physical model of the morphology of block copolymers of
bisphenol-A carbonate dimethyl siloxane

crosslinks and that the matrix was composed of both silicone and polycarbonate blocks. We have offered the following experimental evidence in support of this model.

1. Small-angle x-ray studies indicate that the domains are not pure polycarbonate and that the domains are not dissociated by heating to elevated temperatures.[11]

2. Dielectric and optical studies indicate that the degree of association is dependent on block sizes and composition.[12]

3. We have shown that the mobility of the protons on the phenyl, gem dimethyl, and silicone methyl groups can be detected and measured in swollen samples by high resolution proton magnetic resonance.[13] Our studies indicated that the fraction of the mobile protons in swollen samples depended on the molecular structure, the sample morphology, the swelling liquid, and the degree of swelling.

In order to further elucidate how molecular structure and dynamics, as well as sample morphology, influence the optical and mechanical properties of these materials, we undertook a detailed study of irradiation crosslinked samples. Our primary objectives were to determine the optical anisotropy of the statistical segment at elevated temperatures, and in the swollen state where the blocks are flexible, and to compare this to theory of Stein and Shindo.[6,7]

We show that while the Stein-Shindo theory in contrast to Bullough's theory is highly over-simplified because it neglects the complex internal field effects, which reflect local density, order, and composition, which in turn can give rise to both form dichroism and birefringence, it can be used under the proper experimental conditions to evaluate the optical anisotropy of the statistical segment.

EXPERIMENTAL

The samples, which were used in this study, were supplied by Dr. H. Vaughn. Material characterization data have been presented previously[12] and are given in Table I. Crosslinking was introduced by electron irradiation. The nominal irradiation dose was 30 Mrad. The irradiated samples acquired a yellowish coloration which did not interfere with the measurements.

The stress, strain and birefringence measurements were carried out in the same way as described previously.[8,12]

In order to carry out measurements as a function of temperature, a single glass-walled chamber, which had heating coils around it and which was covered with asbestos, was used. The temperature

was regulated by a Cenco thermoregulator and controlled to ap-
proximately$\pm$2°C of the desired temperature.

Technical grade organic reagents were used for the swelling
studies. The samples were immersed in the swelling agents and
were allowed to come to equilibrium before the birefringence measurements were made. The weight and volume fraction uptake of the
swelling agent was determined by direct weighing and refractometric
measurements respectively.[2] The force, deformation and optical
retardation measurements were made with the sample immersed in the
swelling agent. In order to determine the birefringence and stress
of the strained swollen sample, the width and thickness were measured in place. The optical anisotropy calculations were corrected
for the change in the sample refractive index as indicated by
Treloar.[14]

RESULTS

In Figures 2 and 3 we show the behavior of the SOC as a function of temperature for the samples listed in Table I. From these
curves it is apparent that a transition is occurring within the
samples and this transition is dependent upon the structure of the
sample. In Table II we summarize these transition points and

TABLE I

Sample Characterization Data

Sample #	$\%S$ (a)	$\overline{DP}_s$ (b)	$\overline{M}_n$ (c)	$\rho(\frac{gm}{cm})$	$\underline{n}$
1	64	20	96,000	1.06	1.48
2	65	40	91,600	1.05	1.48
3	65	100	38,400	1.08	1.46
4	80	40	79,000	1.035	1.44
5	83	61	147,500	1.06	1.44
6	75	20	36,000	1.08	1.45

(a) % by weight $S_i(CH_3)_2O$

(b) Average degree of polymerization of polydimethylsiloxane blocks

(c) In figure 1, the digits following the sample number indicate the $\overline{D.P.}$ of the
    silicone blocks and the $\overline{M}_n$ x 10^3

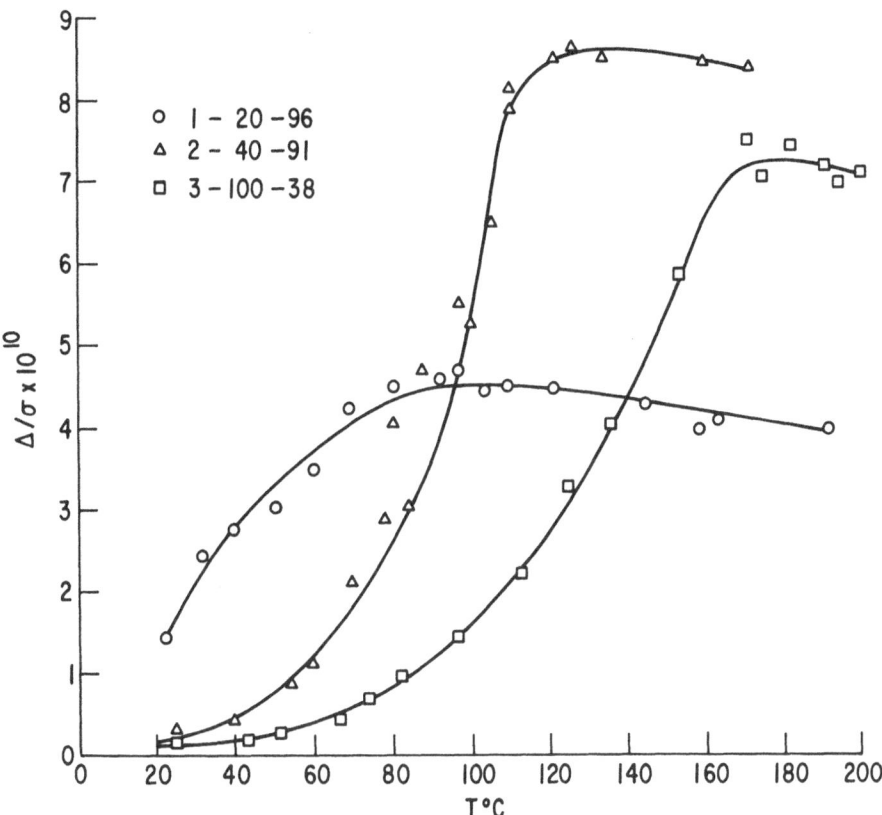

Fig. 2 - The temperature dependence of the stress
optical coefficient for samples 1, 2, and 3

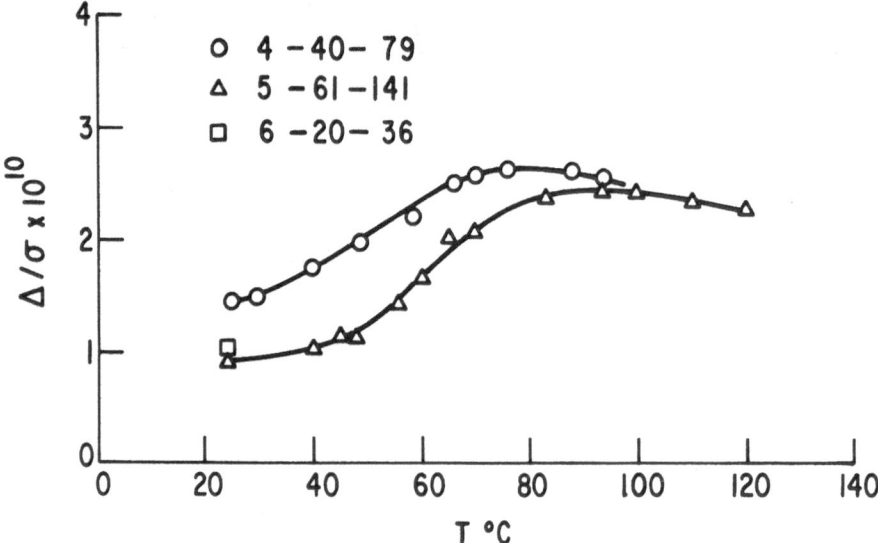

Fig. 3 - The temperature dependence of the stress optical coefficient for samples 4 and 5

TABLE II

Thermal Transition Temperatures

Sample #	S.O.C. Trans., °C	Mechanical[a,b] Trans., °C	Diff. Scanning[c] Calorimetry, °C	Volumetric Trans., °C
4	50-60			--
1	45-60	~50		45
5	55-80			--
2	50-120	~70	59	70
3	65-160	~120	134	90

(a) R. P. Kambour, J. Poly. Sci. 7B, 573 (1969)

(b) M. Narkis & A. V. Tobolsky, J. Macromol. Sci.-Phys., B4(4), 877 (1970)

(c) See Reference 4

compare them to mechanical, volumetric and DSC data. While the absolute value of the temperature at which the transition occurs is dependent on the method of determination, there is a general correlation in their positions. Since considerable evidence has been offered to indicate the presence of domains, the SOC data would suggest a greater ease and better orientation of the bisphenol-A polycarbonate (BPA) blocks. This is supported by the temperature dependence of the dielectric and mechanical loss, which indicates an increase in the mobility of the BPA species.

In Table III we summarize the SOC data at both room temperature and elevated temperature and values of optical anisotropy of the statistical segment.

In order to separate those changes which occur as a result of the thermally induced mobility of those polycarbonate blocks surrounded by silicone from those polycarbonate blocks located in domains, we have measured the SOC in several different swelling fluids. The data for this series of experiments are summarized in Table IV. Neglecting, for the moment, local internal field effects, we expected that fluids, which are good solvents for polycarbonates such as methylene chloride, would cause the domains to dissociate and to increase the molecular mobility of all polycarbonate blocks in the material. This would lead to a larger fraction of the polycarbonate blocks which could be oriented and possibly a higher degree of orientation. On the other hand, liquids such as cyclohexane and silicone fluids which are nonsolvents for polycarbonate would not dissociate the domains but would enhance the mobility of those polycarbonate blocks which are located outside the domains.

Comparison of the MDM and cyclohexane data in Table IV with
the data on dry samples in Table III suggests that cyclohexane
does induce a change while MDM does not.   These findings are in
agreement with NMR studies of swollen samples, which indicate that
BPA mobility is induced by swelling in deuterated cyclohexane but
not in MDM.[13]   The data on the samples swollen with DMP and Aroclor
indicate that these materials are partially soluble in silicones
but we have found them to be better solvents for polycarbonate than
silicones.   We believe that the higher SOC and optical anisotropies
are due to the swelling liquids, i.e. DMP and Aroclor, which may
occur as a result of their geometric and optical anisotropy.

TABLE III

Sample #	T°C	$\frac{\Delta}{\sigma}$ x 10^{-10}	$(b_{\parallel} - b_{\perp})_s$ x 10^{-25}
1	25	2.5	61.8
	70	4.3	125.0
4	25	1.5	38.4
	80	2.6	80.7
2	25	.31	35.96
	123	8.2	282.0
5	25	.1	25.3
	80	2.6	79.8
3	25	.145	17.9
	180	7.43	296.0
6	25	1.0	25.3
Silicone	25	.24	6.2
Rubber	25	.22	6.1
Bisphenol-A	25	2.34[a]	55.0[a]
Polycarbonate	25	.33[b]	7.6[b]
	25	1.95[c]	44.7[c]
	25	5.5[d]	139.0[d]
	200	3.7	138.0

(a) Theoretical value calculated by bond polarizability method

(b) Value obtained at strain below yielding

(c) Value obtained by extrapolation of data at elevated tempera-
    ture to room temperature

(d) Sample swollen in methylene chloride.

TABLE IV[*]

Sample	Rm. Temp.[b]	Trans. Temp.	Methylene Chloride	Carbon Tetrachloride	Cyclohexane	Silicone Fluid	Theoretical[a]
1	62.0	125.0	34.8	102	148	89.0	113
4	38.4	80.7	46.6	131			96.3
2	35.9	282.	27.9	93.2	280	420	113
5	25.3	79.8	35.8	106			96.3
3	17.9	296	24.3	89.6	5.06	18.5	113
6	25.3	--					
Silicone Rubber	6.2		6.1				
Bisphenol-A Polycarbonate		138	139				

[*]Experimental and theoretical values of segmental anisotropy, i.e. $(b_{\parallel} - b_{\perp})_s \times 10^{-25}$

(a) Calculated from Equation (2)

(b) See Table III

DISCUSSION AND CONCLUSIONS

The variation of the mechanical and optical properties of these copolymers as a function of temperature and as a function of the swelling agent are related to basic changes in the molecular order and mobility in the material. Since small-angle x-ray scattering studies indicate the presence of domains to temperatures of 200°C and mechanical, optical and dielectric measurements indicate a transition between room temperature and 160°C, which depends on the structure and composition of the copolymer, we believe that this transition corresponds primarily to a change in the rotational mobility of the species within the domains. On the other hand, swelling with different fluids leads to changes in the mobility of species both in and outside of domains as well as dissociation of the domains. Proton magnetic resonance, small-angle x-ray scattering and dielectric measurements on samples which have been swollen with these different fluids indicate that (1) methylene chloride increases mobility of all polycarbonate blocks and causes domains to dissociate; (2) dimethyl phthalate increases the mobility of polycarbonate blocks in domains but does not cause domains to dissociate; (3) cyclohexane increases the mobility of polycarbonate blocks in matrix but does not affect those within domains; (4) carbon tetrachloride increases the mobility of all polycarbonate blocks; and (5) silicone fluids do not affect significantly the mobility of any of the polycarbonate blocks.

With these points in mind and comparing the SOC data on the dry and swollen samples, it is obvious that the segmental mobility and the degree of dissociation are significant factors. One of the more significant findings is the agreement between the Stein-Shindo theory and the results which are obtained for samples swollen with carbon tetrachloride. On the other hand, the disagreement between the Stein-Shindo theory and samples either dry or swollen with methylene chloride and silicone fluids, which have approximately the same refractive index, serves as an indicator of the necessity for a more complete theory of the stress optical coefficient of block copolymers, which takes into account both sample morphology as well as segmental mobility. Bullough's work is a step in the right direction but must be extended in order to include stress and deformation dependence. Concomitantly, experimental determination of the changes in segmental mobility as measured by dielectric and nuclear magnetic resonance as well as morphological changes as measured by optical scattering such as small- and wide-angle x-ray scattering should be obtained in order to aid in the development of a better theory of the optical properties of block copolymers.

REFERENCES

1.   O. Wiener, Abhandl. Kgl. Sächs. Ges. Wiss. Math-Physik Klasse, 32, 509 (1912)

2.   L. M. Normandin and D. G. LeGrand, J. Poly. Sci. A-2, 7, 231 (1969)

3.   R. K. Bullough, Phil. Trans. Roy. Soc. A259, 397 (1962)

4.   H. A. Lorentz, "The Theory of Electrons" (Leipzig: Teubner, 2nd edition, 1916; reprinted by Dover Publications (New York), 1952)

5.   W. Kuhn and Grün, Kolloid Zeit. 101, 248 (1940)

6.   Y. Shindo and R. S. Stein, J. Poly. Sci. A-2, 7, 2115 (1969)

7.   Y. Shindo, R. S. Stein, and B. E. Read, Macromol. Chem. 118, 272 (1968)

8.   J. F. Henderson, K. H. Grundy, and E. Fischer, J. Poly. Sci. C(16), 3121 (1968)

9.   E. Fischer and J. F. Henderson, J. Poly. Sci. C(26), 149 (1969)

10.   D. G. LeGrand, Polymer Letters 7, 579 (1969)

11.   D. G. LeGrand, Polymer Letters 8, 195 (1970)

12.   T. L. Magila and D. G. LeGrand, Poly. Eng. & Sci. 10, 349 (1970)

13.   D. G. LeGrand, Trans. Rheo. Soc. 15, 541 (1971)

14.   L. R. G. Treloar, "The Physics of Rubber Elasticity", Oxford Univ. Press, 2nd ed., 1967

THE MORPHOLOGY OF IONOMERS

E. P. Otocka

Bell Laboratories

Murray Hill, New Jersey   07974

Ionomers have been studied intensively for the last six years. A variety of different analytical techniques have indicated that these polymers possess unique properties. However, the morphological features of these materials which are responsible for the properties have proven difficult to define.

Extensive mechanical testing has been carried out on ethylene/ acrylic acid and ethylene/methacrylic acid copolymers and their various salts.[1-6] The T_β of the parent polymers, −18 to 27°C depending on acid type and content disappears on neutralization, being replaced by two transitions, one at −20 to 0°C and another at 40 to 70°C. These observations lead to the postulation of a separate "ionic domain" phase associated with the higher temperature transition.[1-3,5]

Electron microscopic search for direct observation of the domains has provided mixed results. In some cases, positive results appear[7] while in other cases equally negative observations are found.[8] The same lack of consistency has been found in observations of the parent acid copolymers.[8,9]

X-Ray diffraction studies yield slightly more consistent results. A new maxima appears in the ethylene/metal methacrylate copolymers,[10] has been observed at much diminished intensity in ethylene/metal acrylate and butadiene/metal methacrylate copolymers and is not found in the methyl methacrylate/metal methacrylate system.[8] This scattering corresponding to a "lattice" spacing of 23-40 Å is relatively insensitive to cation type. A low angle peak, corresponding to.80 Å has been found only for heavy metal counterions in the ethylene/methacrylic acid system.[11] A radial

distribution function achieved by the Fourier Transform of wide
angle data on an ethylene/cesium acrylate sample indicates paired
ion pair complex formation but does not indicate larger cluster
formation.[12]

Continuing studies of relaxation by X-Ray diffraction and
infrared dichroism[13,14] indicate that the orientation function of
both hydrocarbon and ionic segments of ionomers increase as the
material undergoes the higher (40-70°C) temperature transition.
These findings indicate the possibility that the ionic groups may
be associated with the disrupted crystalline portions of the backbone in annealed samples. Evidence for such association of car-
boxylic acid groups in the parent copolymers had previously been
found in a radiolytic study.[15]

As indicated in a recent review, there are a few unambiguous
results in this field, despite the application of advanced analytical methods.[16] Traditionally, major advances in polymer science
have been possible considering the molecules statistically or as
featureless ball and spring models. Significant advances in the
field of ionic polymers appear to await studies which take into
account structural features of the interacting groups.

REFERENCES

1. R. Longworth and D. J. Vaughan, Polymer Preprints, 9, 525
    (1968).

2. W. J. MacKnight, T. Kajiyama and L. McKenna, Polymer Eng. Sci.,
    8, 267 (1968).

3. W. J. MacKnight, et al., J. Appl. Phys., 38, 4208 (1967).

4. E. P. Otocka and T. K. Kwei, Macromolecules, 1, 244 and 401
    (1968).

5. W. J. MacKnight, et al., J. Phys. Chem., 72, 1122 (1968).

6. L. W. McKenna, T. Kajiyama and W. J. MacKnight, Macromole-
    cules, 2, 58 (1969).

7. H. A. Davis, R. Longworth and D. J. Vaughan, Polymer Preprints,
    9, 515 (1968).

8. M. Matsuo, E. P. Otocka, unpublished results.

9. C. L. Marx, J. A. Koutsky and S. L. Cooper, J. Polym. Sci. B,
    9, 167 (1971).

10. F. C. Wilson, R. Longworth and D. J. Vaughan, Polymer Pre-
     prints, 9, 505 (1968).

11. B. W. Delf and W. J. MacKnight, Macromolecules, 2, 309 (1969).

12. R. J. Roe, Polymer Preprints, 12, 730 (1971).

13. T. Kajiyama, T. Oda, R. S. Stein and W. J. MacKnight, Macro-
     molecules, 4, 198 (1971).

14. Y. Uemura, R. S. Stein and W. J. MacKnight, Macromolecules,
     4, 490 (1971).

15. E. P. Otocka, T. K. Kwei and R. Salovey, Die Makromol. Cheimie,
     129, 144 (1969).

16. E. P. Otocka, J. Macromol. Sci-Revs. Macromol. Chem. C, 5,
     275 (1971).

STUDY OF MOLECULAR ORIENTATION IN POLYMERS BY FLUORESCENCE POLARIZATION

G. E. McGraw

Research Laboratories, Tennessee Eastman Company, Division

of Eastman Kodak Company, Kingsport, Tennessee 37662

INTRODUCTION

The morphology of a polymeric system determines to a large extent the ultimate physical properties of the polymer. With such importance placed upon polymer structure, it is not surprising that many physical methods have been used over the years to measure the molecular orientation of polymers. Spectroscopy is one of those tools which has been applied to the characterization of polymers. A spectroscopist can gather three types of information: (1) the band's frequency, (2) the band's intensity and contour, and (2) the band's state of polarization. (Polarization concerns the direction of the electric vector of the radiation with respect to the direction of the particular transition moment in question.) Obviously, a considerable amount of information can be obtained from the first two types of data. This paper discusses the usefulness of the third type of data — measurement of the state of polarization of the interacting radiation.

Perhaps the most well-known use of polarization is the measurement of infrared (ir) or visible dichroism. The intensity of the absorption is dependent upon the angle between the electric vector of the irradiating beam and the direction of the ir or visible transition moment. Maximum absorption of the incident radiation occurs when the electric vector and the transition moment are parallel, and the absorption is zero when the two are perpendicular. Therefore, when the absorption intensity is measured both parallel and perpendicular to some molecule-fixed axis, something can be learned about the direction or orientation of the transition moment.

The fluorescence polarization method is based upon the concept of alignment of electric vectors.[1,2] It differs from conventional dichroic measurements, however, because it utilizes the optical anisotropy of the sample in two ways: the state of polarization of the incident beam upon absorption and the state of polarization of the emitted fluorescent beam are both involved.

EXPERIMENTAL

Methods

The theory of fluorescence polarization has been described[1-3] and two basic equations can be used in studying the theory. Fluorescence intensity (I) is given by

$$I = KWV[\bar{P} \cdot \bar{C})^2(\bar{A} \cdot \bar{C})^2]_{avg}$$

under the conditions that the radiation is absorbed and emitted along the same molecular direction, denoted by the vector $\bar{C}$, and that the fluorescent probe does not rotate appreciably during the lifetime of excitation. K is an instrumental constant, W is the concentration of the fluorescent species, and V is a factor to allow for changes in the illuminated volume of the sample. $\bar{P}$ and $\bar{A}$ are unit vectors in the directions of the electric vectors of the exciting radiation (polarizer) and the fluorescent radiation (analyzer), respectively. Of course, the total intensity is equal to the sum of the contributions from all the fluorescent probes in the sample. For further solution of the molecular orientation problem, a distribution function relating the orientation of the $\bar{C}$ vectors to the macroscopic coordinate axes of the sample may be defined. The intensity expressions which result would then depend upon whether random, uniaxial, biaxial, or conical symmetry or some more complex combination of these were used.

Another parameter which is useful in the study of fluorescence polarization is the degree of polarization (P) normally defined as

$$P = \frac{I_{||} - I_{\perp}}{I_{||} + I_{\perp}}$$

$I_{||}$ represents the intensity with the electric vectors of the polarizer and analyzer parallel to one another; and $I_{\perp}$ represents the intensity with the electric vectors perpendicular to one another. Since the intensity is a function of the angle (ω) between the vector $\bar{C}$ in the sample and the polarization directions, then P is likewise a function of that angle. In practice, the angular distribution of the degree of polarization is plotted from $\omega = 0$ to 2π.

Figure 1 illustrates the angular dependence of the intensity and degree of polarization for a perfect uniaxial orientation in a sample; that is, all of the probes are aligned parallel to the same vertical axis. The intensity of the fluorescence with parallel polarizer and analyzer will be a maximum with perfect vertical alignment of the sample, but it will rapidly fall off as the fourth power of the cosine ω. The pattern will repeat in each of the four

ORIEN-
TATION
AXIS

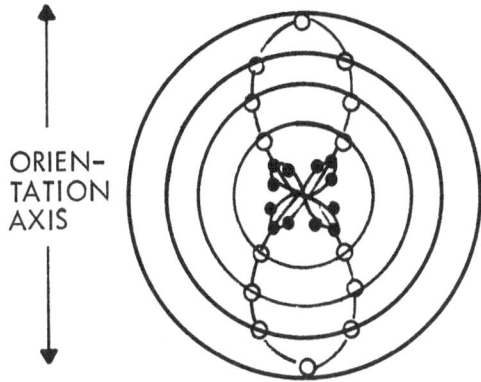

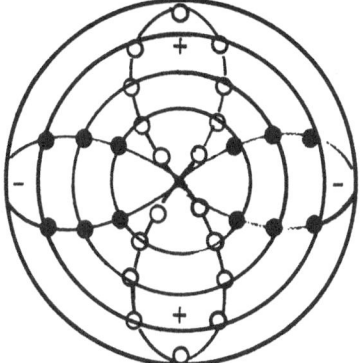

ANGULAR DISTRIBUTIONS OF
POLARIZED COMPONENTS OF
FLUORESCENCE INTENSITY (o) $I_{\parallel}$
AND (•) $I_{\perp}$.

ANGULAR DISTRIBUTION OF THE
DEGREE OF POLARIZATION OF
FLUORESCENCE FOR (o) POSITIVE
AND (●) NEGATIVE VALUES.

Fig. 1.   Perfect Uniaxial Orientation.

quadrants. If, on the other hand, the polarizer and analyzer are crossed,
then clearly the maximum intensity must occur at an angle of $\pi/4$. The
angular distribution of the degree of polarization follows as a consequence of
this angular dependence. P will be a maximum value along the alignment
axis where $I_{\perp}$ is zero; it will go through zero where $I_{\parallel}$ is equal to $I_{\perp}$; then,
it will become negative as $I_{\perp}$ becomes greater than $I_{\parallel}$.

Authors have the option of reporting either one of the intensities or the
degree of polarization. Some authors[4] have chosen to report only one intensity,
such as $I_{\parallel}(\omega)$, but the data reported should not be limited to only the intensity
function. Certain instrumental difficulties, such as parallel diffraction anom-
alies from ruled gratings,[5] will influence each intensity function differently,
and a correction factor must be introduced. The correction can be determined
by exciting the sample with horizontally polarized light and analyzing the
emission again with the electric vectors of the polarizer and analyzer both
parallel and perpendicular to each other.   If there is no effect on the state of
polarization by the fluorometer, the two sets of P's will be identical. In
practice, the sets do differ and a correction for the instrument must be applied
so that the sets become equivalent.

A schematic of the apparatus for measuring the angular distribution of the
polarized fluorescence intensity is shown in Fig. 2. The polymer sample can
be either a solution, a film, or several fibers wrapped in a parallel fashion.
Radiation from a UV source is filtered and polarized before it impinges on the

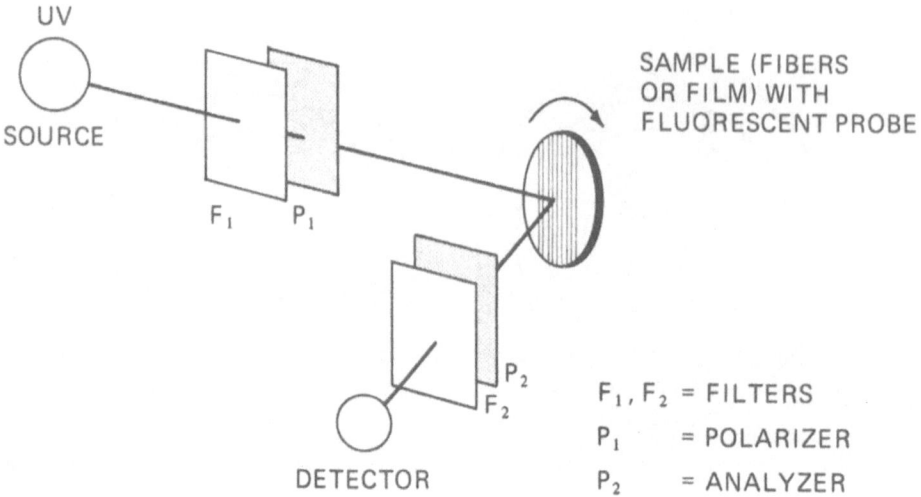

Fig. 2.   Apparatus for Measuring the Angular Distribution of the Polarized
Fluorescence Intensity.

sample and causes fluorescent radiation to be emitted from probes placed
in the sample.   The fluorescent radiation passes through an analyzer and
another filter and is detected by a photomultiplier.   Filters or monochromators
are used so that there is no stray light in the exciting beam and so that the
exciting light does not reach the detector.   By measuring the fluorescence
intensity at various points during the rotations of the polarizer, sample, and
analyzer, we can characterize the orientation of the probes in the sample.

It is assumed that the fluorescent probes assume or imitate the spatial
arrangement of the polymer chains making up their environment.   Figure 3
illustrates this point schematically.   The fluorescent probe shown here, 2,2'-
(vinylenedi-p-phenylene)bisbenzoxazole, has a $\pi-\pi^*$ transition moment
along the long axis of the molecule.[6] These transition moments are shown
as short line segments in the polymer network.   If the polymer chains have no
order, the probes assume a random orientation, and there is no angular dependence of the fluorescent intensity.   However, if the polymer chains have
some type of order as shown in Fig. 3, the probes also reflect that order.
Because of the bulky nature of the fluorescent probes, we assume that they
occur only in the amorphous regions of the polymer.   The model in Fig. 3 is
shown only to convey some polymer order; we are not necessarily proponents
of the simple fringed-fibril model.

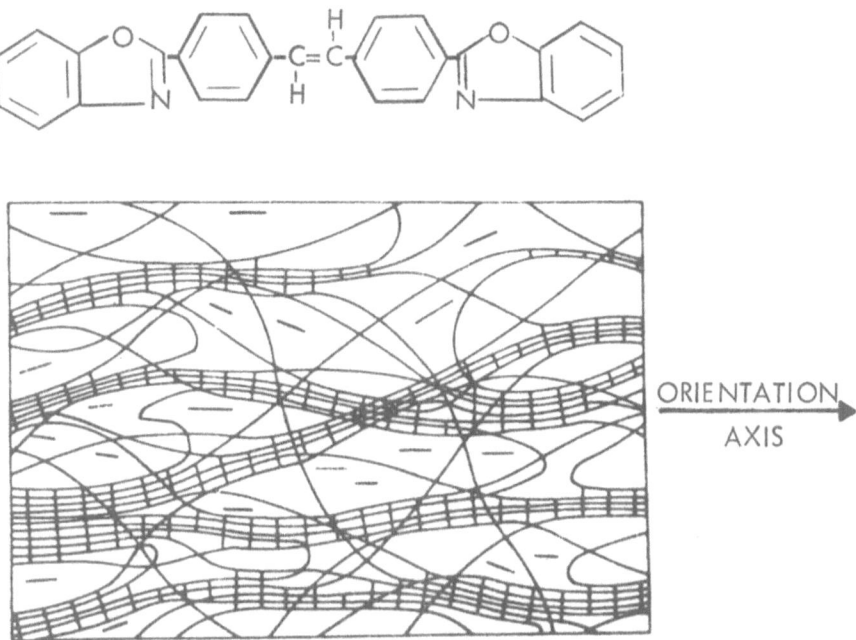

Fig. 3. Schematic of Magnified Polymer Structure Showing the Arrangement of the Probe in Amorphous Regions.

Data Acquisition

We have used several different spectrometers in our laboratory for fluorescence polarization measurements. The preliminary experiments were performed with a commercial Aminco-Bowman spectrophotofluorometer equipped with a Glan-prism polarizing assembly. Later, we purchased an American Instrument Co. Model SPF-125 fluorometer. This instrument consists of two manually controlled grating monochromators: one selects the excitation wavelength, and the other selects the emission wavelength. We chose a grating instrument instead of a filter·instrument to facilitate the study of several fluorescent species having different spectral characteristics. The grating instrument eliminates the need of carefully matching a series of filters. Glan prisms were also used for the polarizer and the analyzer. The optical components of the instrument were arranged so that viewing of the emission was perpendicular to the exciting beam, and the incident beam made an angle of 45° to the surface of the rotating polymer sample.

Initially, the fluorescence intensity was read from the fluorometer point by point and punched onto computer cards for further processing. When this became too cumbersome, we directly interfaced our equipment to a Digital Equipment Corporation PDP 8/S computer. Collection time was reduced from about $1\frac{1}{2}$ hr per sample to about 7 min per sample, and binary tape was directly generated for further data reduction in a larger computer.

We soon found that improvements in design and sample preparation made it no longer necessary to time-average the data. Therefore, instead of tying up a more expensive piece of equipment, we installed a Digitec data acquisition system. Figure 4 shows the fluorometer, a 60-Hz motor to continuously rotate the sample, an operational amplifier, and the data collection system. The data collection system consists of a digital voltmeter and a punch

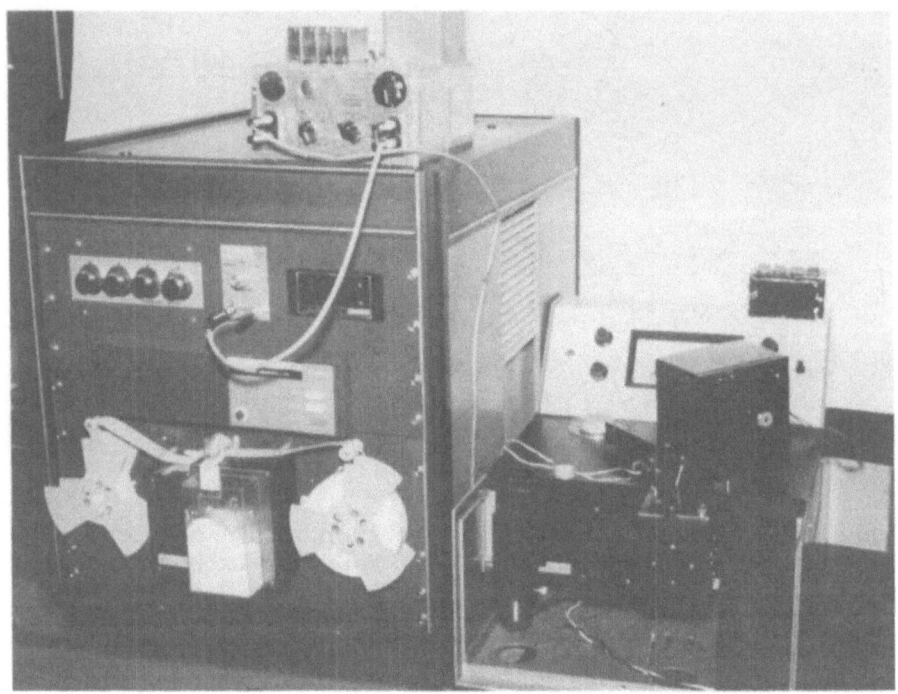

Fig. 4.   SPF-125 Fluorometer With 60 Hz Motor and Digitec Data
          Acquisition System.

controller which punches the data onto paper tape in IBM-odd parity code. A microswitch automatically triggers data collection at a preset angular position, and a manually operated rotary switch bank identifies the sample. The time of data collection with this unit is also about 7 min per sample. The output from the larger IBM computer is shown in Fig. 5. It consists both of a tabulation of the degree of polarization as a function of sample rotation and of a Cal-Comp plot of that data. A fiber sample is also shown to indicate the sample size.

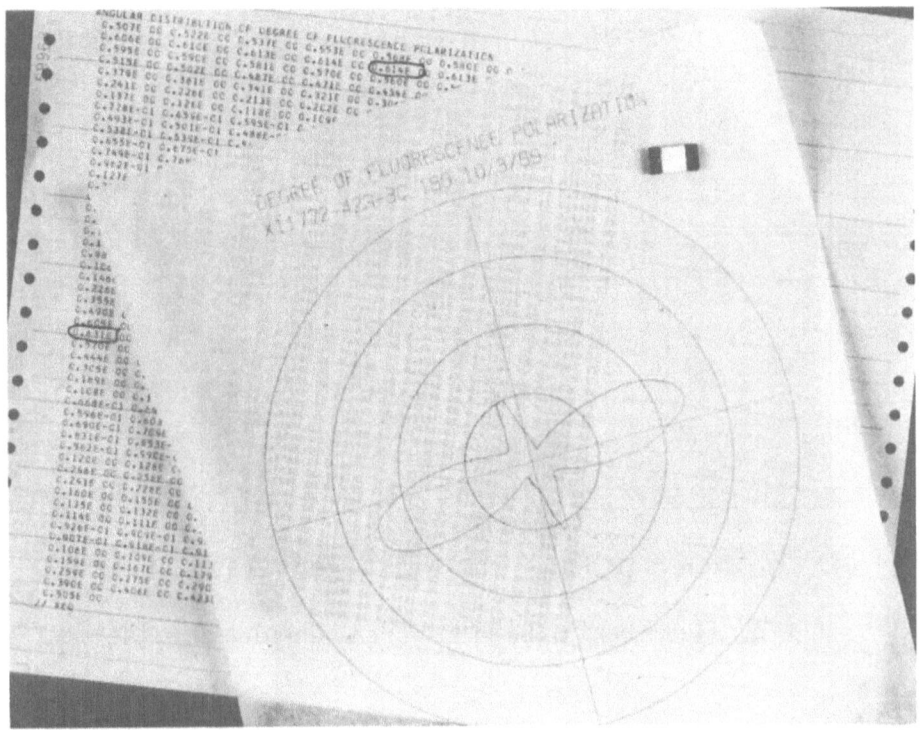

Fig. 5. Computer Output Showing the Angular Distribution of the Degree of Fluorescence Polarization. A mounted sample appears in the upper right hand corner.

At the time we initiated our studies there were no instruments available commercially. However, since that time an instrument has been introduced by Japan Spectroscopic Co., Ltd. and their representative Syntex Analytical Instruments. The JASCO unit utilizes a reference beam and can be used for transmission and scattering measurements; also, it contains a goniometer for three-dimensional adjustment of the sample. On the other hand, it is not computer interfaced and only the intensity data can be recorded; it uses filters which must be matched to the fluorescent probes selected; and it requires the same sampling time as our instrument.

Fluorescent Probe

Most of the work discussed in this paper was done with one fluorescent probe, 2,2'-(vinylenedi-p-phenylene)bisbenzoxazole (Fig. 3). The spectral characteristics of that probe are shown in Fig. 6. The excitation and emission spectra are shown both in methylene chloride and in poly(ethylene terephthalate)

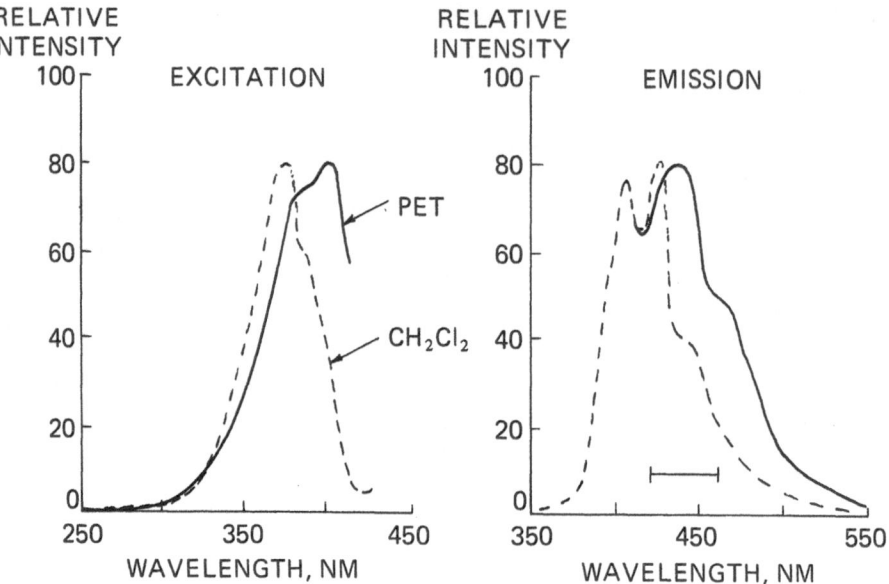

Fig. 6. Fluorescent Probe in Methylene Chloride and PET.

(PET) film. The excitation wavelength was normally 365 nm, and the emission wavelengths selected by a grating and a narrow-bandpass filter were 420 to 460 nm. It was found experimentally that the probe was thermally and photo-lytically stable. Also of concern were fluorescence quenching and polari-zation quenching,[7] whereby a probe may transfer its energy radiationlessly to a neighboring probe having a slightly different orientation. It was found that quenching does not become an important factor with this compound until the fluorescent probe concentration exceeds about 200 ppm.[6] This probe yields sufficiently strong emission for measurements in the range of 50 to 200 ppm.

RESULTS AND DISCUSSION

Tests of Method

For the method to be most effective, it should (1) have sufficient sensitivity to measure changes in polymer orientation and (2) yield information about the amorphous regions of the polymer. Figure 7 displays the results obtained for an as-spun PET fiber containing 200 ppm of the probe which was premixed with the polymer before extrusion. Obviously the orientation is not simply random; there is some degree of orientation along the fiber axis. The sample was melted by placing the fibers in a **press** for 20 sec at 280° C

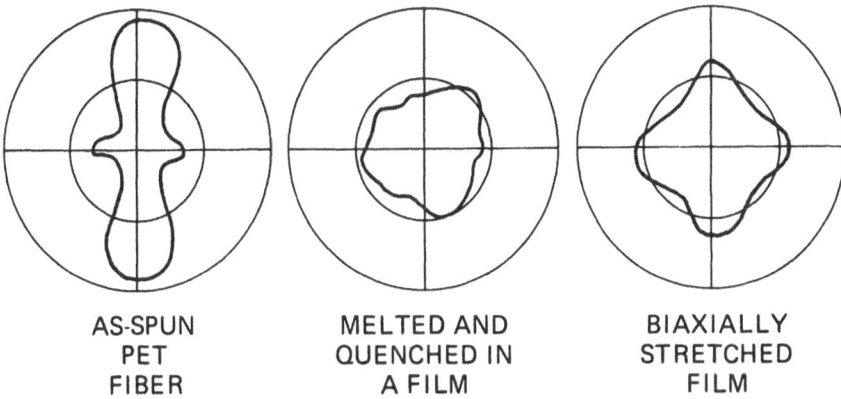

AS-SPUN
PET
FIBER

MELTED AND
QUENCHED IN
A FILM

BIAXIALLY
STRETCHED
FILM

Fig. 7. Angular Distribution of the Degree of Fluorescence Polarization
in PET Samples.

and 50-psi pressure; the sample was then quenched in room temperature water. The resulting degree of polarization plot indicates that the orientation was disrupted and approached the circular behavior of·a random sample. Next, the quenched film was biaxially stretched in a clamp immersed in hot oil, and cooled in air under the load. The pattern again changed and was suggestive of a biaxial orientation. Clearly, there does exist a marked effect of orientation upon the observed fluorescence, and it qualitatively agrees with our knowledge of what a particular sample treatment should produce.

Perhaps the strongest evidence that the method yields information about the amorphous regions of a polymer is the marked change in the degree of fluorescence polarization at the glass transition temperature of the polymeric medium. In addition, simultaneous studies of the crystalline behavior by conventional methods and by the fluorescence method show that the two methods behave differently. Nishijima studied poly(vinyl alcohol) (PVA) by X-ray and fluorescence,[1] Kryszewski studied PVA by light scattering and fluorescence,[8] and we have studied PET by X-ray and fluorescence.[9] In all cases, the fluorescence studies yielded information which was different from the information on crystalline behavior yielded by the other methods. In some experiments, fluorescent probes have been dyed into the polymer after its manufacture. Polymer chemists investigating dye diffusion are very adamant in stating that diffusion of the fluorescent probe into the crystalline regions would be orders of magnitude slower than diffusion into the amorphous regions.

In addition to the method's ability to measure changes in polymer orientation and yield information about the amorphous regions, it is also easily

applied to several polymeric systems. The only requirement is to select a
fluorescent probe of sufficient optical anisotropy which is well dispersed in
the polymer. Rotational depolarization of the probe becomes a problem only
when measurements are being made near or above the glass transition temperature of the polymer under investigation. We have already studied several
different probes in polymers of different types, and other probes and polymers
have been reported in the literature.[1,4,8]

Effect of Orientation

The effect of increasing the molecular orientation in a polymeric sample
can be investigated by the fluorescence method. Figure 8 shows the increase
in the degree of fluorescence polarization along the axis of stretching as a

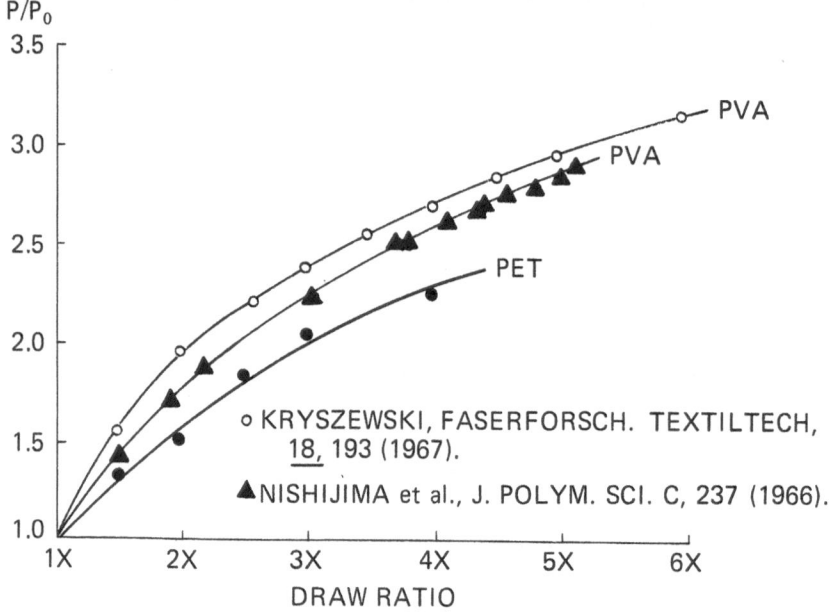

Fig. 8. Effect of Draw Ratio on Fluorescence of Films.

function of draw ratio for two different polymers. P_0 represents the degree of
polarization of the undrafted film. The top curves were obtained independently
for PVA film by Kryszewski in Poland[8] and Nishijima in Japan.[1] Both workers
used sodium fluorescein as the fluorescent probe. There were probably some
differences in the film drafting conditions used by the two workers, and the
film used by Nishijima was about twice the thickness of the film used by

Kryszewski. Our results for PET film[9] with 2,2'-(vinylenedi-p-phenylene)-
bisbenzoxazole as the fluorescent probe are also shown in Fig. 8. The degree
of fluorescence polarization observed gives the extent of orientation of the
amorphous regions of the polymers produced by known mechanical deformations.

Effect of Disorientation

The experiment described in the previous section was aimed at measuring
the increase in orientation of the sample with some externally imposed forced.
It is also possible to investigate the opposite effect by the fluorescence
polarization method. We treated a series of fluorescent-doped PET fibers in
silicone oil at several temperatures and allowed the fibers to shrink freely
during the treatment. Rate studies showed that 10 min at each temperature
was sufficient to obtain full shrinkage. The fluorescence data were then
recorded at room temperature. Figure 9 shows the changes observed in the
angular distribution of the degree of fluorescence polarization for fibers
heated at three temperatures. As the bath temperature increased, a disorientation of the fluorescent probes occurred and approached the circular plot of a

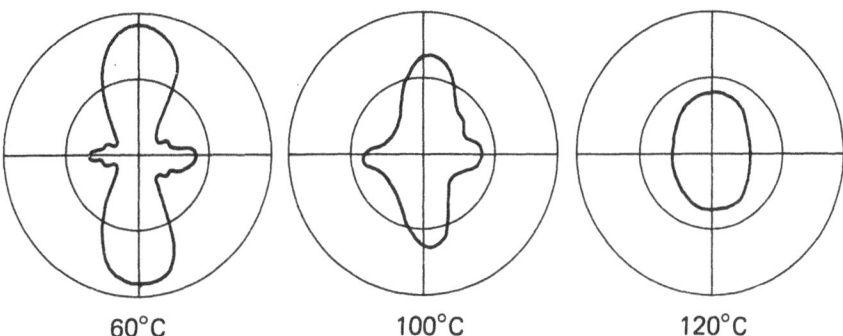

| 60°C | 100°C | 120°C |

Fig. 9.   Angular Distribution of the Degree of Fluorescence Polarization in
As-Spun PET Fibers Heated Unconstrained in Oil.

random distribution. Figure 10 is a plot of the resulting data. The average
degree of polarization (P) along the fiber axis is plotted on the left ordinate.
The PET fiber, initially in a relatively high state of order as a result of spin
drawing, reaches the glass transition temperature at about 70°C. At this
point the chains become mobile and disorientation commences. This disorientation continues until the temperature reaches the crystallization temperature
near 120°C. Here the fiber forms crystalline regions which probably impose
constraints on the remaining amorphous regions.

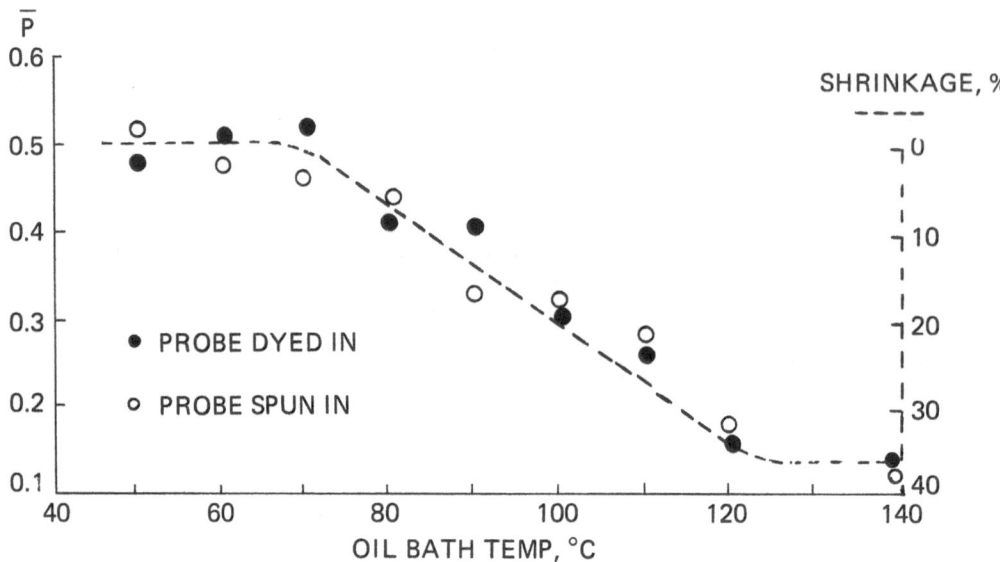

Fig. 10. Change in Average Degree of Polarization and Shrinkage of a
Fiber With Temperature.

An important question which has been asked about the fluorescence
technique is "Do the probes genuinely follow the polymeric environment or
do they behave independently and relax in their own manner?" We attempted
to answer this question by performing an experiment in two ways. In one case,
the fluorescent probes were premixed in the polymer before extrusion, and the
relaxation behavior of the fiber was determined. The behavior of that system
is shown by the open circles in Fig. 10. In the second case, the fiber was
treated in oil, removed, and dyed at 50°C in an aqueous dispersion of the
fluorescent probes. The relaxation behavior of that system is shown by the
solid circles in Fig. 10. The identical results which were obtained, within
experimental error, for the two cases provided very strong evidence that the
polymer is undergoing a change in morphology which then affects the
fluorescent probe's orientation.

The broken line and the right ordinate of Fig. 10 represent the shrinkage
data for the PET fibers. The results indicate a high degree of correlation
between the amorphous orientation as measured by the fluorescence method,
and the amount of fiber shrinkage. This correlation offers further proof that
the probes follow the polymeric environment and perhaps aids in confirming
the hypothesis[10,11] that shrinkage reflects a disorientation of the amorphous
regions in the polymer.

Results from experiments involving an optical measurement, such as
fluorescence intensity should be examined carefully for alternative explanations.
Perhaps the fiber samples had merely crystallized during treatment, and the

resulting crystallites were scattering and depolarizing the fluorescent emission. However, density determinations of the samples showed no measurable crystallization up to a temperature of 120° C. The fibers were also investigated by low angle X-ray and by light microscopy, and no differences were noted. The observed depolarization appears to be a real consequence of the probe's relaxation.

CONCLUSION

The work reported here shows: (1) the fluorescence polarization method can be used to measure orientation, particularly in the unknown amorphous regions of a polymer; (2) the probe is not acting independently but does, in fact, reflect changes in its polymeric environment; and (3) the method can be used to study both orientation and disorientation processes.

Some of the questions which might be asked about a new method have been answered, but not all of them. It would be extremely enlightening, for example, to investigate the forces imposed upon the fluorescent probe by the polymer chains during orientational changes. Reorientation of a short bulky probe should differ significantly from that of a long rod-like probe. We are just beginning studies in this area.

Theoretically, from fluorescence polarization data, the fourth order moments of the orientation distribution function could be determined, and insight into the polymer's morphology could be gained. However, the data were presented in this paper only in a relative or empirical fashion until the additional effects perturbing the observed polarization are more fully understood. For example, the polymer's birefringence may alter the nature of both the incident polarization and the fluorescent emission polarization. The effect of birefringence is minimized when the planes of polarization of the incident and emitted light are along the optical axes of the sample. However, when they are not along these axes, the corrections may be large. For example, at an angle of 45° the correction factor[4] for the parallel intensity of a 30-μ-thick PET film approaches 30% of the observed value even with a birefringence of only 0.02. Clearly, when dealing with a correction of such a magnitude, we must proceed cautiously.

Another factor which may perturb the observed polarization is the effect of scattering and refraction of the light at the polymer-air interface. We recently studied this by comparing the fluorescence polarization of 2,2'-(vinylenedi-p-phenylene)bisbenzoxazole in an amorphous PET film for samples in air (n = 1.000) and for samples immersed in water (n = 1.333) and glycerol (n = 1.473). The last two media have indices of refraction which approach that of PET film (n = 1.588). The observed degree of polarization did not vary significantly among the three media; therefore, it appears that surface scattering effects may not greatly perturb the observed polarization.

A third area of importance which we are presently investigating is the electronic behavior of the probe. The absorption and emission transition moments may not be coincident, as is commonly assumed. For the simple case of trans-stilbene, even when the absorption is highly dichroic, the emitted light is not highly polarized. [12] This result would demand a more complete theory for interpreting fluorescence polarization data than has previously been offered.

Another complicating feature is the possibility of vibronic interaction. If the probe is excited with light of too short a wavelength or if the emission is monitored at too long a wavelength, vibrational levels excited in the electronic states may affect the polarization characteristics.

Once these factors which may perturb the observed polarization have been clearly understood, we will have an extremely powerful tool with which to investigate the distribution functions and thus obtain a firm knowledge of the polymer's morphology. The full significance of the fluorescence polarization method can only be determined by comparing it with other methods, such as polarized Raman or wide-line NMR, which are also capable of yielding information on the moments of the distribution function up to the fourth order. [13-15]

REFERENCES

1. Y. Nishijima, Y. Onogi, and T. Asai, in U. S.-Japan Seminar in Polymer Physics (J. Polym. Sci. C, 15), R. S. Stein and S. Onogi, Eds., Interscience, New York, 1966, p. 237.

2. C. R. Desper and I. Kimura, J. Appl. Phys., 38, 4225 (1967).

3. R. S. Stein, J. Polym. Sci. A-2, 6, 1975 (1968).

4. J. Seki, Sen-i Gakkaishi, 25, 16 (1969).

5. R. F. Chen and R. L. Bowman, Sci., 147, 729 (1965).

6. G. E. McGraw, J. Polym. Sci. A-2, 8, 1323 (1970).

7. R. S. Knox, Physica, 39, 361 (1968).

8. M. Kryszewski, Faserforsch. Textiltech., 18, 193 (1967).

9. G. E. McGraw, unpublished work.

10. D. Patterson and I. M. Ward, Trans. Faraday Soc., 53, 1516 (1957).

11. P. R. Pinnock and I. M. Ward, Trans. Faraday Soc., 62, 1308 (1966).

12.   R. H. Dyck and D. S. McClure, J. Chem. Phys., 36, 2326 (1962).

13.   R. Roe, J. Polym. Sci. A-2, 8, 1187 (1970).

14.   R. S. Stein and B. E. Read, Appl. Polym. Symp. 8, 255 (1969).

15.   G. L. Wilkes in "Fortschritte der Hochpolymerenforschung" ("Advances in Polymer Science"), Vol. 8, H. J. Cantow, et al., editors, Springer, Berlin, 1971, p. 91.

TRANSMISSION MULTIPLE BEAM INTERFEROMETRY OF POLYMER FILMS

M. B. Rhodes

Chemistry Department, University of Massachusetts
Amherst, Massachusetts

When an experimental technique is applied in a manner
differing from previous applications there are certain logical
counterparts to the actual experimental application. These begin
with an analysis or classification of the subject field in order
to gain the proper perspective on the specific technique so that
any necessary modifications of the technique will have a significant direction. Such an analysis is usually accompanied by a
careful consideration of the basic theory in which all assump-
tions, approximations, and limitations are evaluated for their
applicability to the new situation. Therefore this communication
will be divided into three general parts. The first part will
deal with a broad classification of interferometry and where in
the overall scheme one finds the specific techniques that are to
be applied to polymeric films. The second part will include
brief comments on the theory, the differences that distinguish
one specific method from another, concepts to be considered upon
application to polymers, and reference to previous investigations.
Finally, the third part will present some of the preliminary
results obtained when multiple beam techniques are applied to
polymer films.

CLASSIFICATION

Any discussion of a classification system of optical interferometry must begin with a definition of terms. Interference is
the interaction of light waves such that the resulting intensity
may be lesser or greater than that of any individual interacting
wave. When radiation travels more than one pathway between
origin and detection, this interaction in the form of interference

113

becomes possible. Whether or not interference occurs depends on
the coherence of the radiation, since if the radiation is incoherent then the intensities rather than the amplitudes will add
when the radiation combines. Coherence and interference are the
theoretical and experimental aspects of the same phenomenon.
The visual result of an interference effect is usually a set of
alternating light and dark bands, known as interference fringes.[1-3]

The two traditional methods for the classification of interferometric systems, (1) by the number of interfering beams and
(2) by the method used to separate the beams, have been extended
by Tolansky to include (3) chromatism and (4) fringe location.[3]
While these four characteristics may not essentially yield a
comprehensive classification scheme, they do represent four
important and distinguishing qualities that are associated with
all types of interferometry. Thus any discussion of an interferometric experiment describes the experiment within the framework
of these four characteristics.

1. Number of Interfering Beams

Multiple beam methods are obviously by their name, methods
employing many beams for the interference process in contrast to
any method that makes use of just two beams. The normal two beam
interferometric methods use one beam as the sample beam and the
second as the reference beam, with the two beams traveling
different paths before recombining. The path difference between
the sample and reference beam is made visible by the fringe
pattern with its varying degrees of constructive and destructive
interference. There are two alternate means by which these
separate paths can be experimentally obtained. In the first way,
the reference wave travels in such a manner that it is not
influenced by the object. While in the second, both the reference
and sample beams enter and pass through the sample. Krug, Rienitz
and Schulz discuss the relative merits of these basic approaches,
explaining that one of the important advantages found in systems
employing a reference wave that is influenced by the object, is
the ease and simplicity of the experimental set up.[4] This one
advantage usually more than offsets the increased difficulty
associated with the interpretation of the fringe pattern.

Michelson restricted the term "Interferometer" to any system
or arrangement that would separate a beam of light into two parts
and then allow them to recombine to produce interference. At
present this terminology has been extended, simply by continued
usage, to include all methods that divide a beam into any number
of parts which eventually recombine for interference. The Fabry-
Perot interferometer and the commonly observed interference

effects in thin films are examples of this broader definition of
interferometry.[5]

The advantages of using multiple beam techniques have been
discussed.[6,7] However specific these advantages are, they
essentially depend on the extreme sharpness of the typical
multiple beam fringe. Tolansky reports that a multiple beam
fringe, under optimum conditions, can be one twenty fifth the
width of a conventional two beam fringe. Sharp fringes permit
very high precision in any quantitative analysis involving fringe
displacements or variation. Multiple beam fringes are capable
of measuring vertical direction dimensions in the order of a few
angstroms, thereby placing these interferometric methods in the
same sized domains as the methods of electron microscopy.
However, when the lack of optimum experimental conditions result
in deterioration of a multiple beam method, the resulting fringe
characteristics become similar to that expected for the conventional two beam methods. Consequently although these experiments
will be all identified in name as multiple beam methods, the
fringes will not necessarily appear as such. This results from
nonideal conditions inherently associated with the use of
polymeric materials.

2. Wave Division

Coherent radiation can be obtained by division of the light
wave in one of two ways. The first is by a wave front division
in which a point source having wavefronts of a similar phase
progressing in slightly different directions, has each wavefront
further separated by simple optics before recombining for interference. The second is by amplitude division in which the light
is divided by the use of a beam splitter, the exact nature of
which leads to a further classification. In these multiple beam
investigations, amplitude division is accomplished through use
of the semi-reflecting surfaces of metallic silver deposited on
microscope glass slides.

3. Chromatism

Chromatism refers to the type of radiation used for the
interference. In these investigations both monochromatic and
polychromatic radiation was found to be applicable, although only
the results from white light fringes will be illustrated.

4. Fringe Location

Different types of interferometric systems will result in different locations for the interference fringes. This study utilized microscope optics to view the sample with the fringes observed in the plane of the specimen.

THEORY AND APPLICATION

Methods for producing multiple beams have been described.[6-8] Light of wavelength λ, at an incidence angle φ, enters a sample of thickness t and refractive index μ. The multiple beams are generated when the incident radiation undergoes continued reflection within the sample that is contained between silver coated glass slides. The success of the multiple beam technique depends primarily upon the high reflection coefficient of the silver deposit. The quality of the silver coating is itself a function of proper preparation of the glass surface prior to the vacuum deposition of the silver, and to this end Tolansky's recommendations should be explicitly followed.

The variation in the interference fringe intensity is described by the Airy formula,[6-8]

$$I = \left(\frac{I_{max}}{1 + F \sin^2 \pi \dfrac{2 \mu t \cos \varphi}{\lambda}} \right)$$

and R is the reflection coefficient in

$$F = \frac{4R}{(1-R)^2}$$

Although this expression is exact only under very well defined conditions, it is still a valid approximation for a wide range of nonideal experimental conditions. As such, an examination of the functional behavior of the parameters in this expression reveals; (a) the need for the reflection coefficient, R, to be as high a value as possible in order to increase the rate at which the fringe maximum intensity decreases between orders, (b) the origin of the fringe intensity variation to be in the $\sin^2$ term and (c) that the parameters λ, μ, t, and $\cos \varphi$ either individually or in combination produce this variation.

RADIATION	CONSTANT PARAMETER	FRINGE TYPE	NAME
Monochromatic λ constant	t	Equal inclination	Fabry-Perot
	φ	Equal thickness	Fizeau
White λ variable	φ	Equal $\mu t/\lambda$	FECO (Equal Chromatic Order)
	t	Equal $\dfrac{(\mu t \cos \varphi)}{\lambda}$	White light Fabry-Perot

TABLE 1. A classification of fringe types based on the expression $n\lambda = 2\,\mu t \cos \varphi$

The interrelationship of these four parameters expressed as $n\lambda = 2\mu t \cos \varphi$, leads to the detailed classification system seen in Table 1.[6] This is the same as that given by Tolansky except for the addition of the term μ, which is essential for the polymeric systems because the value of the refractive index in these cases is greater than one.

Figure 1 is a diagram to illustrate the production of either Fizeau or FECO fringes. Radiation is parallel (the angle of incidence, φ, is $90°$) to a sample volume defined by ACDF. The diagram shows beams CD and BE, two of the many possible reference beams undergoing internal reflection at the highly reflecting silvered surfaces and then interfering with the sample beam AF at F. The internally reflected intensity of successively displaced reference beams drops off very rapidly unless the reflection coefficient is extremely high and absorption is low. The more beams that undergo interference at specific points, the sharper the resulting interference fringes. A reflection coefficient of 0.94 has fringes that are characterized by a half width that is one fifth the half width of fringes when the reflection coefficient is 0.70.[6]

Only two of the four fringe types listed in Table 1 were investigated, the monochromatic Fizeau fringes and the white light FECO fringes. The FECO fringes produced results of more significance and of greater potential for polymeric studies. Fizeau fringes while also suitable for polymeric systems have a specific and limited applicability because they are less versatile and give poorer fringe definition when conditions of extreme optical hererogeniety exist within the polymer sample. Although no results of Fizeau fringe studies will be presented here, a few comments on the distinction between the Fizeau and FECO methods are necessary. The basic theory for Fizeau fringe production incorporates a wedge angle θ into the geometry as illustrated in Figure 1. The existence of this wedge angle is not always experimentally applicable and especially not in the case of the polymer samples where the polymer is sealed between silvered glass slides that are essentially parallel. As a consequence of this parallel sided geometry, one observes no conventional wedge fringes but only Fizeau contour fringes, of constant μt, associated with the internal sample morphology. However a consideration of the wedge angle provides a simple means of understanding why the multiple beam method breaks down. Therefore one can make a logical extension towards understanding any limitations when the method is applied to polymers.

Even when the angle θ is reduced to a zero value, the successive reflections become farther and farther displaced from the interference site, as one can observe from Figure 1. It can be calculated that an air sample one micron thick uses a linear

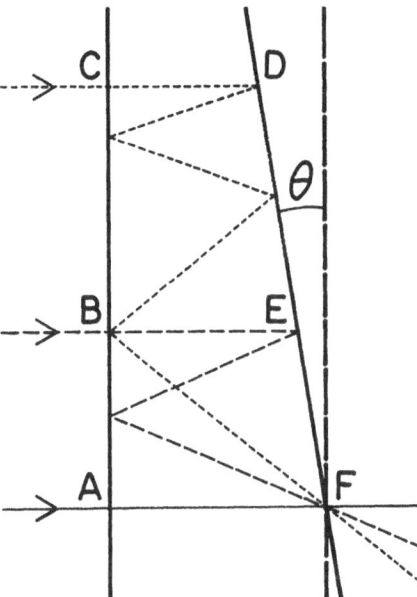

Figure 1. Diagram of a sample volume contained between ABC and DEF showing two reference beams at B and C and the sample beam at A.

distance of approximately forty microns in which to generate some
sixty reflections. When this distance is of nonuniform composition,
as would be the case in polymers, then the optical (and sometimes
geometrical) heterogeniety of this region will govern the complexity
of the interference at sites such as F.

A finite wedge angle can quantitatively describe the phase
lag associated with each successive internal reflection. This
phase lag increases with the increasing linear distance over which
the multiple reflections are generated. Thus a distance will exist
that will ultimately yield a phase lag of the proper magnitude to
effectively destroy contributions to the Airy expression. The
phase lag is the primary cause for the reduction in fringe quality
and in polymeric materials it does not originate from any well
defined experimental factor of geometry but rather from the
intrinsic refractive index variation characteristic of the polymer
crystallite orientations. There exists a wide variation in this
type of optical heterogeniety due to the morphological differences
among polymers, preparations, and even small sample domains within
a given polymer preparation. This modification of the light as it
repeatedly traverses the sample volume does not affect the interpretation of the results, at least to a first approximation.[4]

It would be unrealistic to attempt a comprehensive literature
review in this communication. Therefore references will be
restricted to include material most pertinent to the application
of multiple beam interferometric methods to polymeric systems.
The book by Krug, Rienitz and Schulz on interference microscopy
offers an adequate background discussion with accompanying literature references for the general subject of multiple beam methods.[4]
However, the extensive development and application of these methods
is due to Tolansky whose publications on the study of the micro-
topography of surfaces and films by multiple beam techniques also
includes the application of such techniques using transillumination
for samples of mica.[3,6] Recently there appeared a short communication by Israelachivi in which the author explained the
simultaneous determination of thickness and refractive index on
thin films by a consideration of the characteristics of the odd
versus the even fringe orders.[9] In view of the inherent complications associated with any analysis of polymer multiple beam
fringes, the report by Glauert on the distortion of multiple beam
fringes and subsequent interpretation by Hunter and Nabarro, is
of the utmost importance.[10,11] Any elaboration of these multiple
beam methods for polymers will have to give serious thought to
the extent of sporadic fringe modulation arising from the polymer
crystallite orientation. Courtney-Pratt used FECO fringes for
measuring the thickness of adsorbed monolayers and found the
results to be in agreement with X-ray data.[12] Verma and Reynolds
used multiple beam Fizeau fringes to study the spiral growths on

stearic acid crystals and Wunderlich applied these same techniques to the investigation of polyethylene single crystals.[13,14] Faust has made three very important contributions to the literature on multiple beam interferometry. One is a theoretical discussion in which he considers, among other things, the influence on the final interference image of very small structural entities within the sample.[15] The other two publications include theory and applications of FECO fringes (channeled spectra) for the determination of refractive index and refractive index variations in solid samples.[16,17]

EXPERIMENTAL

1. Instrumentation

Figure 2 is a schematic diagram of the optics suitable for either Fizeau or FECO fringe observation and recording. The dual light source is a monochromatic mercury lamp for the production of Fizeau fringes and for the calibration of the wavelength for the chromatic orders, which is interchangeable with a high intensity white light source for the production of the FECO fringes. This well collimated (parallel) illumination is normal to the sample. The Fizeau fringes can be observed with just the optics of a transmission microscope. For the observation of the FECO fringes, the image of the sample is focused from the microscope on the slit of the spectrograph. The prism resolves the various chromatic orders and they are viewed through the spectrograph eyepiece at which time the wavelength values can be read off the calibrated drum or the fringes photographed for subsequent measurement. These investigations were done on a Vickers metallurgical microscope using a Hilger Quartz spectrograph.

2. Sample Preparation

Samples were prepared by melt sealing the polymer between freshly silvered microscope slides. The silvering technique was that described by Tolansky and included the recommended glass cleaning procedure. Sample thickness varied between five and twenty five microns. After the sample was sealed between the silvered slides, the preparation could be used for crystallization studies on a Kofler hot stage mounted on the microscope. Once the silver deposition was protected by the polymer sample, essentially from exposure to air, it proved remarkably resistant to deterioration effects at elevated temperatures. Only FECO fringe patterns from polyethylene oxide and polypropylene, will be illustrated, but similar results and generalizations apply also to polystyrene, polybutene, and polyethylene.

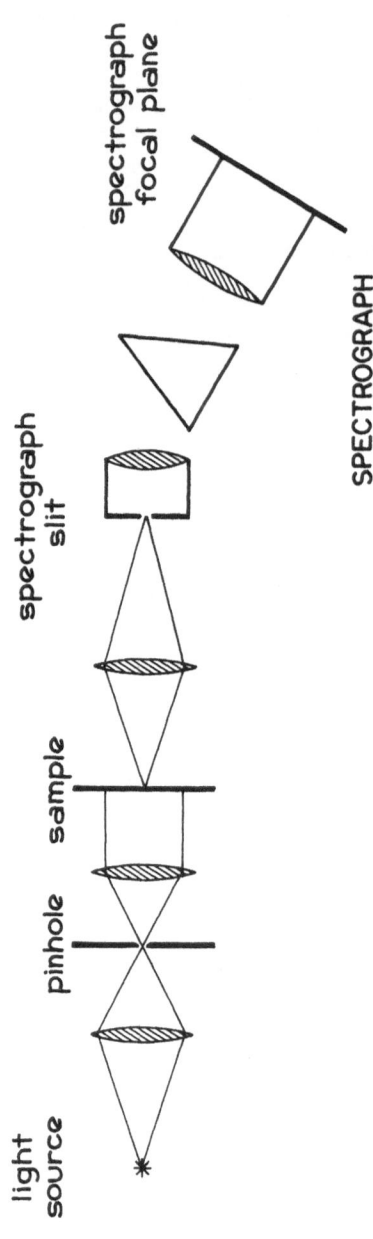

Figure 2. Diagram of the optics for the production of either Fizeau or FECO fringes.

3. Experimental Measurements

The fringes of equal chromatic order that are observed in the focal plane of the spectrograph represent regions within the sample where there is a constant $\mu t/\lambda$ value. The separation between fringe orders is a function of the sample thickness and the separation of a fringe doublet is a function of the birefringence of the specific morphology associated with the image on the spectrograph slit. These two important quantitative values can be calculated from the basic relationship, $n\lambda = 2\mu t \cos \varphi$.

The separation between orders measured by $\Delta\nu$ permits a calculation of thickness. Since $\cos \varphi = 1$ in these investigations,

$$n = 2\mu t/\lambda = 2\mu t\nu$$
$$\Delta n = 2\mu t\Delta\nu$$

Then since the separation between two adjacent fringes results in $\Delta n = 1$,

$$\Delta\nu = 1/2\mu t$$

The separation of any doublet measured as $\Delta\lambda$ permits a calculation of the birefringence. Since

$$n\lambda = 2\mu t$$
$$n\Delta\lambda = 2\Delta\mu t,$$
$$\lambda/\Delta\lambda = \mu/\Delta\mu$$
$$\Delta\lambda = \lambda\Delta\mu/\mu$$

An average refractive index value for the polymer sample is required for the calculation of the thickness and the birefringence. This was obtained either from the literature, by use of the microscope Becke line method or a refractometer. Frequently a characteristic morphological feature within the polymer thin films permitted a comparison of the apparent thickness, t/μ, with the interferometric thickness, μt. The latter value was obtained from the separation of the fringe orders, while the t/μ value was obtained from readings on the calibrated fine focus of the microscope as the specimen feature was focused for the upper surface and then for the lower surface. Whenever such determinations could be made, it was possible to evaluate t^2 and μ^2 by multiplication and division of the values for these two expressions of sample thickness. Such comparisons gave values that confirmed the value for the refractive index obtained by one of the other means. This technique is based on the Tolansky light slit method and when used together with the FECO fringe method provides for the simultaneous determination of μ and t in very thin films.[18]

Figure 3.   Transmission fringes of equal chromatic order
from a sample of mica (courtesy of Professor Tolansky).

Since relative relationships were more important than absolute values in these preliminary experiments, no attempt was made to accurately include dispersion effects into the calculations. Instead fringe measurements were restricted to the spectral region where error in the approximate value of μ would be minimal. Or else, when a quantitative comparison of two spectra was made, the comparison was confined to two similar wavelengths.

4. Fringe Patterns

Figure 3 illustrates the appearance of the FECO fringes from a thin sheet of mica. As was previously mentioned, the doublet separation, sometimes barely resolvable, is a measure of the mica birefringence. Any curvature of the fringes reflect variation in the optical thickness. Each fringe is a constant $\mu t/\lambda$ value and if either μ or t varies, a change in the value of λ will compensate. A fringe curvature towards the red region of the spectrum indicates increasing optical thickness and when in the case of mice, μ is constant, this becomes an increase in actual sample thickness.[6] Although the FECO fringes from polymeric materials may be somewhat more complex in their appearance than these shown for mica, the same major characteristics will be observed. Recorded in the fringes of equal chromatic order is a thickness and birefringence profile of the sample area that has been imaged on the spectrograph slit.

Figure 4 illustrates two spectra from different regions of a polyethylene oxide spherulite at room temperature. The most notable feature is the variation in the second member of the fringe doublet. This is in contrast to the mica fringes in which the intensity and separation of the doublet members was uniform. The FECO fringes for the polymer reflect the variation in the birefringence values associated with the specimem morphology imaged on the slit. There was no such birefringence variation in the mica sample since the orientation of the optic axes remained constant over the imaged area in spite of the thickness variation. The upper spectrum in Figure 4 arises from a region very close to the center of the spherulite. Here the doublet variation is suggestive of a type of periodicity. However, in the lower spectrum, which is representative of the outermost regions of the spherulite, the variation in the second member of the doublet is very random with respect to both intensity and separation. This extreme variability is interpreted as resulting from the irregular growth and orientation of the crystallites in the peripheral regions of the spherulite. The birefringence values calculated from such patterns range from a minimum of .001 to a maximum of .026.

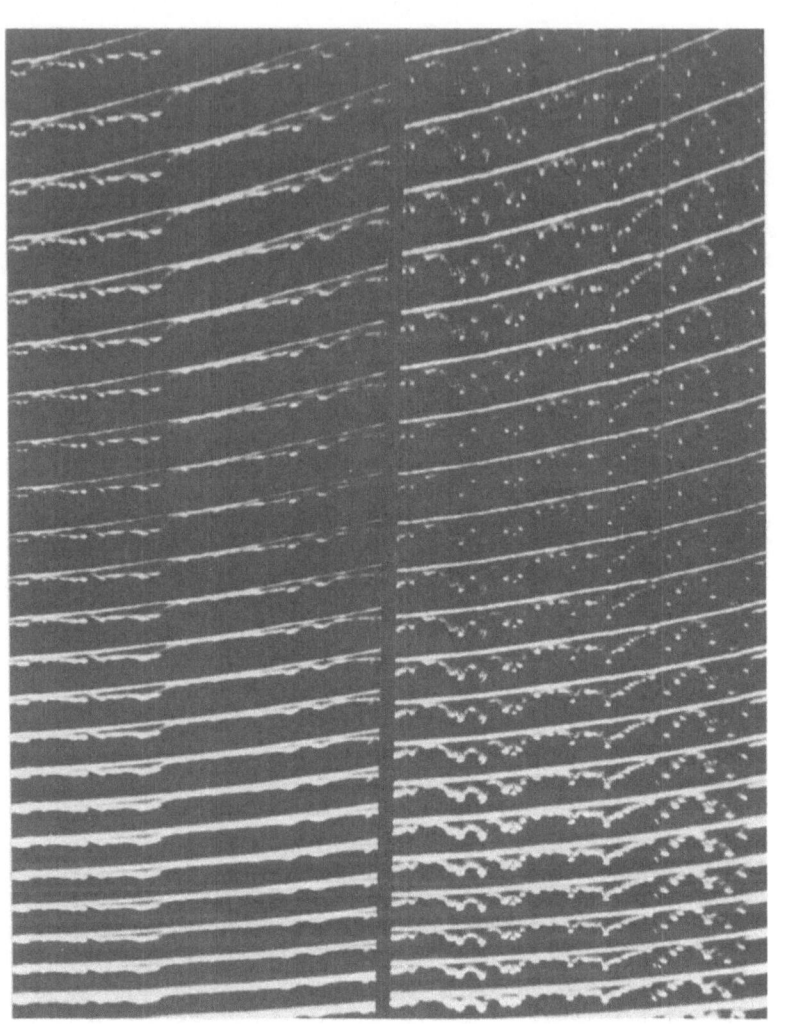

Figure 4. Fringes of equal chromatic order from two different regions of a polyethylene oxide spherulite at room temperature.

The nuclear region of spherulites is frequently associated with a complex overgrowth, appearing during the later stages of crystallization or growing during the cooling process. Figure 5a FECO fringes of a polypropylene spherulite growing at 140°C are contrasted with fringes in 5b from the same spherulite after the polymer has cooled to room temperature. At 140°C, the small uniform separation of the fringe doublet in the exact center of the spherulite identifies a very low birefringence when compared to other regions of the growing spherulite. This is characteristic of spherulites growing at temperatures close to the melt temperature. Comparing spectra from 5a and 5b, one can calculate that the central region has undergone a birefringence change from an initial value of .003 at 140°C to a final value of .070 at room temperature. In addition to this birefringence change associated with the spherulite center, the entire spherulitic and nonspherulitic birefringence profile has been altered. The single fringe that is characteristic of the melted polymer in 5a becomes a closely spaced doublet at the room temperature conditions, indicative of a low birefringence region with no specific resolvable orientation in the fringe pattern that can be assigned to the small sized crystalline entities. The noticeable curvature observed in the 5b fringes indicate a gradual increase in the optical thickness of the spherulite. From these observations alone it cannot be determined what part of this increase, if any, results from changes in μ. One should also note that the fringe quality in 5a is much better than that in 5b. The fringe broadening in the room temperature spectrum results from the increased phase lag experienced by the multiple reflections when the optical homogeniety decreases as the melted state transforms into the final crystalline state.

Figure 6 illustrates another characteristic of spherulitic central regions. These are FECO fringes arising from a polypropylene spherulite at room temperature centrally imaged on the spectrograph slit. The only difference between 6a and 6b is the use of an analyzer in the microscope optics for fringe recording in 6b. The extinction properties of the doublet in 6b indicate that the polarization characteristics of the spherulitic profile are not uniform across the spherulite. Specifically, the central region with its overgrowth does not go to extinction with the outer regions of the spherulite. Also there is no abrupt change from one type of orientation to another but rather a gradual transition. This gradual change is frequently demonstrated by varying degrees of extinction over small regions in one of the doublet members indicating the existence of many and varied crystallite orientations in the central region of a spherulite as a result of certain crystallization conditions.

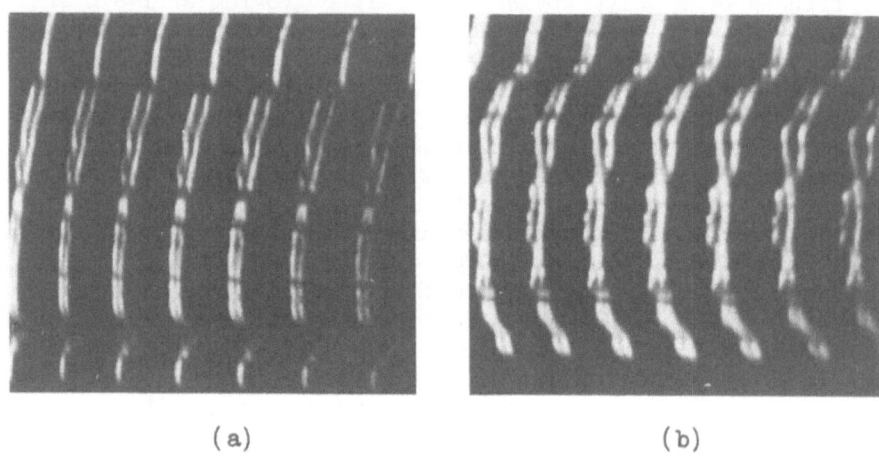

(a)   (b)

Figure 5. Fringes of equal chromatic order from a polypropylene
spherulite, (a) growing at 140°C and (b) after the sample has
cooled to room temperature.

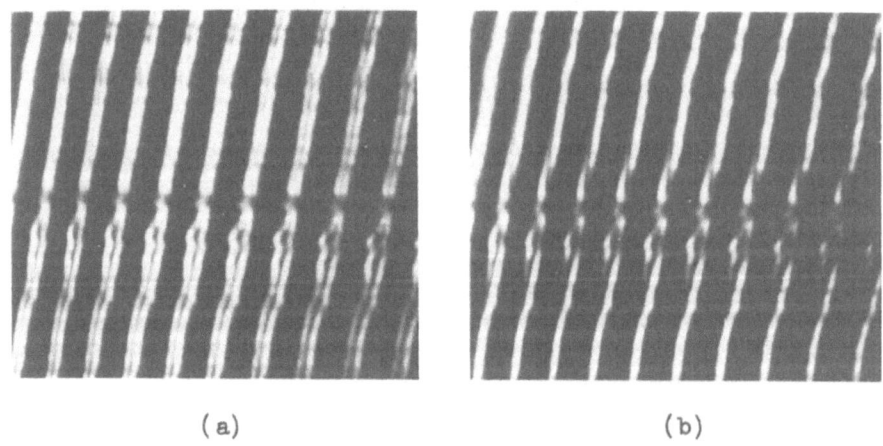

(a)   (b)

Figure 6. Fringes of equal chromatic order from a polypropylene
spherulite at room temperature, (a) no analyzer and
(b) with an analyzer.

Figure 7 contrasts the FECO fringes from different regions of a polypropylene spherulite growing at 145°C. The spectrum of 7a is characteristic of an equatorial side region of the spherulite while 7b is from a vertical central section of the same spherulite. Since μ was found to remain constant in this instance, the curvature of the fringes in 7a represent a spherulite thickening of 0.3 to 0.4 microns towards the equator. These two spectra illustrate the problem arising when it becomes necessary to identify the exact boundary between spherulite and melt. Such a boundary is usually characterized by a fringe discontinuity in addition by the obvious distinction between the birefringent or doublet fringe and the single melt fringe. However, confusion often arises when there is no abrupt transition between these two distinct types of fringes. Thus careful identification of fringes with morphology becomes necessary when a fringe pattern demonstrates an unusual fringe broadening, like a poorly resolved doublet, together with more than one discontinuity. Examples of this type of fringe behaviour can be observed in the spherulite upper boundary region in 7a and in the spherulite lower boundary region of 7b. It has usually proven extremely difficult to correlate the exact morphological spherulite boundary observed with bright field or polarization optics with the FECO fringe discontinuity. This suggests that an optical boundary does not always identify with a conventional microscopic morphological boundary.

Confusion of the boundary is further enhanced by observing the fringe pattern of the polypropylene spherulite growing at 140°C shown in Figure 8. In this instance an analyzer in the microscope optics places at extinction one of the doublet fringe members associated with the spherulite birefringent material. The single isotropic melt fringe is still observed. The most striking feature of this FECO fringe pattern is the extension of the spherulitic one-membered fringe into the melt region and the gradually decreasing birefringence value. Observations such as this make it difficult to uniquely characterize a growing spherulitic boundary.

Although this communication represents only limited experiments to determine the applicability of multiple beam interferometry to polymer systems, the results obtained from these preliminary studies are encouraging. The observations from these multiple beam methods have all been substantiated by the holographic techniques of time differential interferometry using a holographic microscope. Future experimental investigations using multiple beam methods should be carefully planned to consider the consequences of the more fundamentally inherent limitations and approximations. There are certain aspects of the theory that should be seriously evaluated with a view to using heterogeneous

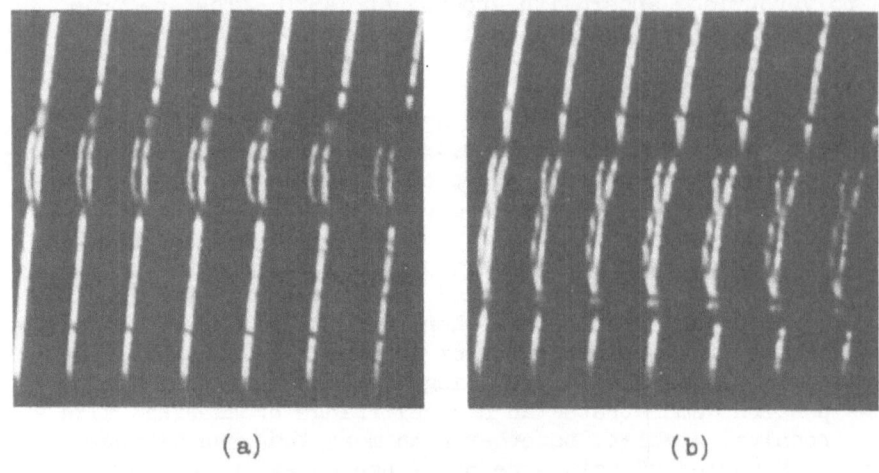

(a)   (b)

Figure 7. Fringes of equal chromatic order from two regions of
a polypropylene spherulite growing at 145°C, (a) the
equatorial side region and (b) the vertical central region.

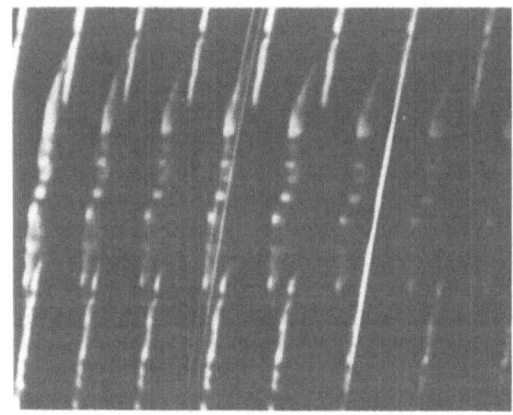

Figure 8. Fringes of equal chromatic order, with an analyzer
in the microscope optics, for a polypropylene spherulite
growing from the melt at 140°C. (Calibration lines can be
observed superimposed on the fringes.)

samples and expecting valid interpretations from the fringe patterns. This would be a significant evaluation in regard to two important features of spherulitic growth and crystallization. The first concerns the complex changes that originate in the central region of the spherulite sometimes long after the spherulites have become volume filling and sometimes while the spherulites are still growing. These changes are manifestations of the secondary crystallization process and can be studied by any of several interferometric techniques. However, the multiple beam methods offer the advantage of precise quantitative measurements for specific localized changes. The second feature of spherulitic crystallization, one that almost necessitates a unique experimental approach, concerns the mechanisms operating within the spherulite and within the adjacent melt region as the spherulite boundary becomes established. Again, such changes are best evaluated by interferometric methods and the multiple beam methods offer an advantage in their extreme sensitivity to optical path changes and to low degrees of order or organization. Consequently, as a result of these studies it is concluded that an unusual quantitative description of spherulitic growth in polymer thin films can be conveniently obtained with multiple beam interferometric methods.

These experimental investigations were performed in Professor Tolansky's laboratory at Royal Holloway College and grateful acknowledgement is made to Professor Tolansky for his assistance, hospitality and patience, also to Mr. Michael Thyer, the staff of the Physics Department, and the many friends at the College.

REFERENCES

1. W. H. Steel, "Interferometry", Cambridge University Press, Cambridge, 1967.

2. H. Lipson and S. G. Lipson, "Optical Physics", Cambridge University Press, Cambridge, 1969.

3. S. Tolansky, "An Introduction to Interferometry", Longmans, Green and Co. Ltd., London, 1955.

4. W. Krug, J. Rienitz and G. Schulz, "Contributions to Interference Microscopy", (Translated by J. Home Dickson) Hilger and Watts Ltd., London, 1964.

5. W. Ewart Williams, "Applications of Interferometry", Methuen and Co. Ltd., London, 1930.

6. S. Tolansky, "Multiple Beam Interferometry of Surfaces and Films", Oxford University Press, London, 1948.

7. M. Born and E. Wolf, "Principles of Optics", 3rd Edition, Pergamon Press, Oxford, 1964.

8. F. Jenkins and H. White, "Fundamentals of Optics", McGraw-Hill Book Co., New York, 1967.

9. J. Israelachvili, Nature, 229, 85 (1971).

10. A. M. Glauert, Nature, 168, 861 (1951).

11. S. C. Hunter and F. R. N. Nabarro, Phil Mag., 43, 538 (1952).

12. J. S. Courtney-Pratt, Proc. Roy. Soc., 212 A, 505 (1952).

13. A. R. Verma and P. M. Reynolds, Proc. Phys. Soc., 66, 414 (1953).

14. B. Wunderlich, "Interference Microscopy of Crystalline Linear High Polymers", Technical Report NSF Research Grant GP-74, 1963.

15. R. C. Faust, Proc. Roy. Soc., 211 A, 240 (1952).

16. R. C. Faust, Proc. Phys. Soc., 65 B, 48 (1952).

17. R. C. Faust, Proc. Phys. Soc., 67 B, 138 (1954).

18. O. S. Heavens, "Optical Properties of Thin Solid Films", Dover Publications, New York, 1965.

INFRARED STUDIES OF CHAIN FOLDING IN UNIAXIALLY

ORIENTED POLYHEXAMETHYLENE ADIPAMIDE *

J. L. Koenig and Masaaki Itoga

Case Western Reserve University

Cleveland, Ohio

SUMMARY

The fold content of uniaxially drawn polyhexamethylene adipamide (nylon 66) is measured as a function of annealing conditions by an infrared tilting method. Unoriented, rapidly quenched and isothermally crystallized samples were also studied as a function of annealing temperature. The incremental changes in fold content and crystallinity were compared as the samples were sequentially annealed under restraint and relaxed conditions at increasing temperatures.

INTRODUCTION

It is generally accepted that bulk crystalline polymers, as well as polymer single crystals, consist of a folded-chain structure. Also for oriented polymers, if they are annealed after stretching (1) or crystallized from oriented melt (2), a well-developed fold structure is observed where lamellae are basically perpendicular to the draw direction. Furthermore, for as-drawn fibers, a longitudinal periodicity of the order of 100 $\overset{\circ}{A}$ has been observed by means of small-angle X-ray diffraction, electron microscopy with an iodine-staining or nitric acid-etching technique (3) and dark field microscopy (3). This periodicity in drawn polymers is comparable to the thickness of single crystals and depends on

----
* Published in the Journal of Macromolecular Science-Physics, B6(2), 309-326 (1972). Reprinted by permission.

the drawing temperature of the bulk polymer. Based on these
observations, Peterlin (4) has proposed a folded-chain model for
drawn fibers. The deformation experiments with lamellar structures show a phase transformation, twinning, chain tilting, and
slipping. The deformation leads to a breakup of the original
lamellae into small blocks, which may be incorporated into fibrils
(5). The lamellae in the drawn state are highly irregular; but,
once they are annealed, regular lamellae are reformed (1).

For low crystallinity polymers, such as polyesters and polyamides, the existence of amorphous regions seems to complicate
the fiber structure and folding mechanism during annealing.
Dismore and Statton (6) assume an extended paracrystalline chain
structure in as-drawn fibers. On annealing, these chains melt
and recrystallize as folded-chain lamellae producing an increase
in the small-angle scattering intensity, thermal shrinkage, and
the fluid -like fraction in the nuclear magnetic resonance (NMR).

Bell and Dumbleton (7) suggest a mixture of folded-chain
lamellae and extended-chain crystals in drawn nylon 66 from their
observation of two melting phenomena at intermediate draw ratios.
The melting fraction shifts toward the extended type at higher draw
ratios. Refolding phenomena have also been observed in the annealing of extended-chain crystals of solid-polymerized polyoxymethylene (8). Amano et al. (8) have shown that the extended-chain
crystals transform to folded-chain lamellae on annealing far below
the melting point, without any change of the c-axis orientation. At
the same time, a large amount of the twinned crystals melts first
and then recrystallizes on the extended form creating "shish-kebabs".

These different models of fiber structure and the annealing
behavior of drawn polymers can be tested by studies of regular
fold formation. One of the most sensitive tools for detection of
changes in fold structure and content is infrared spectroscopy.

Infrared studies of chain folding were originated by Koenig
and Witenhafer (9), who showed that the 1304- and 1352-cm^{-1} bands
assignable to the gauche structure of polyethylene are sensitive to
the number and kinds of folds. Koenig and Hannon (10) found a
unique fold band for polyethylene terephthalate (PET) at 988 cm^{-1}
from their selective degradation and annealing studies of single
crystals.

Furthermore, Koenig and Agboatwalla (11) have assigned the
1224- and 1329-cm^{-1} bands to the unique regular fold conformation
in nylon 66 crystals, by the analogous selective degradation study

of single crystals. Their results have been confirmed by Cannon and Harris (12) and also by studies by cyclic model compounds (13). According to their results, single crystals exhibit various amounts of regular folding depending on the supercooling during the preparation. The irregular folds regularize to the unique conformation on annealing. When annealed at 258°C the samples contain the same number of regular folds based on a reduced long period regardless of their preparation. These results indicate that all of the irregular folds have been regularized at this annealing temperature. The structure of the fold in bulk-crystallized nylon 66 is the same fold structure as in single crystals as both samples exhibit the 1224- and 1329-cm^{-1} bands. Annealing of melt-quenched films, which initially contained little regular folding, increases the crystallinity and forms regular folds. Stretching of the annealed films distorts the regularity of the folds as evidenced by the disappearance of the 1224- and 1329-cm^{-1} bands, but subsequent annealing allows the folds to recover their regularity.

Recently, a quantitative analysis of the fold content of oriented PET fibers has been reported, and the absorbances of the fold bands were a function of the orientation and annealing (14) effects.

The changes in folding for oriented and unoriented samples during annealing are still obscure because of lack of quantitative data. Hence, in this study, quantitative measurements of the spectral absorbances of oriented nylon 66 were made.

EXPERIMENTAL

1. Sample Preparation

Nylon 66 pellets (Zytel 101, DuPont), after being dried at 150°C for 12 h under vacuum, were melted for 15 min at 280°C, while being sandwiched between Teflon-coated aluminum foils and pressed to 0.5 to 3-mil film in a heated ram press. Then, the sample was plunged rapidly into a silicone bath, which had been adjusted to a predetermined temperature ranging from 25° to 250°C. After being kept in the bath for 15 min, this sandwiched film was left to cool in the air at room temperature. Strips of 1 x $\frac{1}{2}$ in. were uniaxially oriented to 2.0 to 4.0 times the original length with a hand stretcher in water at 50°C and dried at room temperature for 3 days under vacuum. The draw ratio was calculated from the extension of the marked distance (2.0 mm) on initial films. No crystallinity was induced during the drying as measured by the 936- and 1140-cm^{-1} infrared absorbance.

The annealing heat treatments were done successively by dipping a sample in silicone oil at predetermined temperatures for 15 min. The constant length samples were clamped in the hand stretcher with a fixed distance (constraint-type annealing). The relaxed samples were placed directly in the silicone oil. After the heat treatments, the samples were placed in another silicone bath at room temperature, then washed in benzene to remove silicone oil on the surface, and dried under vacuum at room temperature. The thermal shrinkage during the relaxation-type annealing was obtained from the change in the marked distance on the sample with respect to the as-drawn length (corrected for a dimensional recovery during storage).

All samples were stored under vacuum until used.

2. Instruments

Infrared Spectroscopy

All spectra were recorded on the Perkin-Elmer 521 grating spectrometer from films of 0.5 to 1.0 mil thickness. For uniaxially oriented samples, the film was tilted with respect to the incident beam, on a special tilting device, at various angles between 0 and 45 deg. in order to obtain the degree of orientation and the true absorbance free from molecular orientation. A detailed description of the infrared tilting method has been reported (15).

The absorbances at 936 (crystalline band), 1140 (amorphous band), and 1224 and 1329 cm^{-1} (fold bands) were determined, but only the 1329 cm^{-1} band was subjected to fold analysis, since this was characterized better than the 1224 cm^{-1} band. The base lines for the 936-, 1224-, and 1329-cm^{-1} bands were drawn in the same way as in the previous report (11). The base line for the 1140 cm^{-1} band was a tangent between low absorption points in the vicinity of the 1125 and 1150 cm^{-1}.

Differential Thermal Calorimetry

The Perkin-Elmer differential scanning calorimeter was utilized at a heating rate of 80°C/min. For oriented samples, the aluminum holder containing 3/16 x 3/16 in. film was pressed more than usual to prevent contraction during heating. This was confirmed below the melting temperature by checking a dimension before and after heating.

X-Ray Small-Angle Scattering

Long period measurements were carried out on the Rigaku Denki Rota Unit RU-3 and small-angle diffractometer at a scanning rate of $6°C/hr$, with slit gaps of 0.2 mm and sample detector distance of 200 mm.

3. Internal Mass Standard for Crystallinity and Fold Content Determination

Nylon 66 is reported to have a crystalline band at 936 cm^{-1} and an amorphous band at 1140 cm^{-1} (16, 17), but no internal thickness band is available to normalize observed absorbances for our work. It is necessary to establish an internal parameter, representing a sample thickness or mass in the beam area. Based on Starkweather and Moynihan's experimental result (16) that absorbances per thickness of those bands are linear as a function of density and, therefore, proportional to the volume fraction of crystalline region, the following equations can be derived (assuming uniform thickness):

$$A_{936} = k_c' X_c \rho t \tag{1}$$

$$A_{1140} = k_a' (1 - X_c) \rho t \tag{2}$$

where k_c and k_a are the specific extinction coefficients; X_c = fraction of crystallinity; ρ = density; t = thickness; and

$$\frac{A_{1140}}{A_{936}} = \frac{k_c}{k_x} (\frac{1}{X_c} - 1) \tag{3}$$

But

$$X_m = \frac{A_{936}}{k_c \rho t} \tag{4}$$

Eliminating X_m from the above two equations,

$$\frac{A_{1140}}{A_{936}} = C_1 (\frac{W}{A_{936}}) - C_2 \tag{5}$$

where W is the mass of sample in the beam area. The A_{936}, A_{1140} are observed absorbances at 936 and 1140 cm^{-1} (the extrapolated value of $A(\beta) \cdot \cos \beta$ for oriented samples). The data for samples with a variety of thermal histories are shown in Fig. 1, where the films were carefully weighed to obtain W. There is a linear correlation between A_{1140}/A_{936} and W/A_{936}. One can obtain the following experimental equation:

$$\left(\frac{A_{1140}}{A_{936}}\right) = 0.056 \left(\frac{W}{A_{936}}\right) - 0.208 \tag{6}$$

The sample mass utilized in the next sections always refers to this calculated value.

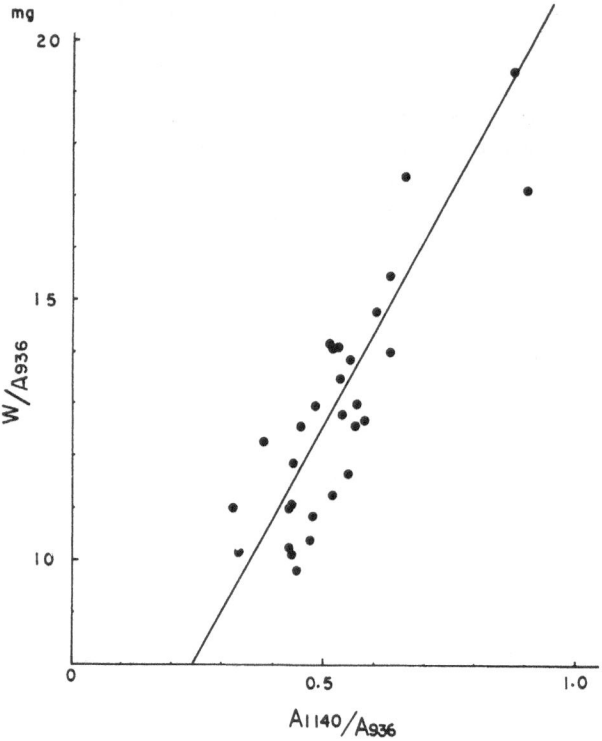

FIG. 1. Relationship between amorphous-crystalline absorbance ratio and reciprocal of crystallinity for nylon 66.

The term A_{936}/W is proportional to crystallinity, or $C_1 A_{936}/C_2 W$ is equal to the percent crystallinity. The percent crystallinity, calculated by the above equation for samples throughout this experiment, ranges from 25 to 40%, which seems to be consistent with the previous spectroscopic data (17). However, these values are particularly sensitive to the values of C_1 and C_2, compared with the sample mass estimation.

We have assumed a simple two-phase model with the densities of the crystalline and amorphous phases being constant. Within the experimental range of these experiments, the peak extinction coefficients are found to be independent of the sample history.

RESULTS

1.   Studies of Randomly Oriented Samples Crystallized At Various Temperatures

Two characteristic quantities, A_{1329}/W and A_{936}/W, proportional to the regular fold content and crystallinity, respectively, are plotted against isothermal crystallization temperature for melt-crystallized, randomly oriented samples in Fig. 2.

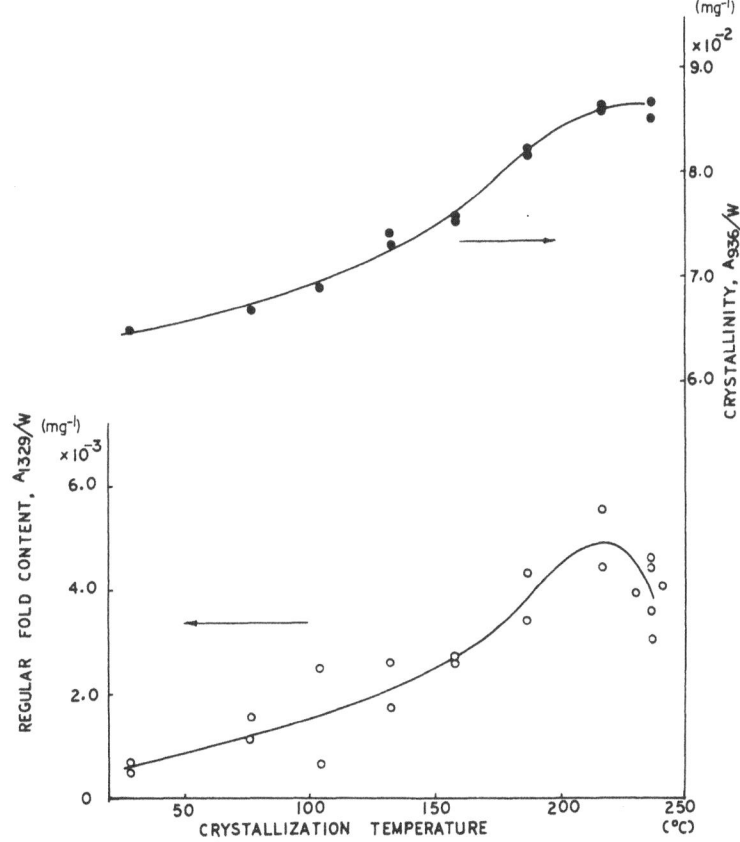

FIG. 2. Dependence of crystallinity and regular fold content on crystallization temperature for isothermal crystallization from 280°C melt.

Both quantities increase with crystallization temperature to about
230°C and then the regular fold content decreases abruptly, while
the crystallinity levels off. This leveling off of crystallinity at the
highest crystallization temperature is felt to be caused by short
crystallization period due to a relatively long induction period
and slow crystallization rates at these high temperatures (18).
The long period data, obtained from the X-ray-small angle scattering for the above samples, are shown in Table 1.   The long
period increases particularly above 235°C, which coincides with
the decrease of regular fold content in Fig. 2.

TABLE I

Long Period of Nylon 66 in Various States

	Long period ($\overset{\circ}{A}$)
A. Random orientation	
1. Melt-crystallized at 105°C	94
2. Melt-crystallized at 159°C	103
3. Melt-crystallized at 215°C	103
4. Melt-crystallized at 235°C	104
5. Melt-crystallized at 250°C	113
B. Uniaxial orientation	
6. As-drawn from (1) (x3.0, 50°C water)	86
7. As-drawn from (2) (x3.0, 50°C water)	86
8. Annealed from (6) at 220°C	97
9. Annealed from (7) at 220°C	101
10. Annealed from (7) at 230°C	109

The regular fold content and crystallinity are plotted for samples,
originally cooled to 28°C and annealed at 150 to 250°C for 15 min,
in Fig. 3.   The time dependence of fold formation during annealing
for quenched random samples has already been studied (11).   The
regular fold formation and increase in crystallinity occur rapidly
during the initial stage of crystallization at times less than 15 min
and then remain almost unchanged.   The annealing period in this

experiment corresponds to the early stages of this plateau.

Crystallinity and regular fold content begin to increase at annealing temperatures around 180°C. This temperature is a little lower than the temperature at which the crystallographic transition from triclinic to pseudohexagonal packing occurs, but higher than that of the motional narrowing in the NMR (19, 20). The regular fold content in Fig. 3 increases monotonously with annealing temperature, approaching the same maximum value as the melt-crystallized sample but the crystallinity is much higher at the highest annealing temperatures. The long period is expected to show almost the same temperature dependence as found for the isothermal crystallizations (11).

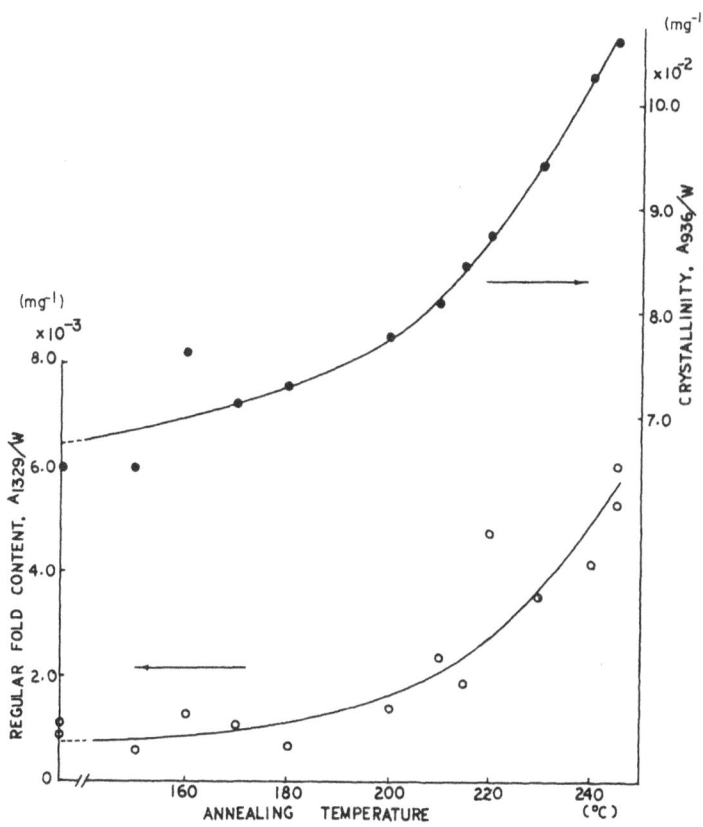

FIG. 3. Dependence of crystallinity and regular fold content on annealing temperature for quenched, random samples.

2. Studies of Uniaxially Oriented Samples Annealed at Various Temperatures

The temperature dependences of crystallinity and regular fold content during the 15 min successive annealing of uniaxially oriented samples are shown in Fig. 4 with free shrinkage and in Fig. 5 when restrained to constant length. The data in Fig. 4 are shown as a reference line in Fig. 5. These samples were originally crystallized at higher temperature, such as 235°C, and

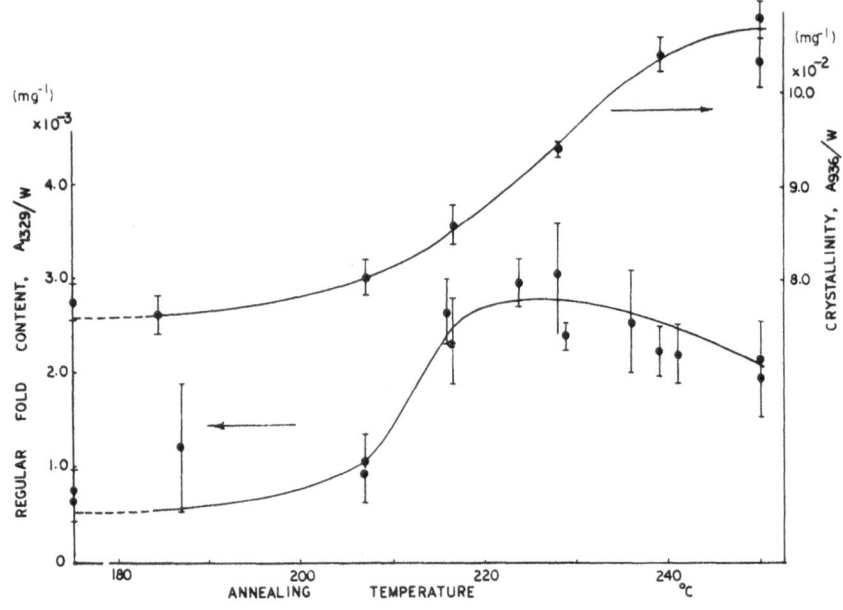

FIG. 4. Dependence of crystallinity and regular fold content on annealing temperature for uniaxially oriented samples with high initial crystallinity (annealing with free contraction).

stretched to 3.0 to 3.5 times in 50°C water. They were liable to fracture, showing a sharp neck and opaque appearance during stretching, compared with the low crystallinity samples mentioned below.

It is to be noted that, comparing the data for the as-drawn samples in Figs. 4 and 5 with those for the corresponding initial unoriented samples in Fig. 2, the uniaxial orientation causes a drastic decrease in regular fold content, with almost no decrease in crystallinity. The subsequent annealing leads to the increase in both quantities, but the regular fold content at its maximum is half or one-third of those of the corresponding nonoriented samples.

The temperature dependences of the fold content and crystallinity are quite different depending on the annealing conditions and whether the samples are allowed to shrink. For a sample that is allowed to relax during annealing, the regular fold content shows a much larger increase at lower temperature and a maximum at

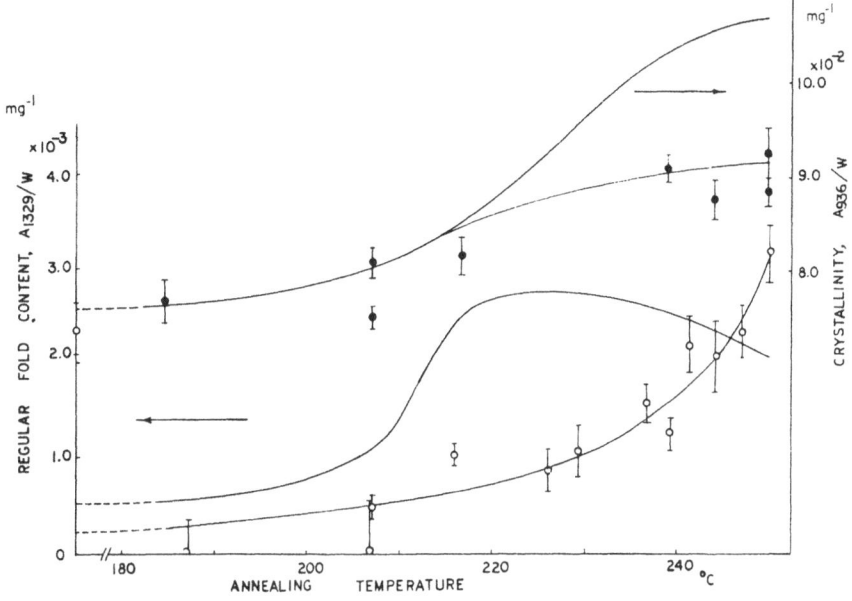

FIG. 5. Dependence of crystallinity and regular fold content on annealing temperature for uniaxially oriented samples with high initial crystallinity (annealing with constant length).

approximately 220°C. For a constrained sample the regular fold content is much smaller at lower temperatures but increases gradually with increasing temperature without any maximum or saturation. The latter behavior apparently coincides with Dismore and Statton's small-angle scattering data (6). The temperature range for the abrupt increase in the fold content of the relaxed sample is higher than the crystallographic transition region (19, 20). The crystallinity is larger for the relaxed sample, particularly at higher annealing temperatures. A similar result has been obtained for uniaxially oriented PET annealed before stretching (21).

Similar data for samples originally of low crystallinity, obtained by quenching the melt to 50°C, are shown in Figs. 6 and 7. In contrast to the samples with high crystallinity, stretching

was much easier, and these samples apparently drew homogene-
ously.   Also, the crystallinity was increased by drawing, suggesting stress-induced crystallization, but there is apparently no

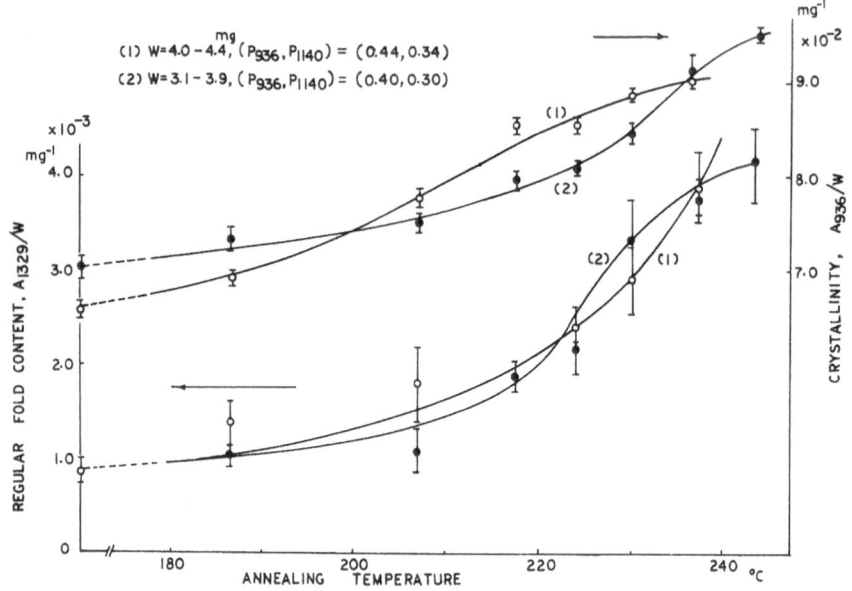

FIG. 6. Dependence of crystallinity and regular fold content on annealing
temperature for uniaxially oriented samples with low initial crystallinity
(annealing with free contraction).

change in regular fold content.   Both quantities increase by
annealing, and the maximum values are a little larger than in the
sample of high crystallinity.   There also seems to be slightly
lower crystallinity and regular fold content for the constrained
samples, but not so distinct a difference as for the high crystallinity samples.   Figure 8 shows data for samples with low
crystallinity, drawn with various draw ratios and constrained
during annealing.   The data for the quenched random sample in
Fig. 3 are shown as a reference line.   With increasing draw
ratio, the regular fold content deviates more from the results for
the quenched, random sample.   Furthermore, at the lower draw
ratio, the regular fold content deviates more from the results for
the quenched, random sample.   Furthermore, at the lower draw
ratio, the same leveling-off-type dependence on annealing temperature was or is observed, whereas at the higher draw ratio, no
saturation is observed.

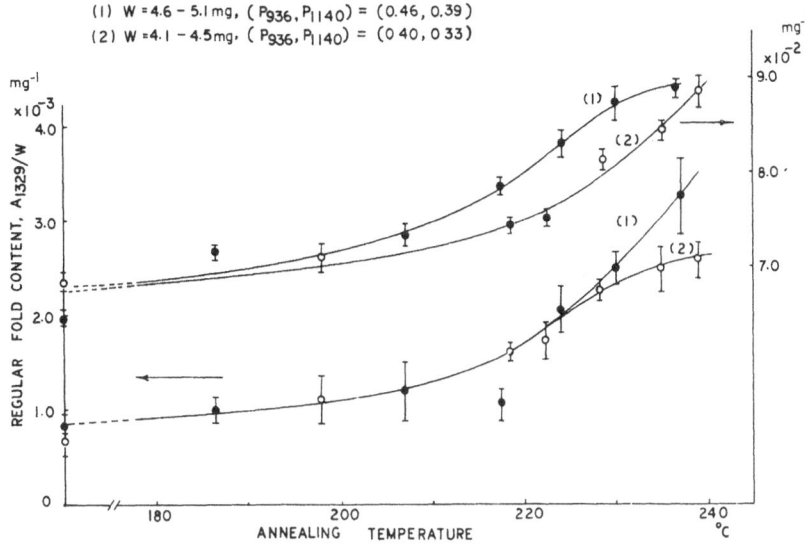

FIG. 7. Dependence of crystallinity and regular fold content on annealing temperature for uniaxially oriented samples with low initial crystallinity (annealing with constant length).

The temperature dependence of the fold content for annealing of oriented samples can be classified into two types-saturation
type and continually increasing type.   Samples exhibiting saturation of the fold content with annealing temperature always have
lower crystallinity than samples showing continued development of
fold content for the same annealing condition.   The differences
are also dependent on the sample mass and the degree of orientation of the sample; the samples of larger mass (thicker film) or
higher orientation tend to have higher crystallinity and to exhibit
the continually increasing fold formation with increase in annealing temperature.   At any rate, drawing conditions and the crystallinity of the original, unoriented samples have major influences
on the ability of the sample to form folds for the uniaxially oriented
samples.

DISCUSSION

A more convenient form of the data is the relationship between crystallinity and regular fold increase (see Figs. 9 through
12).   The results for the annealing of the quenched random
samples, obtained from Fig. 3, are shown in each figure as a

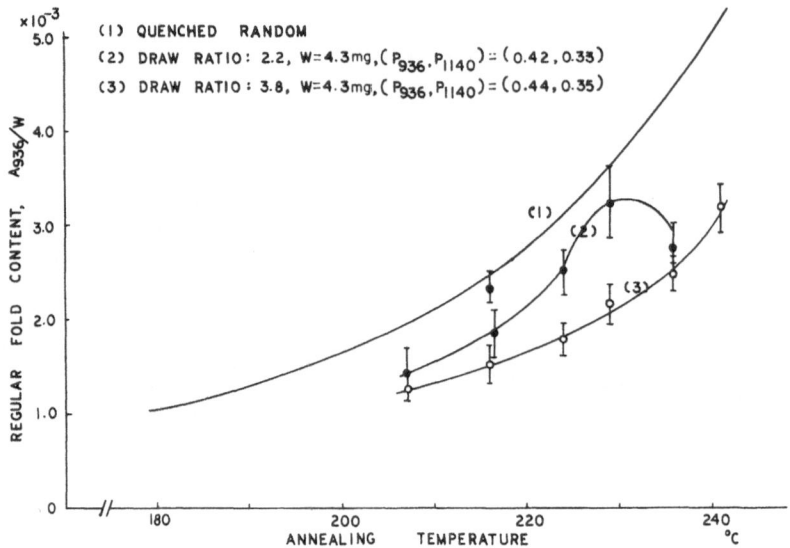

FIG. 8. Dependence of regular fold content on annealing temperature for uniaxially oriented samples with low initial crystallinity and various draw ratios (constrained during annealing).

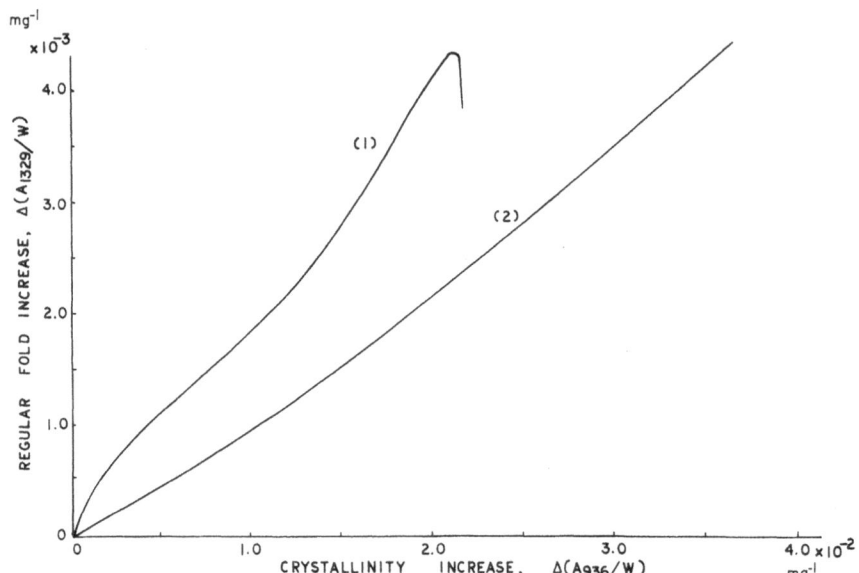

FIG. 9. Relationship between crystallinity increase and regular fold increase during annealing or melt crystallization. (1) Isothermal crystallization from 280°C melt (from Fig. 2); (2) annealing of quenched, random samples (from Fig. 3).

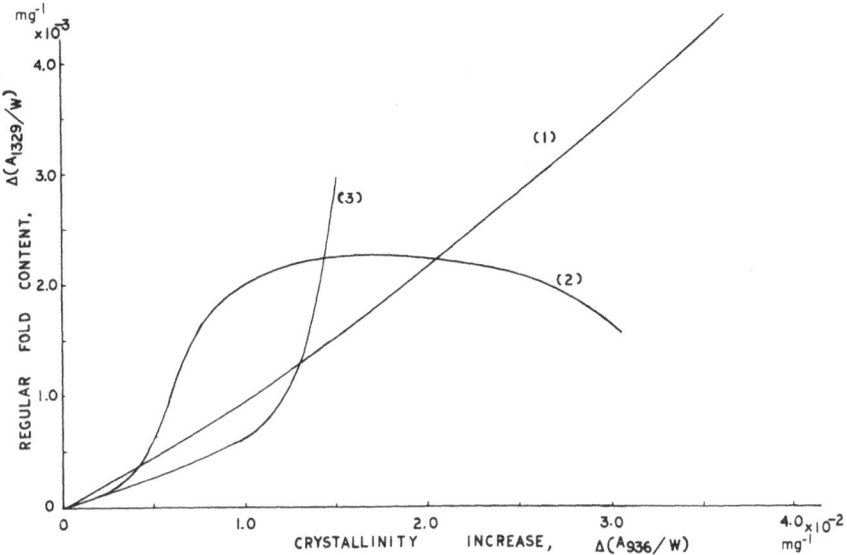

FIG. 10. Relationship between crystallinity increase and regular fold increase for annealing of uniaxially oriented samples with high initial crystallinity. (1) Quenched random (from Fig. 3), (2) relaxed during annealing (from Fig. 4), (3) constrained during annealing (from Fig. 5).

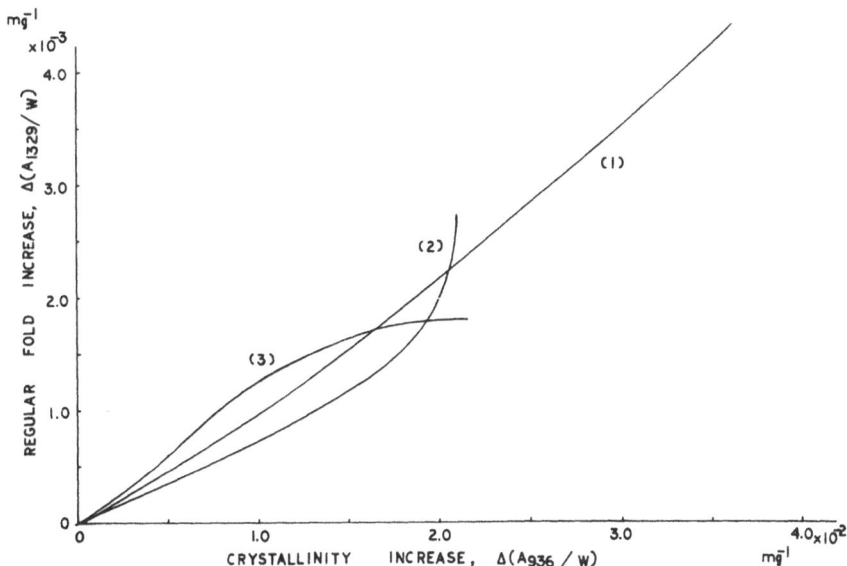

FIG. 11. Relationship between crystallinity increase and regular fold increase during annealing of uniaxially oriented samples with low initial crystallinity. Annealing under constraint. (1) Quenched random (from Fig. 3), (2) corresponds to (1) in Fig. 7, (3) corresponds to (2) in Fig. 7.

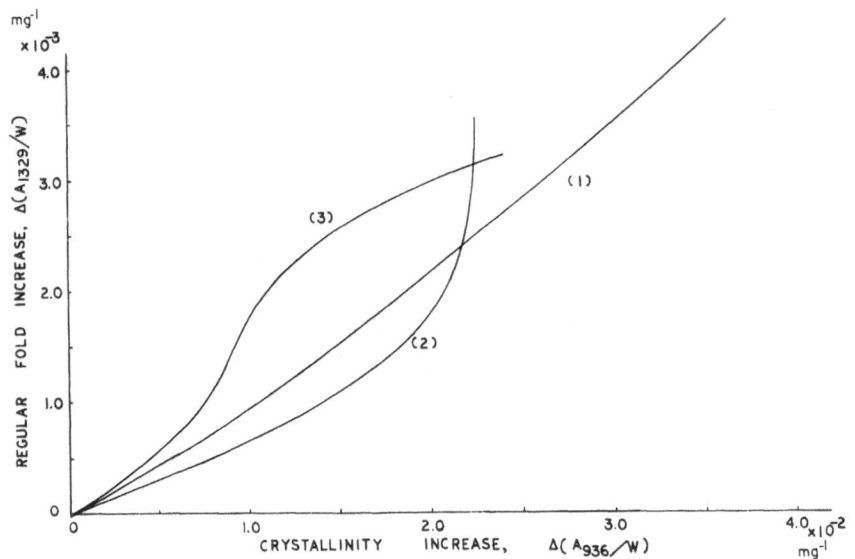

FIG. 12. Relationship between crystallinity increase and regular fold increase during annealing of uniaxially oriented samples with low initial crystallinity. Annealing with free shrinkage. (1) Quenched random (from Fig. 3), (2) corresponds to (1) in Fig. 6, (3) corresponds to (2) in Fig. 6.

reference.   For the melt-crystallized samples in Fig. 9, the crystallinity and regular fold content for the sample quenched at 30°C, as shown in Fig. 2, has been taken as a control or initial value to calculate the differences (increments) from those for higher crystallization temperatures.   For other curves, the initial values are those of corresponding nonannealed samples.

The annealing of the quenched, random samples in Fig. 9 shows almost linear or slightly greater fold increase over a long range of these two quantities. The findings for the melt-crystallized samples (Fig. 9) are quite different from those of other samples.   They are characterized by a larger increase of fold content increments relative to the crystallinity and an abrupt decrease at the final stage.

The annealing of the uniaxially oriented samples with high initial crystallinity gives rise to different behavior, depending on annealing conditions, as shown in Fig. 10.   The fold content for the relaxed samples increases linearly at first, then turns upward, and finally decreases with respect to the crystallinity increase. On the other hand, for the constrained samples, there is a longer initial linear portion, and then the fold increment increases

continually without any final saturation or decrease.   For all of
the quenched and drawn samples (Figs. 11 and 12) the results are
also linear, at first, with almost equal slopes which are also
nearly equal to that of the quenched, random sample. Then, the
fold increments increase similarly, but the relaxed samples tend
to show a larger increase.   From the above results, it can be
summarized that these increments of fold content relative to the
crystallinity show essentially three successive portions.

1. Initially, the increase in regular fold content is linear
    with increase in crystallinity.   The slopes are nearly
    equal for all the samples except the isothermally
    crystallized samples.
2. After the initial linear portion, the increase in regular
    fold content becomes larger relative to the increase in
    crystallinity.
3. Finally, in some cases, the increase in regular fold
    content is smaller relative to the increase in crystal-
    linity.   This final portion is not always observed for
    all samples.

These three types of behavior are indicative of different
structural changes in the samples occurring at different annealing
temperatures and their understanding should contribute to our
structural knowledge of the folding process.

REFERENCES

(1)   P. Ingram, H. Kiho, and A. Peterlin, Polymer, 7, 135
       (1966).
(2)   K. Kobayashi, cited in P.H. Geil, Polymer Single Crystals,
       Interscience-Wiley, New York, 1963.
(3)   A. Peterlin, P. Ingram, and H. Kiho, Makromol. Chem.,
       86, 294 (1965).
(4)   A. Peterlin, Kolloid-Z., 216-217, 129 (1967).
(5)   A. Peterlin, J. Polymer Sci., C9, 61 (1965).
(6)   P. F. Dismore and W. O. Statton, J. Polymer Sci., C13,
       133 (1966).
(7)   J. P. Bell and J. H. Dumbleton, J. Polymer Sci., (A-2)7,
       1033 (1969).
(8)   T. Amano, E. W. Fischer, and G. Hinrichsen, J. Macromol.
       Sci., B3, 209 (1969).
(9)   J. L. Koenig and D. E. Witenhafer, Makromol. Chem., 99,
       193 (1966).

(10)   J. L. Koenig and M. J. Hannon, J. Macromol. Sci., B1, 119
       (1967).

(11)   J. L. Koenig and M. C. Agboatwalla, J. Macromol. Sci., B2,
       391 (1968).

(12)   C. G. Cannon and P. H. Harris, J. Macromol. Sci., B3, 357,
       (1969).

(13)   P. D. Frayer, J. L. Koenig and J. B. Lando, J. Macromol.
       Sci., B3, 329 (1969).

(14)   W. O. Statton, M. J. Hannon, and J. L. Koenig, J. Appl.
       Phys., 41, 4290 (1970).

(15)   J. L. Koenig and M. Itoga, Appl. Spectr. 25, 355 (1971).

(16)   H. W. Starkweather and R. E. Moynihan, J. Polymer Sci.,
       22, 363 (1956).

(17)   I. Sandeman, and A. Keller, J. Polymer Sci., 19, 401
       (1958).

(18)   J. H. Magill, Polymer, 3, 43 (1962).

(19)   W. P. Slichter, J. Polymer Sci., 35, 77 (1958).

(20)   W. P. Slichter, J. Appl. Phys., 26, 1099 (1955).

(21)   J. L. Koenig and S. W. Cornell, J. Polymer Sci., C22, 1019
       (1969).

(22)   D. R. Buchanan and J. H. Dumbleton, J. Polymer Sci.,
       (A-2), 7, 113 (1969).

MECHANISMS FOR REGULARIZATION OF CHAIN FOLDS DURING ANNEALING OF POLYHEXAMETHYLENE ADIPAMIDE*

J. L. Koenig and Masaaki Itoga

Case Western Reserve University

Cleveland, Ohio

SUMMARY

The regular fold content of polyhexamethylene adipamide depends on the initial crystallization conditions, the degree of orientation, and the annealing time and temperature. At low an annealing temperatures, the regular fold content increases linearly with crystallinity and arises from lamellar crystallization of isolated amorphous or interlamellar regions. At intermediate annealing temperatures, the increase in regular fold content greatly exceeds the crystallinity increase. The excessive increase in folds in this temperature range arises from regularization of loose loops and melting and recrystallization from extended to more folded type of crystals. At the higher annealing temperature, the crystallinity shows an increase relative to the fold content, and this implies that the increase in fold period occurs at the expense of folding.

INTRODUCTION

It can be assumed that a uniaxially drawn or randomly oriented semicrystalline polymer consists essentially of lamellar-type structures (1). In the drawn state, these lamellae are highly distorted and are linked together by tie molecules (interlamellar links).

---

*Published in the Journal of Macromolecular Science-Physics, B6(2), 327-342(1972). Reprinted by permission.

The fold surfaces of crystals probably consist of (1) adjacent entry-type regular fold, as often suggested for single crystals, (2) irregular loops, (3) interlamellar links, and (4) chain ends (cilia).

In addition, there may be separate amorphous or paracrystalline regions. Upon annealing these polymers, considerable numbers of regular folds are formed and, as shown in the previous paper (2), the annealing behaviors as a function of temperature are different, depending on the annealing, drawing, and initial crystallization conditions. These regular folds probably arise from several different sources including crystallization by chain folding from interlamellar regions and independent amorphous regions, regularization of loose-loop-type folds or the transformation from extended-type to fold-type crystals. The regularization of loose-loop folds can occur by two different processes, i.e., reorganization on the lamellar surface or melting and fold-type recrystallization of imperfect lamellae. The purpose of this paper will be to attempt a mechanistic interpretation of the infrared results reported previously and of the present results.

RESULTS

Table 1 shows the change in molecular orientation in the crystalline portion (P_{936}) and amorphous portion (P_{1140}) of polyhexamethylene adipamide during annealing at successively increased temperature. In every case, whether the samples are constrained or relaxed, with samples of initial high crystallinity or of low crystallinity, there is no conspicuous change in molecular orientation neither in the crystalline nor amorphous region as a result of the thermal treatment. The annealing of uniaxially oriented polyethylene terephthalate shows a slight decrease in crystalline orientation (3). At the same draw ratio there is a higher degree of orientation in both regions for the samples with high initial crystallinity.

The thermal shrinkage during successive annealing, allowing free shrinkage, is plotted against annealing temperature in Fig. 1. The samples with high initial crystallinity show much larger thermal shrinkage. It has been observed that thermal shrinkage or shrinkage force has a good correlation with both small-angle scattering intensity and infrared absorption intensity of a fold band (4, 5). Reviewing the infrared data of the relaxed samples in the previous paper (2) from this point, a coincidence

TABLE 1

Change in Molecular Orientation during Successive
Annealing of Uniaxially Oriented Samples[a]

Sample No.:	1	2	3	4
Crystallization temp. (from 280° C melt):	50°C	50°C	216°C	235°C
Draw ratio:	×3.5	×3.8	×2.9	×3.5
Annealing condition:	Free shrinkage	Constraint	Free shrinkage	Constraint
I P_{936} (crystalline)				
As-drawn	0.52 ± 0.03	0.41 ± 0.02	0.46 ± 0.04	0.50 ± 0.05
187°C	0.40 ± 0.02	0.44 ± 0.02	0.54 ± 0.02	0.49 ± 0.01
207°C	0.47 ± 0.05	0.43 ± 0.01	0.49 ± 0.02	0.48 ± 0.05
216°C	0.45 ± 0.02	0.39 ± 0.03	0.49 ± 0.04	0.43 ± 0.03
224° C	0.42 ± 0.02	0.42 ± 0.01	0.46 ± 0.04	0.50 ± 0.02
229°C	0.46 ± 0.02	0.45 ± 0.02	—	0.44 ± 0.04
236°C	0.43 ± 0.02	0.42 ± 0.01	0.50 ± 0.06	0.56 ± 0.04
241°C	0.49 ± 0.03	0.44 ± 0.01	0.52 ± 0.03	0.52 ± 0.02
II P_{1140} (amorphous)				
As-drawn	0.41 ± 0.02	0.33 ± 0.01	0.49 ± 0.09	0.41 ± 0.07
187°C	0.29 ± 0.04	0.28 ± 0.02	0.57 ± 0.06	0.46 ± 0.04
207°C	0.40 ± 0.12	0.36 ± 0.03	0.43 ± 0.09	0.50 ± 0.07
216°C	0.37 ± 0.04	0.29 ± 0.01	0.49 ± 0.06	0.41 ± 0.03
224°C	0.32 ± 0.02	0.28 ± 0.01	0.30 ± 0.04	0.42 ± 0.03
229°C	0.35 ± 0.04	0.32 ± 0.04	—	0.49 ± 0.06
236°C	0.35 ± 0.07	0.33 ± 0.03	0.41 ± 0.11	0.57 ± 0.07
241°C	0.47 ± 0.03	0.30 ± 0.03	0.49 ± 0.04	0.52 ± 0.05

[a]$P = 1$ (perfect uniaxial orientation), $\frac{1}{3}$ (random orientation), 0 (planar orientation).

between the fold content and the thermal shrinkage data will be noticed. It is to be noted, however, that such a correlation should be confined to oriented samples having the same crystallization history. The long period data from X-ray small-angle scattering for the uniaxially oriented samples was reported in the previous paper. The long period is reduced by drawing and increases again during annealing. The values of the long period for the annealed samples are comparable to those for randomly oriented samples crystallized at the same temperature. The long period for annealed uniaxially drawn nylon 66 has already been studied (6, 7), and it is a linear function of the reciprocal of supercooling, $1/(T_m - T)$, where T_m and T are the melting point and annealing temperature, respectively.

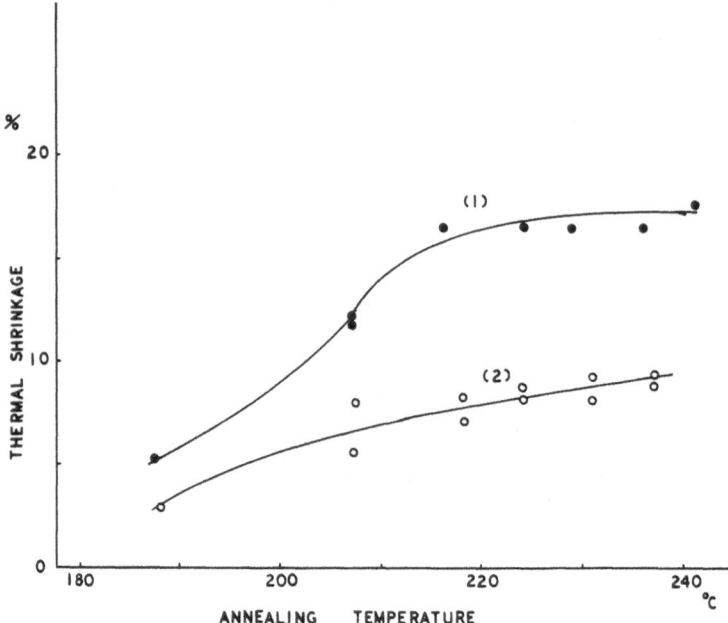

FIG. 1. Thermal shrinkage during successive annealing of uniaxially
oriented samples. (1) Crystallized at 216°C and drawn ×2.9 in 50°C water;
(2) quenched to 50°C and drawn ×3.5 in 50°C water.

        The differential scanning calorimeter (DSC) thermograms
are shown in Fig. 2.   They were obtained at high heating rate and
are corrected for sample mass differences.   The as-drawn sample
with high initial crystallinity shows a much sharper melting peak
and the highest melting point, whereas the quenched and stretched
sample shows a broader peak and lower melting region.   For
samples drawn to low draw ratios, i.e., 2.3 or less, another
small peak was sometimes observed on the lower shoulder of the
main peak, apparently in agreement with Bell and Dumbleton's
observation of two melting peaks (6).   The quenched, random
sample shows an exothermic peak, presumably due to cold crystallization, as well as a broad melting peak.   These results should
be useful in our efforts to interpret the different chain-folding
processes.

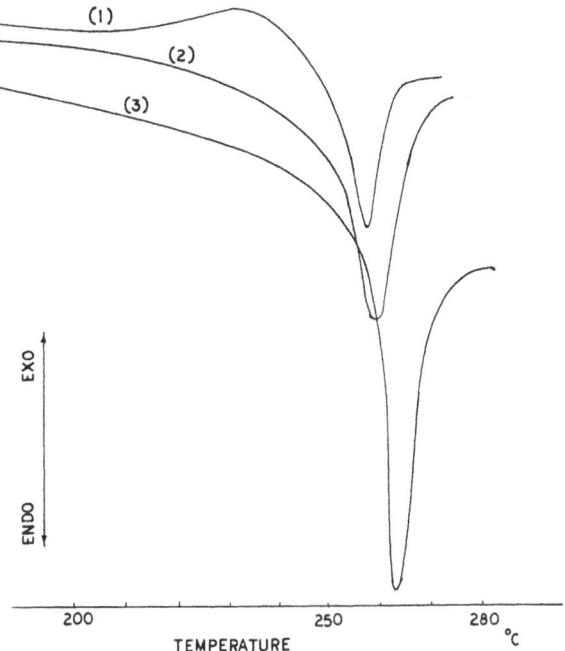

FIG. 2. Differential scanning calorimeter thermogram of random and uniaxially oriented samples. (1) Quenched to 30°C (random); (2) quenched to 30°C and drawn ×3.5 in 50°C water; (3) crystallized at 236°C and drawn ×3.5 in 50°C water. Heating rate, 80°C/min.

DISCUSSION

In order to discuss the possible sources of regular folds it would be helpful to relate crystallization with fold formation. This can be treated as follows. As mentioned above, the molecular chains emerging from the crystalline lattices terminate at the surface with regular folds, irregular folds, interlammelar links, or free chain ends (cilia). One fold or one interlamellar link involves two emerging chains, whereas each cilium corresponds to a single emerging chain. But one chain in a crystal has two ends which we assume will be on the surface. Thus, if N_o is the total number of segmental chains composing the cross section of crystals in a unit weight of polymer, the following relation can be obtained:

$$2N_o = 2N_{rf} + 2N_{if} + 2N_t + N_e \tag{1}$$

where N_{rf} is the number of regular folds, N_{if} the number of irregular folds (loose loops), N_t a number of interlamellar links
(tie molecules), and N_e is a number of cilia in a unit weight of
polymer.

The infrared absorption of 936 cm^{-1} is proportional to
crystallinity, which corresponds to the total number of chain
segments in crystals.   Therefore

$$N_c = \frac{A_{936}}{K_c \cdot W} \tag{2}$$

$$= N_o \cdot \frac{L}{\ell_o} \tag{3}$$

where N_c is a total number of segments in crystals, L and ℓ_o
are the crystalline thickness and the chemical repeat distance of
nylon 66, respectively, and W and K_c are polymer weight and
extinction constant.   From Eqs. (1) and (3),

$$N_c = \frac{L}{\ell_o} (N_{rf} + N_{if} + N_t + \frac{1}{2} N_e) \tag{4}$$

Furthermore, the absorbance at 1329 cm^{-1} is proportional to the
total number of regular folds in a unit volume of polymer. Thus

$$N_{rf} = \frac{A_{1329}}{K_f \cdot W} \tag{5}$$

and, from Eqs. (2), (4), and (5),

$$\frac{A_{1329}}{W} = \frac{K_f}{K_c} \cdot \frac{\ell_o}{L} \cdot \frac{A_{936}}{W} - K_f(N_{if} + N_t + \frac{1}{2}N_e) \tag{6}$$

Since, we are interested in the changes in the regular folding
process, by differentiating this equation (neglecting higher terms),
we obtain

$$\Delta(\frac{A_{1329}}{W}) = \frac{K_f \ell_o}{K_c L} \cdot \Delta(\frac{A_{936}}{W}) + \frac{\ell_o K_f}{2K_c} \cdot \frac{A_{936}}{W} \cdot \Delta(\frac{1}{L})$$

$$- K_f \cdot \Delta(N_{if} + N_t + \frac{1}{2} N_e) \tag{7}$$

Particular cases can be studied from this result.

For well-formed single crystals, there will be neither ir-regular loops ($N_{if} = 0$) nor interlamellar links ($N_1 = 0$). For high molecular weight polymer the number of cilia will be negligibly small compared with the number of folds. Since crystallinity is extremely high the lamellar thickness may be almost equal to the long period obtained by X-ray small-angle scattering. Therefore, for this special case, Eq. (6) can be simplified to

$$\frac{A_{1329}}{A_{936}} = \frac{\ell_o K_f}{LK_c} \tag{8}$$

According to previous data (8) for nylon 66 single crystals, $A_{1329}/A_{936} = 0.10$, $L = 58$ A, and also $\ell_o = 17.2$ A. Then

$$\frac{K_f}{K_c} = 0.34 \tag{9}$$

Therefore, Eq. (7) can be arranged as follows:

$$\Delta\left(\frac{A_{1329}}{W}\right) = \frac{5.8}{L} \cdot \Delta\left(\frac{A_{936}}{W}\right) + \frac{5.8 A_{936}}{W} \cdot \Delta\left(\frac{1}{L}\right)$$

$$- K_f \cdot \Delta\left(N_{if} + N_t + \frac{1}{2} N_e\right) \tag{10}$$

The crystallinity increase, $\Delta(A_{936}/W)$, and lamellar thickness in-crease, ΔL, are always positive for single crystals [$\Delta(1/L)$ is negative], with respect to increasing annealing temperature. Consequently, from Eq. (10) the following criteria can be obtained with regard to crystallinity increase and regular fold increase.

Case A: linear relationship between increase in fold con-tent and crystallinity. If the regular fold increase, $\Delta(A_{1329}/W)$, is linear as a function of crystallinity increase, $\Delta(A_{936}/W)$, with a reasonable slope which corresponds to L, the lamellar thickness, then

$$\Delta L = \Delta\left(N_i + N_t + \frac{1}{2} N_e\right) = 0 \tag{11}$$

The regular fold increase in this case arises only from formation of new folded-chain crystals from amorphous regions. Crystallization of amorphous chains without any other lamellar reorganization is probably the preferred process at relatively low temperature.

Case B: excessive nonlinear relationship between increase in fold content and crystallinity. If the regular fold increase, $\Delta(A_{1325}/W)$, is larger than expected from the crystallinity increase, $\Delta(A_{396}/W)$, there must be additional sources for regular fold formation to the crystallization mentioned. In such a case, possible causes are the regularization of loose loops and the transformation from the extended-type to the folded-type crystals. The contribution of folding back of a free chain end is minor because of its short length (9). In a strict sense, the negative effect of lamellar thickening must also be considered and the following relation can be obtained:

$$- K_f \; \Delta \, (N_i + N_t + \frac{1}{2} N_e) \; \rangle \; -5.8 \; \frac{A_{936}}{W} \; \Delta \, (\frac{1}{L}) \tag{12}$$

This implies that the positive contribution to the regular fold formation of the decrease of loose loops or interlamellar links exceeds the lamellar thickening. This behavior may be expected at intermediate temperatures.

Case C: reduced nonlinear relationship between increase in fold content and crystallinity. This is the opposite case of B, i.e.,

$$- \frac{5.8 \; A_{936}}{W} \; \Delta(\frac{1}{L}) \; \rangle \; - K_f \; \Delta(N_i + N_t + \frac{1}{2} N_e) \tag{13}$$

The negative effect of lamellar thickening exceeds the positive effects, such as crystallization and fold regularization. Since lamellar thickening is much more conspicuous at higher temperatures, this behavior may be expected at a high annealing temperature.

Thus, examining the relationship between crystallinity increase and regular fold increase in this manner, possible mechanisms for regular fold formation and annealing processes can be analyzed in more detail. As seen in the previous paper the changes in fold content compared to crystallinity are linear at first, then turn upward with the fold content showing excessive increases relative to the crystallinity, and finally some of the curves level off with little change in fold content with crystallinity. In the initial linear region, the slopes are almost equal indicating crystallization of amorphous regions. The lamellar thickness, calculated from the initial slope, is 45-80 Å, which is

of the same order as the long period of nylon 66 single crystals
(8) (60-70 Å), obtained from X-ray small-angle scattering data.
Therefore, the initial linear portion can be interpreted as regular
folding, caused by fold-type crystallization from independent
amorphous regions or amorphous interlamellar regions.   Furthermore, because of the agreement between the calculated and
measured values of the fold period, this crystallization during
annealing is assumed to yield almost perfect regular folds as in
single crystals.   This will be possible, in terms of the nodule
structure, where amorphous polymers exist as small clusters,
relatively independent of each other and, on annealing, merge to
form lamellae (11).

The regularization of loose loops and the transformation
from extended-type to fold-type crystals, which may occur without
appreciable crystallinity increase, must be considered in addition
to the simple crystallization.   The regularization of loose loops
may involve surface regularization without melting of lamellae,
whereas the transformation from extended - type to fold-type
crystals will be possible only by melt recrystallization.   In terms
of annealing temperature, such phenomena appear above 210°C
and at lower temperatures for samples annealed free to relax,
regardless of initial crystallinity.   At such a temperature, molecular chains in crystalline lattices may also have considerable
mobility, such as twisting (12), which is required for fold regularization.

The final region of the curve represents the region where
little increase in regular folds occurs relative to the increase in
crystallinity.   This suggests the negative effect of lamellar
thickening is predominant.   The calculation of the lamellar thickness increase for the uniaxially oriented sample with initial high
crystallinity and the melt-crystallized samples, by assuming L
as 70 Å at the maximum fold content and neglecting
$\Delta(N_i + N_t + \frac{1}{2} N_e)$, shows 15-25 Å of ΔL, which is comparable
to the data estimated by Tsvankin's method.

Thus, for all samples, random or oriented, at low annealing temperatures crystallization from amorphous regions may
initiate an increase in regular fold content.   Then, at slightly
higher temperature the regular fold formation without appreciable
increase in crystallinity begins due to such features as the regularization of loose loops or the crystalline transformation from
extended to fold type occurs in addition to the crystallization.

This process gives rise to a large increase in regular fold content without much change in crystallinity. Finally, at high annealing temperatures for some samples, the negative effect of lamellar thickening becomes predominant, and apparently no fold increase is observed.

Let us reexamine the data from this viewpoint. The melt-crystallized samples show an exceptionally large increase in regular fold content relative to the crystallinity, followed by an abrupt loss in fold content at the highest crystallization temperature. The small, regular, fold content for samples formed at lower crystallization temperatures is due to irregular fold surfaces as well as to low crystallinity or large amounts of interlamellar links. But, with increasing crystallization temperature, the fold regularity is improved just as in the case of single crystals (8). At the highest annealing temperatures, the regularity will be complete and the influence of lamellar thickening is larger, producing the decrease in regular fold content.

The relationship between crystallinity increase and fold increase during the annealing of quenched, random samples shows an initial linearity, and, then, a slightly excess amount of folding relative to crystallinity over a long temperature range, indicating that crystallization from amorphous regions or interlamellar links causes the initial regular fold increase and that regularization of irregular folds, which is expected from the above discussion on the melt-quenched samples, occurs over a wide range of increase in crystallinity. For the annealing of uniaxially oriented samples with initial high crystallinity, the difference is distinct between the constrained and relaxed samples. The relaxed sample shows a shorter initial stage, the second stage terminates earlier, and the final saturation stage is observed. The uniaxially oriented samples with initial low crystallinity do not show such a difference; however, the relaxed samples tend to show higher improvement of surface regularity. Higher draw ratio seems to give rise to a longer initial linear part, and disappearance of the final stage, since it is not seen in constrained samples.

Mechanisms for Regular Fold Formation

In addition to the regular folds obtained by crystallization from the amorphous regions, there is an incremental number of regular folds formed from other sources and not accompanied by an appreciable increase in crystallinity. There are at least two possible sources for these regular folds--(1) the regularization of loose-loop folds and (2) the melting and regular fold-type recrystallization of imperfect lamellar or extended-chain crystals.

The regularization of loose-loop folds, which implies a reorganization of the lamellar surface without melting, requires a lamellar structure in the original, unannealed samples. In a cold-drawn polymer, crystalline lamellae may be highly distorted and the interlamellar links be under considerable stress. If such a polymer is relaxed at sufficiently high annealing temperatures, these chains will be relaxed and lamellar distortion will be removed (13). Then, the loose loops on the surface will be regularized, fulfilling the thermodynamic requirement (14). However, if the sample is constrained during annealing, this crystalline perfection process will be restricted, since the retractive force of interlamellar links increases with increasing temperature due to the entropy effect. The unique fold conformation probably will not be formed on these distorted lamellar surfaces under stress since the process will go in the direction of relieving the stress (15).

The larger thermal shrinkage of the samples with higher degree of orientation and higher initial crystallinity indicates larger stresses in the interlamellar links. However, this thermal shrinkage process is completed at 220°C, which suggests that the residual stress and crystalline distortion is already removed below this temperature. Slichter has observed the crystallographic transformation from the ordinary α for β type to pseudohexagonal γ type below 210°C and a motional narrowing in the nuclear magnetic resonance (NMR) at much lower temperatures (16). He suggests that, in the γ form, there is a vigorous molecular motion within a crystal, such as local skeletal twisting between hydrogen bonds (12). This molecular motion will allow the removal of stress and lattice distortion.

Buchanan and Dumbleton (10), in their annealing experiments on uniaxially oriented nylon 66 fibers with free shrinkage, have observed an increase in crystalline perfection, which is completed at 230°C. They report an increase in the average crystallite size perpendicular to the fiber axis and suggest a transformation from irregular (oblique) crystallites to more regular (rectangular) crystallites. From the above results, the perfection of irregular or distorted lamellae is expected to occur far below the ordinary melting temperature of nylon 66 when annealed free from constraint.

The free energy of a crystal consists of three contributions: perfect crystalline portions, defects, and surfaces (17). Thus, small and/or imperfect crystals have reduced melting temperatures. However, if crystalline lamellae are sufficiently large and perfect, any loose loops will regularize through surface regularization rather than through the melt recrystallization. The DSC thermogram in Fig. 2 shows a much sharper and higher melting peak for the uniaxially oriented sample with high initial crystallinity, indicating larger crystals and less defects, in spite of larger lattice distortion.

For the melting and subsequent regular fold-type recrystal lization mechanism, an expression can be written, based on the assumption that the rate-controlling step in this process is the breaking of the intermolecular bonds with a Maxwell distribution. In our terminology,

$$\Delta(\frac{A_{1329}}{W}) = C \exp(-\frac{\Delta E}{RT})$$

where ΔE is an activation energy for folding, and R is the gas constant. From the kinetic point of view, this expression corresponds to isothermal crystallization from a supercooled melt where the transport term is the rate-controlling step (18).

This relationship of the increase in regular folds is plotted against the reciprocal of the annealing temperature for a quenched, random sample in Fig. 3. A linear relationship is observed and from the slope of the line, the activation energy is calculated to be 19 kcal/mole. From the temperature dependence of spherulite growth for polyamide melts, an activation energy of 18 kcal/mole is obtained (19). The activation energy measured from viscous flow experiments for polyamides is 21 kcal/mole. The agreement between these activation energies suggests that the viscous transport of amorphous chain segments is the rate-determining step in fold formation. The logarithm of the increase in regular folds vs. the reciprocal of the annealing temperature for the uniaxially oriented samples with high initial crystallinity annealed under restraints is shown in Fig. 4, and with low initial crystallinity and higher molecular orientation in Fig. 5. The activation energy for the oriented sample with high initial crystallinity is 25 kcal/mole which is higher than for the unoriented sample, indicating that additional energy is required when the sample is oriented and annealed under stress. The uniaxially oriented

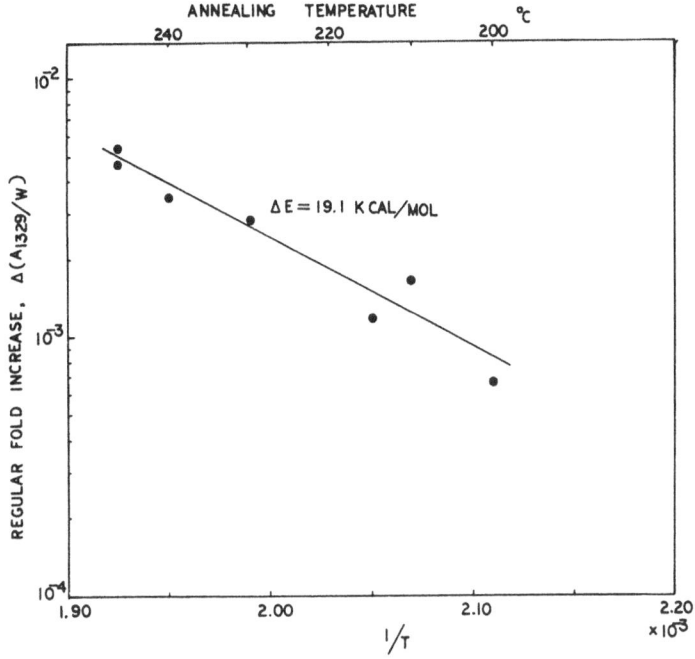

FIG. 3. Logarithmic plots of regular fold increase against reciprocal of annealing temperature (quenched, random samples).

sample with low initial crystallinity has an activation energy of 23 kcal/mole which is also slightly higher than for the unoriented sample.

Thus, the quenched, random samples and the uniaxially oriented samples with high initial crystallinity annealed under restraint or with low initial crystallinity and relatively higher orientation follow this approach, and their activation energies are comparable to the activation energy for segmental transport. This result suggests that regular fold formation in these samples proceeds through melt recrystallization. Thus, for the quenched, random sample at low annealing temperature, crystallization with regular chain folding will occur in the isolated amorphous regions and the interlamellar regions. As the annealing temperature becomes sufficiently high, the irregular and imperfect lamellae with loose loop or regular folds will melt and recrystallize with formation of more regular folds.

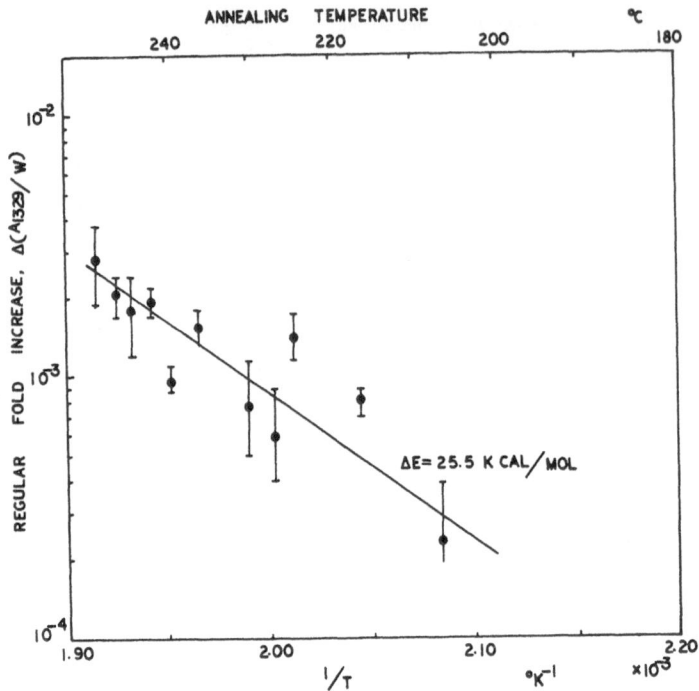

FIG. 4. Logarithmic plot of regular fold increase against reciprocal annealing temperature (uniaxially oriented samples with high initial crystallinity, annealed under constraint).

However, the uniaxially oriented samples with high initial crystallinity, relaxed during annealing, or with low initial crystallinity and lower orientation do not obey this approach. They show much larger regular fold content at low annealing temperatures, and saturation or decrease in fold content at higher temperatures. These results suggest an alternative mechanism - surface regularization.

The surface regularization mechanism may be associated with molecular rearrangement by bond rotation, instead of segmental transport as in melt recrystallization, and requires a smaller activation energy. Therefore, if crystals are sufficiently large and perfect, and free from distortion, the surface regularization tends to occur much faster or at lower temperatures than the melt recrystallization. The uniaxially oriented samples with high initial crystallinity contain larger crystals

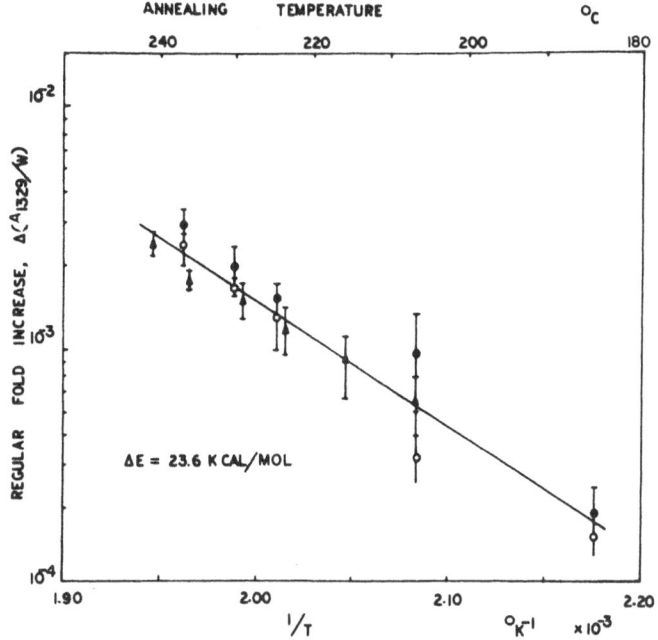

FIG. 5. Logarithmic plot of regular fold increase against reciprocal annealing temperature (uniaxially oriented samples with low initial crystallinity). (●) Relaxed $(P_{936}, P_{1140}) = (0.44, 0.34)$, $W = 4.0\text{-}4.4$ mg; (▲) constrained $(0.43, 0.35)$, 4.3 mg; draw ratio 3.8; (○) constrained $(0.46, 0.39)$, 4.6-5.1 mg.

or less defects. Annealing may remove the lattice distortion below 220°C, when the sample is relaxed during annealing, as seen in Fig. 1 with thermal shrinkage data. The dependence of regular fold content on annealing temperature is in good agreement with these shrinkage data. Thus, the regular folds formed in this sample, relaxed during annealing, seem to result from surface regularization, in contrast with the annealing under constraint where the fold regularization seems to occur through melt recrystallization. In the as-drawn state, the lamellae in this sample may be relatively large and perfect, but highly distorted. If the sample is relaxed during annealing, this distortion will be removed and the loose loops on lamellar surfaces will regularize without melting of the lamellae. However, if the sample is constrained, the recovery of this distortion is restricted, and loose loops may regularize through melt recrystallization.

In the same manner, fold regularization for the uniaxially oriented samples with low initial crystallinity and low molecular orientation is assumed to result from surface regularization, in contrast to melt recrystallization for the highly oriented samples. In this case, drawing conditions and initial structure seem to have a much more important effect on the temperature dependence of regular fold content. However, the relaxation type of annealing still tends to yield a larger fold content. All sources of regular fold formation, such as lamellar crystallization from amorphous or interlamellar links and lamellar regularization, are favorable for annealing in the relaxed stated.

CONCLUSIONS

The comparison of regular fold content with crystallinity during annealing reveals that the regular fold content increases linearly with crystallinity at lower temperatures with similar slopes for every sample. At intermediate temperatures the regular fold content increase exceeds the crystallinity increase and, in some cases, at higher temperatures, the fold increase shows a saturation or decrease. The initial linear portion implies that the increase in regular fold content arises from lamellar crystallization of isolated amorphous region or inter-lamellar links. The second additional increase results from the regularization of loose loops or the melting and recrystalliza-tion from extended-type to folded-type crystals. The final saturation or decrease implies that the increase in fold period exceeds the increase in regular fold content.

The annealing of oriented samples with free shrinkage always results in a higher regular fold content than with constant dimension. Loose loops in uniaxially oriented samples with initial high crystallinity may regularize through melt crystalliza-tion, when they are annealed under constraint, whereas relaxa-tion during annealing may give rise to regular folds through surface reorganization after perfection of distorted lamellae. The regular fold formation in quenched drawn samples is subject to a larger influence of drawing and initial quenching conditions.

REFERENCES

(1)   P. Ingram, H. Kiho, and A. Peterlin, Polymer, 7, 135
       (1966).
(2)   J. L. Koenig and M. Itoga, J. Macromol. Sci., B6(2), 309,
       (1972).
(3)   J. L. Koenig and S. W. Cornell, J. Polymer Sci., C22, 1019
       (1969).
(4)   P. F. Dismore and W. O. Statton, J. Polymer Sci., C13,
       133 (1966).
(5)   W. O. Statton, M. J. Hannon and J. L. Koenig, J. Appl. Phys.
       41, 4290 (1970).
(6)   J. P. Bell and J. H. Dumbleton, J. Polymer Sci., (A-2) 7,
       133 (1966).
(7)   D. R. Buchanan and J. H. Dumbleton, J. Polymer Sci.,
       (A-2) 7, 113 (1969).
(8)   J. L. Koenig and M. C. Agboatwalla, J. Macromol. Sci.,
       B2, 391 (1968).
(9)   A. Keller and D. J. Priest, J. Macromol. Sci., B2, 479
       (1968).
(10)   D. R. Buchanan and J. H. Dumbleton, J. Polymer Sci.,
       (A-2) 7, 113, (1969).
(11)   G. S. Y. Yeh and P. H. Geil, J. Polymer Sci., B1, 235, 251
       (1967).
(12)   W. P. Slichter, J. Polymer Sci., 35, 77 (1958).
(13)   I. L. Hay and A. Keller, Intern. Union Pure Appl. Chem.,
       Toronto, 1968.
(14)   T. Kawai and T. Goto, J. Polymer Sci., (A-2) 4, 521 (1966).
(15)   J. L. Koenig and M. J. Hannon, J. Macromol. Sci., B1, 119,
       (1967).
(16)   W. P. Slichter, J. Appl. Phys., 26, 1099 (1955).
(17)   B. Wunderlich, Polymer, 5, 125 (1964).
(18)   J. D. Hoffman, Soc. Plastics Engrs. Trans., p. 316 (1964).
(19)   B. B. Burnett and W. F. McDevit, J. Appl. Phys., 28, 1101
       (1957).

ORIENTATION EFFECTS IN BIAXIALLY STRETCHED POLYPROPYLENE FILM

George C. Adams

E. I. du Pont de Nemours & Co., Inc.

Film Department, Wilmington, Delaware

SYNOPSIS

Unoriented, quenched sheet of polypropylene has been reheated and biaxially stretched sequentially in two perpendicular directions on a pantograph stretcher for stretch ratios up to 6X6 over a twelve degree stretch temperature range. Orientation of (040) in the film plane, called population 1 crystals, predominates at and above 4X4. At 5X5 a second crystal orientation occurs which is orientation of (110) in the film plane called population 2 crystals, found also in similarly prepared polyethylene. A method of 040 peak resolution enables calculation of the fraction of population 2 crystals with stretch ratio and temperature. A proposed mechanism of formation of population 2 crystals is intralamellar slip along (110). Correlation of the amount of population 2 crystals with the area under the stress-strain curve obtained during stretching, as well as line breadth measurements, support such a mechanism. The generation of population 2 crystals occurs concurrent with no significant change in c-axis orientation which orients in the film plane below 4X4. The significance of biaxial over uniaxial stretching to detect this structural change is emphasized.

INTRODUCTION

A major effort on orientation effects due to uniaxial stretching of polypropylene film and fiber has been made by Samuels [1]. Orientation of the type a- and c-axis of spherulitic polypropylene film is quantitatively studied by x-ray diffraction

and a model of deformation based on intralamellae slip is presented
which is similar to studies on polyethylene [2].

Despite its industrial importance, there are few papers on
orientation changes due to biaxial stretching. The process of
biaxial stretching of quenched, crystallized and reheated sheets
is difficult due to the need for elaborate equipment to stretch
in two perpendicular directions simultaneously at a controlled
temperature. A unique and simple method of polyaxial stretching
on a small scale of polypropylene has been reported in a series
of five papers by Okajima et al. [3]. Gross orientation changes
were studied by refractive index measurements in the three principal directions of the sample. A possible representation of the
elementary deformation process is presented in Part V of the
series. While refractive index values perpendicular to the film
surface are the principal measurement made, diffraction photographs
show a "uniplanar" orientation of the chain axis and an increas-
ingly greater "selective uniplanar" or (040) planar orientation
with stretch ratio.

The elementary deformation process discussed by Okajima
relates to refractive index changes the following modes of
deformation: (a) principally amorphous orientation, (b) lamellar
tilting and breaking of lamellae, and (c) chain unfolding from
lamellae with changes in rate of refractive index change with
increasing sample area due to biaxial stretching.

In all of this work a (040) planar orientation occurs in the
plane of the film, which is the biaxial or polyaxial plane of
stretching. Such orientation occurs to a greater extent for more
nearly biaxial as opposed to uniaxial stretching. A (110) planar
orientation is not detected.

Another paper by Uejo [4] reports on orientation of biaxially
stretched polypropylene with different amounts of restraint and
subsequent stretching in a direction perpendicular to the first
direction of stretching. The pole figure technique was applied
to these studies. In addition to the (040) planar orientation
(b-axis along the film normal) there is a (110) planar orientation.
These two orientation types are particularly significant to the
present work, but a systematic study of the development of (110)
planar orientation is lacking.

The present paper constitutes a study of the development of
planar (040) and (110) orientation as found in simultaneous
biaxially stretched polypropylene film that has been quenched from
the melt, reheated and biaxially stretched. X-ray photographs and
quantitative wide angle x-ray scans were used to explore the
effects of biaxial stretch ratio and stretch temperature on the
two orientation types mentioned.

EXPERIMENTAL

Sample Preparation and Stretching Procedure

Rexall 413S polypropylene resin was pressed with rapid quenching into ice-water from the melt to 0.025" thick sheet. Pieces measuring two inches square were simultaneously biaxially stretched to equal elongation in two directions in a T. M. Long pantograph stretcher. The stretching rate was approximately 10,000%/min. Stretch ratios of 4X4, 5X5, 6X6 and 7X7 were studied at stretch temperatures of 142°, 148° and 154° with selected samples stretched at 160°C. The stretching produced film from 0.001 to 0.002 inches thick. For diffraction work a stack to make about 0.020 inches thick was prepared by the aid of a light oil spray between film pieces.

Wide Angle X-Ray Scans

Quantitative x-ray diffractometer intensity scans enabled determination of crystal orientation with respect to the film normal. Incremental tilting at 3.6° intervals of the sample in the accessible region in transmission was performed. Corrections were made for background, incoherent intensity, polarization and changes in sample volume according to the usual procedure. A data point was taken in reflection at ω equals 90° which served as a basis of a fill-in of data points by interpolation in the inaccessible region of the scan (due to interference from the sample holder). Orientation of the c-axis with respect to the film normal was calculated utilizing orientation function values determined on the (110) and (130) peaks and utilizing the equation supplied by Wilchinsky [5]. The a*-axis is defined as the direction in the unit cell perpendicular to the b- and c-axes and this orientation function was obtained from the orthogonality relationship [6]. Orientation of the a*-, b- and c-axes is expressed as the average cosine squared of the angle between the film normal and the principal direction of the axis studied. Its value is 0.0, 0.333 and 1.0 for axis orientation perpendicular to, random about and parallel to the film normal direction [6].

Method of Calculation of Percent of the Two Populations

Two crystal populations differing by virtue of different orientations result from this work. The percent of each population of crystals is calculated by resolution of area under the intensity vs. tilt angle ω curve for the 040 peak as outlined in Figure 1. The angle ω is defined as the angle between the normal to the diffraction plane and the plane of the film. The intensity at tilt angle ω constitutes input to a computer program that calculates a corrected intensity I_i(corr.)

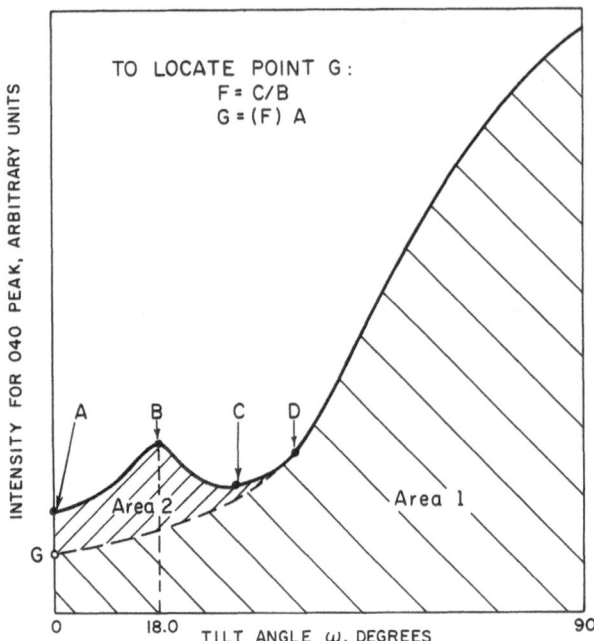

FIG. 1.   Schematic representation of method to obtain percent of
population 2 crystals

$$I_i(\text{corr.}) = I(\omega_i) \times \cos(\omega_i)$$

where ω_i is incremented in intervals of 3.6°. The intensity is
thereby[1] corrected to be the number of poles at each tilt angle.
Area 1 and area 2 are calculated from the corrected intensities,
the increment in angle and appropriate intensity differences.
        The percent of population 2 crystals, P_2, is then calculated
by the equation

$$P_2 = (100) \ \frac{\text{Area 2}}{\text{Area 1} + \text{Area 2}}$$

        It becomes apparent that some population 1 crystals contribute
to intensity where population 2 crystals do also. No attempt was
made to eliminate this effect from the method of calculation.
Rather points A, B and C of Figure 1 were chosen to obtain area
2 = 0 when B = C in which case G = A.   It will be seen that the
method is sufficiently accurate to show reasonable variations in
population 2 percentage with stretch temperature and ratio.

Low Angle Photographs

A Warhus camera was used at a sample to film distance of
17 cm. at 16 hours exposure on 0.050 inches samples thickness.
The beam was directed parallel to the film plane on a stack of
several layers of film pieces. Microdensitometer traces provided 2θ identification of the peak maximum. Indistinct patterns
were obtained with the beam directed perpendicular to the film
surface (through direction). The through direction photograph on
unstretched cast sheet shows a distinct long period intensity
maximum of 145Å spacing and no orientation.

Crystal Size from Line Broadening

Relative crystal size was determined by slow (1/2 degree per
minute) strip chart recordings of intensity vs. 2θ angle for biaxially stretched film. Because of the different orientations
assumed for crystals of (110) and (040) planar orientation, it is
possible to orient the film sample so that a scan through the
appropriate 2θ range provides line broadening data (β values) for
each of these peaks on each crystal orientation. No instrument
correction has been applied so that only relative crystal size is
obtained, the size being smaller the larger the half height width.
A Du Pont curve resolver enables resolution of adjacent 110 and
040 peaks.

Differential scanning calorimetry (DSC) scans were performed
on a Du Pont Model 900 Thermal Analyzer with a scanning rate of
20°C/min. Sample weights were 10 mg in a crimped sample holder.

RESULTS

Wide Angle X-Ray Photographs

Several crystal forms of polypropylene have been reported [6a],
but only the monoclinic form is found in the present study. This
form has unit cell dimensions given by Natta [7]. The chain axis
makes an angle of about 10° with the normal to the base plane or
plane made by the a- and b-axes.

With the x-ray beam directed parallel to the film plane, the
photographs of Figure 2 are obtained for stretching at 148°C. The
relatively indistinct diffraction spots at a 4X4 stretch ratio
become sharp at a 6X6 stretch ratio. Examination of the photographs with a model of the a*-b axes plane of the monoclinic unit
cell reveals two crystal orientations which are designated populations 1 and 2 defined as follows:

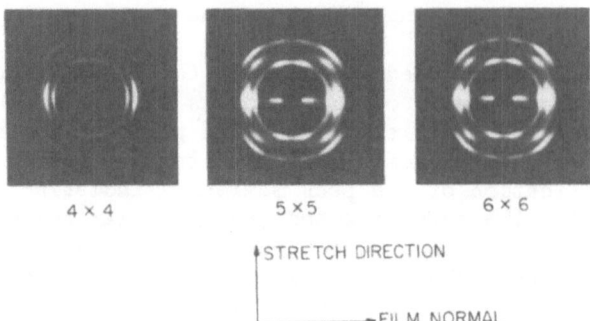

FIG. 2. Wide angle x-ray photographs of biaxially stretched
polypropylene film

Population Number	Crystal Plane in Film Plane	Referred to as
1	(040)	(040) planar orientation
2	(110)	(110) planar orientation

A (110) planar orientation was first reported by Uejo [4] who did
not explore its generation with stretch ratio or temperature and
relationship to (040) planar orientation. The latter orientation
has been reported by several authors in the past [3,5,8]. The

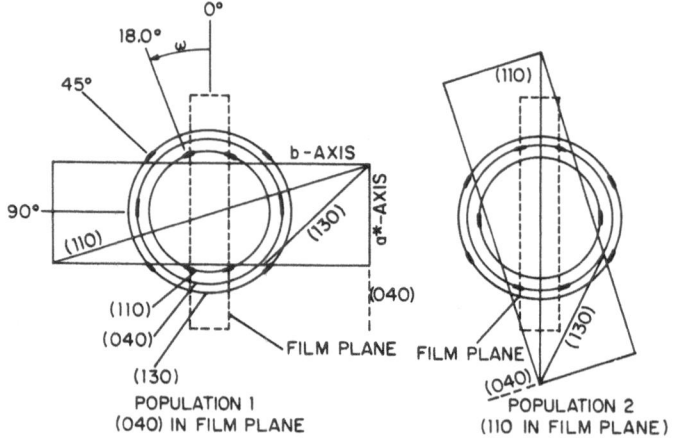

FIG. 3. Orientation of the unit cell for the two crystal populations in biaxially stretched polypropylene

superposition of the a*-b plane of the polypropylene monoclinic
unit cell and the x-ray photographs of Figure 2 enables a visual
identification of the 16 diffraction spots comprising the 110,
040 and 130 peaks as shown schematically in Figure 3. This figure
indicates that a 72.0° rotation about the c-axis of the plane containing the a*-b axes accounts for the diffraction spots observed.
This angle as read from the photograph agrees well with that predicted from the relationship between layer-line spots on a flat
plate photograph and the Bragg equations.

The dual crystal population discussed in the present work is
distinct from and unrelated to a-axis and c-axis oriented crystals
reported by Samuels in stretched fibers [1]. The latter arises
from either a-axis or c-axis orientation along the stretch directions in amounts dependent on stretch ratio.

Since it is not evident from the photographs which, if either,
of the two populations of crystals forms first with stretch ratio,
a quantitative identification of their formation is obtained from
the omega scan.

The Omega Scan

The omega scan is so named because from such a scan the diffraction intensity variation is obtained for diffraction plane
normals oriented between the plane of the film ($\omega = 0°$) and the
film normal ($\omega = 90°$). As discussed in the Experimental Section,
intensity values are obtained for the region $64.8° < \omega < 0°$ in
transmission with one data point obtained in reflection at $\omega = 90°$.
Interpolation of intensity between these values results in the
complete intensity scans as shown in Figure 4 for the 110 and 040
peaks for stretching at 142°. The 040 peak at $\omega = 90°$ arises from

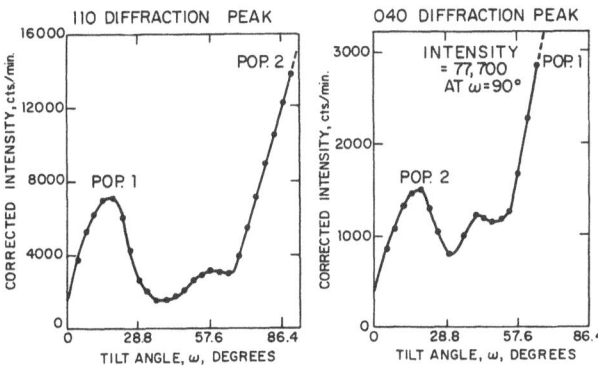

FIG. 4. Typical omega scan plots for 6X6 biaxially stretched polypropylene; plots are for 110 and 040 diffraction peaks showing two
crystal orientations

an (040) planar orientation which also gives rise to the 110 peak
at ω = 18.0°. This crystal orientation is designated population 1,
see Figure 3. Population 2 crystals result in the intense 110
peak at ω = 90° and the 040 peak at ω = 18.0°. A 79.2° rotation
of the unit cell about the c-axis exists between these two orientations resulting in an (040) normal peak maximum at ω = 18.0°.

The intensity variation with tilt angle omega for stretching
at 142°, 148° and 154° is shown in Figure 5.

An obvious effect of biaxial stretching on 040 and 110 peaks
is to change the intensity of each of the peaks for each of the
two crystal populations while less significant changes occur in
c-axis orientation which remains preferentially in the film plane
at all stretch ratios and temperatures. In addition, it was clear
early in the work that a change in fraction of crystals of each
population may be occurring and that a method of calculation of
this fraction could be performed by some method of resolution of
peak areas in an omega scan plot. Such a resolution procedure was
developed and used in the manner illustrated in Figure 1.

The variation in fraction of population 2 crystals with
stretch temperature at several stretch ratios is shown in Figure 6.

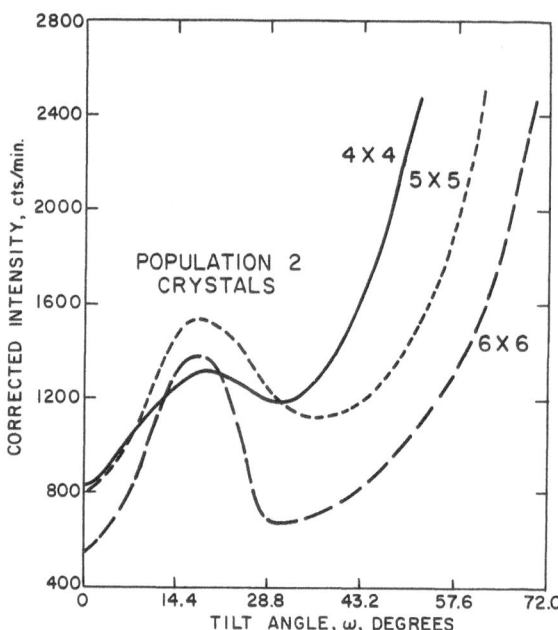

FIG. 5. Detailed omega scan plot in vicinity of peak for population
two crystals for 040 peak for biaxially stretched polypropylene

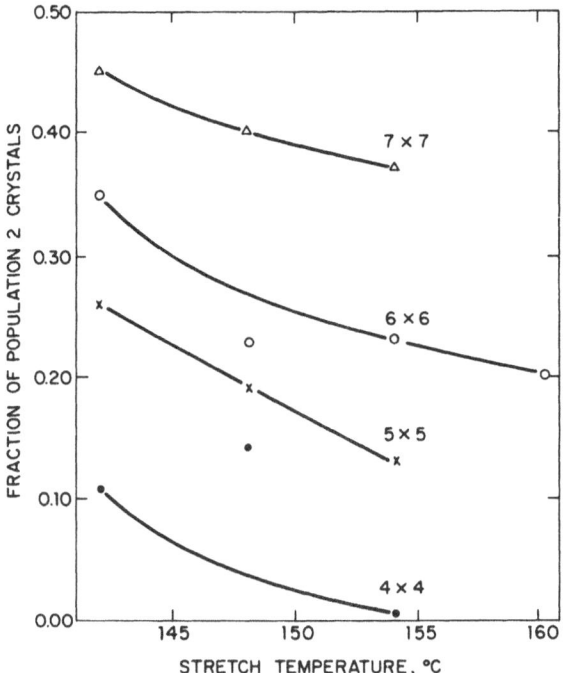

FIG. 6.   Variation of fraction of small crystals with stretch
temperature and stretch ratio for biaxially stretched polypropylene

It is apparent that population 2 crystals are generated with increasing stretch ratio and decreasing stretch temperature. Up to
40% of population 2 crystals can be generated by 7X7 biaxial
stretching at 154° by this method of calculation. For stretching
at 154° a ratio above 4X4 is necessary to form any population 2
crystals. The morphological implications of this condition to
form such crystals is discussed in the Discussion Section.

Line Broadening Measurements

It is not apparent from Figure 6 if population 2 crystals form
at the expense of population 1 crystals, that is if crystals with
a planar (110) orientation result from a reorientation of crystals
with an (040) orientation. If such is the case one might consider
the relative size of each crystal population.

That each of the two populations can be detected separately
at the appropriate 2θ value can be seen in Figure 7. A strong
040 peak at 2θ equals 16.5° occurs when the film is oriented such
that diffraction is occurring at ω = 90°. The line broadening of
the 110 peak of population 2 crystals and the 040 peak of population 1 crystals may thereby be obtained. Similarly, 110 line

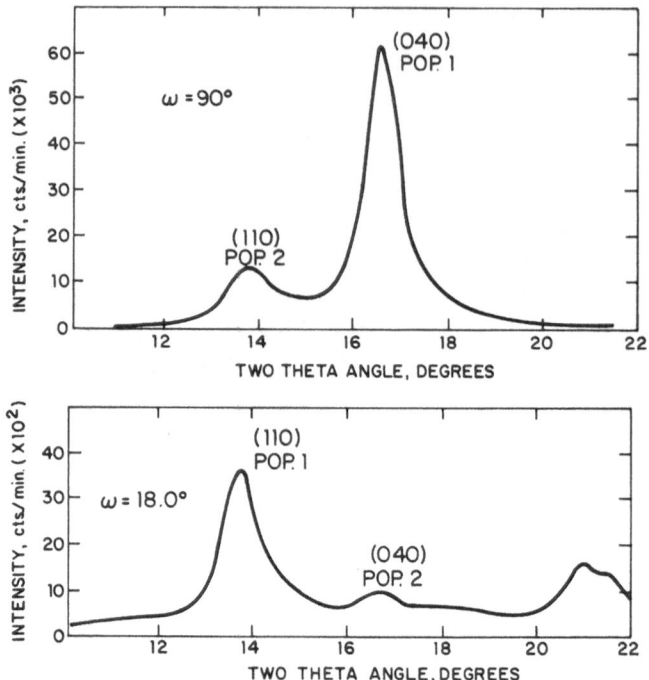

FIG. 7. Typical radial scan plots for two tilt angles for 5X5
biaxially stretched polypropylene

broadening of population 1 crystals may be obtained when 2θ is
scanned at $\omega = 18.0°$ as shown in Figure 7. It is apparent from
these scans that at neither of these tilt angles does the 130
peak ($2\theta = 18.6°$) show diffraction effects.

Figures 8-9 show variation of half height width values with
stretch temperature for the prominent peaks available for study.
The 040 peak at $\omega = 18.0°$ was adjacent to the strong 110 peak and,
in addition, too weak for study. Hence, changes in crystal size
along the b-axis were not obtained. Figure 8 shows values for the
110 peak for each of the two populations. Greater line broadening
is exhibited by population 2 crystals, but the difference converges
to zero at the highest stretch temperature studied. Since no
attempt was made to obtain an instrument correction line broadening value, no crystal size determinations were performed. Suffice
to say that, because of the inverse relationship between crystal
size and half height width values, experimental data is at hand
which says crystal size is smaller and/or internal defects greater
for population 2 crystals.

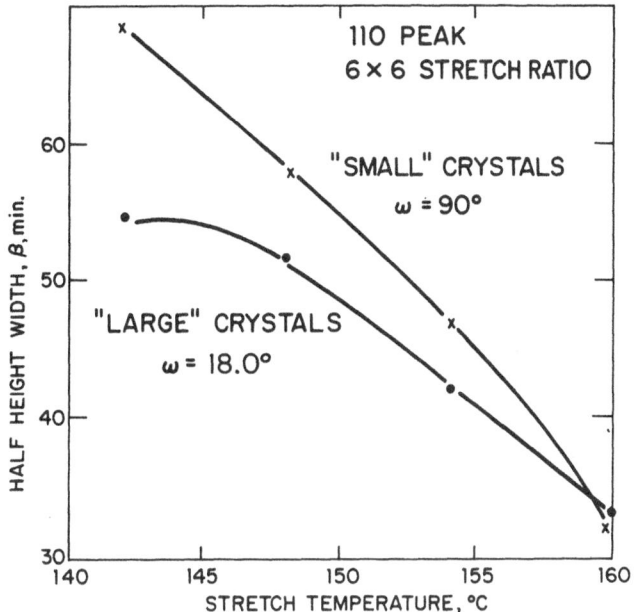

FIG. 8. Half height width of polypropylene (110) diffraction peak
for biaxially stretched film showing differences in crystal size
and perfection for the two populations of crystals

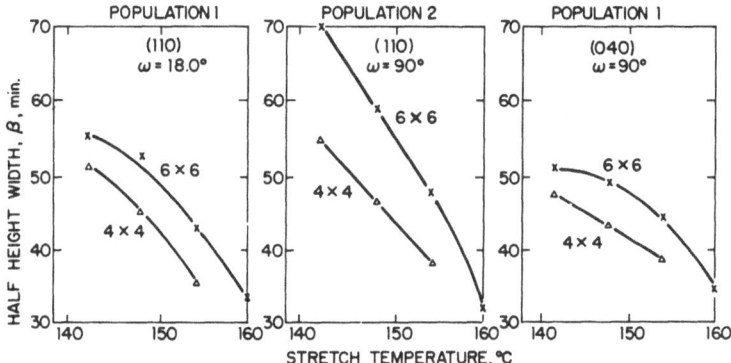

FIG. 9. Change in half height width of diffraction peak for biaxially stretched polypropylene for each of the two populations
of crystals

Chain Axis Orientation

The procedure used to obtain the chain or c-axis direction is outlined in the Experimental Section. Figures 10, 11, and 12 show $< \cos^2\phi_{a,fn} >$, $< \cos^2\phi_{b,fn} >$ and $< \cos^2\phi_{c,fn} >$ values for the stretch temperature and ratio conditions of this study. A value below 0.333 indicates a preferential orientation perpendicular to the film normal. Over the stretch ratio studied there is very little change from a nearly perpendicular chain or c-axis orientation with respect to the film normal with both stretch ratio and stretch temperature. While negative values are prohibited such values for the 4X4 stretch ratio are considered due to experimental error associated with each of the values for 110 and 130 peaks used in the calculation. Samuels found a similar effect of no change in c-axis orientation above 250% elongation of polypropylene fibers [9].

It should be apparent that the chain axis orientation changes but remains preferentially oriented in the film plane over the stretch ratio range where a significant increase occurs in the fraction of a (110) planar orientation, i.e. fraction of population 2 crystals. Because the emergence of a (110) planar orientation results only from a rotation about the chain axis of the monoclinic unit cell, c-axis orientation parameter is not the parameter

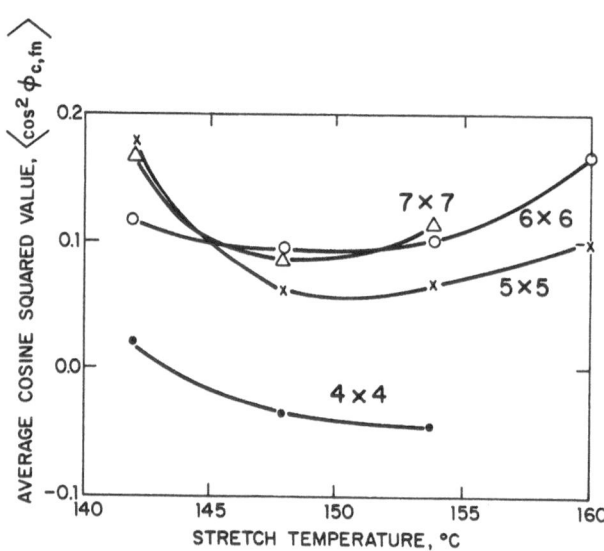

FIG. 10. Average cosine squared of angle between c-axis of crystal and the film normal for biaxially stretched polypropylene film

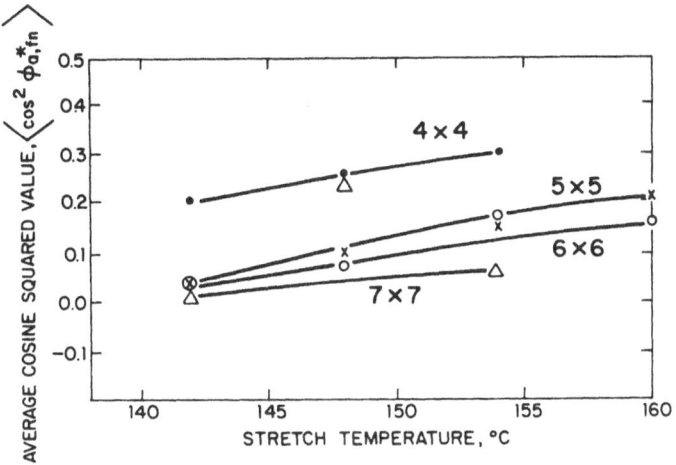

FIG. 11. Average cosine squared for angle between a*-axis and
film normal for biaxially stretched polypropylene film

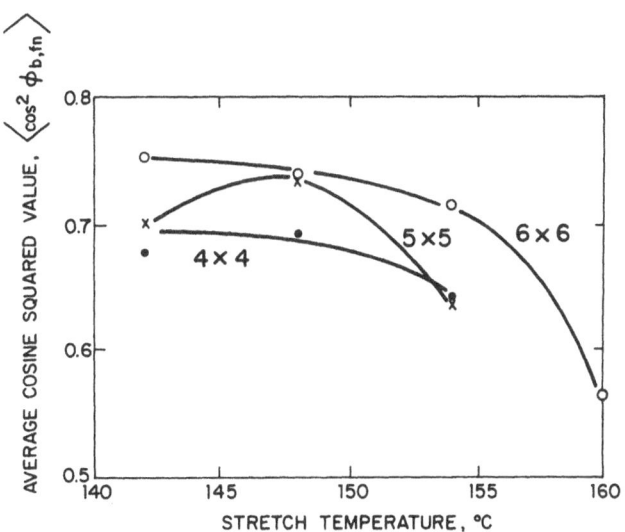

FIG. 12. Average cosine squared of angle between b-axis and film
normal for biaxially stretched polypropylene film

indicative of the structural change occurring in biaxially
stretched polypropylene. It is for this reason that the parameter
developed from the plots of intensity vs. omega angle, the fraction of population two crystals, takes on special significance.

Figures 11-12 show the a*-axis also orients predominantly
in the plane of the film and the (040) normal or b-axis orients
predominantly perpendicular to this plane. Discussion of the
origin of an (040) planar orientation is provided later. The
average cosine squared values reported are average values calculated without consideration of the presence of two crystal
orientations.

Calorimetry Results

The DSC scans (Figure 13) show a general trend to greater
distribution in melting points with stretch ratio and some suggestion that separate lower temperature melting peaks are develop-
ing. It cannot be stated with certainty whether the lower melting
peaks result from population 2 crystals except that such an interpretation is in agreement with the greater line broadening values
for these crystals.

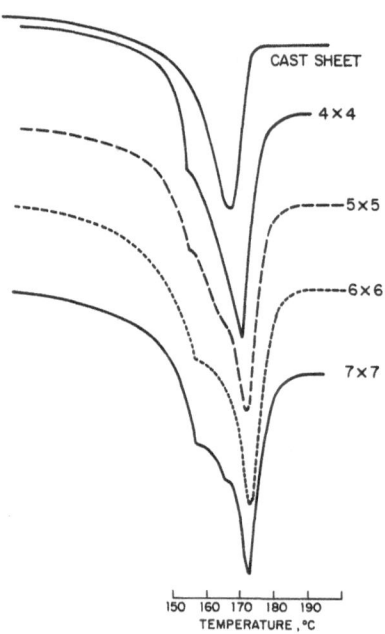

FIG. 13. DSC scans for several stretch ratios for biaxially
stretched polypropylene film

Low Angle Photographs

The unstretched cast sheet shows a peak maximum at a spacing
of 145Å. Photographs on biaxially stretched film with the x-ray
beam directed parallel to the film plane (Figure 14) show an increasing intensity at higher angles in the equatorial region of
the photograph than that of the unstretched cast sheet (film plane
vertical). The implication is that smaller structure gives rise
to the diffraction. The patterns are elongated in the direction
perpendicular to the film plane. Microdensitometer traces show
maxima which change in relative intensity but not in 2θ angle
(Figure 15). Low angle photographs taken with the beam perpendicular to the film plane show diffraction at lower angles than in the
cast sheet with no evident intensity maxima.

Work of Stretching from Stress-Strain Curve

This study on biaxially stretched polypropylene film shows
the chain or c-axis orients in the film plane and remains essentially unchanged over a biaxial stretch ratio range wherein a sizeable
increase in (110) planar orientation occurs. Therefore, the
structural parameter developed in this work, that of the fraction
of population 2 crystals, reveals the significant structural change
occurring during biaxial stretching.
It is possible to record the stress-strain curve during the
stretching operation. An interesting nearly linear relationship
is found between the fraction of population 2 crystals and the

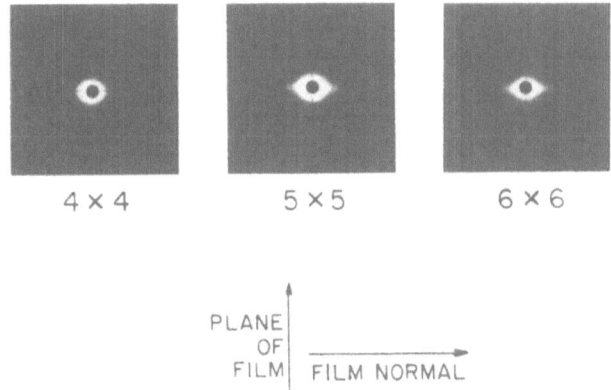

FIG. 14. Low angle x-ray photographs taken with beam directed
perpendicular to film plane for biaxially stretched polypropylene
film

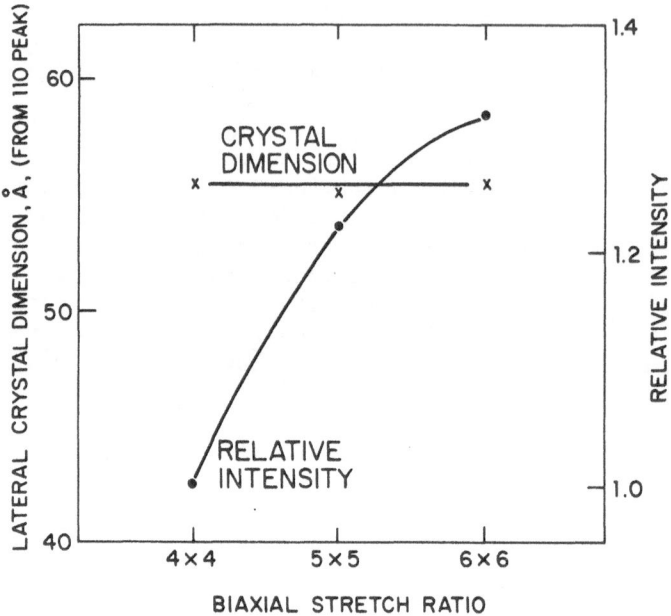

FIG. 15.   Lateral crystal dimension and relative intensity determined from microdensitometer scans of low angle x-ray photographs

area under the stress-strain curve as shown in Figure 16.   This
area is a measure of the total work performed on the unstretched
sample and higher stretch ratio or lower stretch temperature result in greater work. The structural change that occurs during
stretching is dependent on the work input to the sample as is
evident from Figure 16. A certain minimum work of about 1000
units is necessary before any structural change occurs.

DISCUSSION

The significant and different finding from earlier studies
of the present work is the formation of a dual crystal orientation
in biaxially stretched polypropylene film. The evidence from the
wide angle photographs and the omega scans indicates that the
initial (040) planar orientation gives rise with stretch ratio to
a (040) and (110) planar orientation. While an (040) planar
orientation in polypropylene film has been reported on numerous
occasions in the literature, the (110) planar orientation has been
observed only once [4]. No quantitative investigation of the dual

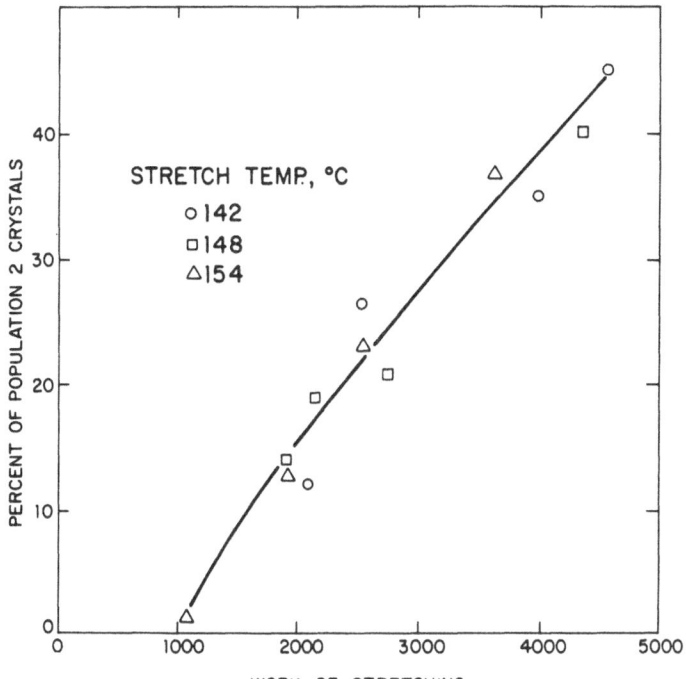

FIG. 16.   Percent of small crystals with work of stretching for biaxially stretched polypropylene film

crystal orientation has been made and, when first observed in a qualitative manner, a reference to the need for further work was made in order to clarify the types of orientation observed in biaxially stretched polypropylene film [4].

The quantitative nature of the present work reveals that (a) the first and predominant crystal orientation occurring at all stretch ratios is an (040) planar orientation which is designated population 1 crystals and (b) above a stretch ratio of 4X4 there is an additional orientation which is a (110) planar orientation. It is clear that at a low stretch temperature (142°) and high stretch ratio (6X6) up to 40% of population 2 crystals is present. Since the latter crystal orientation is reported as a percentage of that of the two crystal orientations, the inference is that population 2 crystals result from a structural change occurring on population 1 crystals.

The formation of population 1 crystals comprising an (040) planar orientation is taken as the counterpart of a (200) planar orientation formed in biaxially stretched linear polyethylene reported earlier [10].   The reason for this orientation type in each of these polyolefins is not understood, but it is conceivable that it results from a tendency for the long crystal dimension in the direction perpendicular to the chain axis to orient into the

biaxial plane of stretching. In polypropylene single crystal work
reported by Morrow [11], the long crystal dimension of the lath
shaped crystals is the a-axis. It is the direction of fastest
growth. Similarly in the biaxially stretched films, an orientation
of the a-axis and of the chain or c-axis in the film plane results
in orientation along the film normal of the axis mutually perpendicular with the a- and c-axes (the b-axis). This is manifested
in an (040) planar orientation as illustrated in Figure 3 for
population 1 crystals.

In polyethylene single crystal work, the diamond shaped crystals with (110) fold planes do not lead to as convenient a ratio-
nale for (200) planar orientation based on relative crystal dimensions along the three crystal axes. Perhaps from the reasoning
herein supplied, it can be deduced that for melt crystallized
polyethylene the long crystal dimension in the lateral direction
is the b-axis.

It is noteworthy to observe that both for biaxially stretched
polypropylene and linear polyethylene a (110) planar orientation
is obtained in biaxially stretched film. Again, the origin of
this orientation cannot be stated with certainty. However, in
keeping with findings on uniaxially stretched polyolefins, an
intralamellar slip mechanism is conceivable such that lamellar
fragmentation occurs along (110) with stretch ratio. Following
thoughts mentioned earlier, the new fragment resulting from such
cleavage has a long dimension along 110 and, hence, this plane
will orient into the biaxial stretching plane by rotation about
the c-axis which does not change orientation. This gives rise to
population 2 crystals. By this mechanism population 2 crystals
result from and at the expense of population 1 crystals. Accordingly, the amount of population 2 crystals has been reported as a
percentage of the two crystal populations. A schematic representation of orientation, fragmentation and reorientation is given
in Figure 17.

The suggestion that lamellar fragmentation occurs during biaxial stretching is consistent with the correlation of fraction
of population 2 crystals with the work of stretching as measured
by the area under the stress-strain curve and shown in Figure 16.
That a certain minimum work of stretching is necessary before
lamellar fragmentation occurs is explained by crystal c-axis rotation into the film plane and by an increase in interlamellar
stress through tightening of tie molecules. These processes
occur up to a 4X4 biaxial stretch ratio and require about 1000
units of work before the structural disruption by intralamellar
slip along (110) occurs. At higher stretch ratios the work of
stretching is expanded on fragmentation a measure of which is
obtained from the extent of (110) planar orientation as indicated
above.

Except for the work previously reported on biaxially stretched
linear polyethylene, no dual crystal orientation has been reported

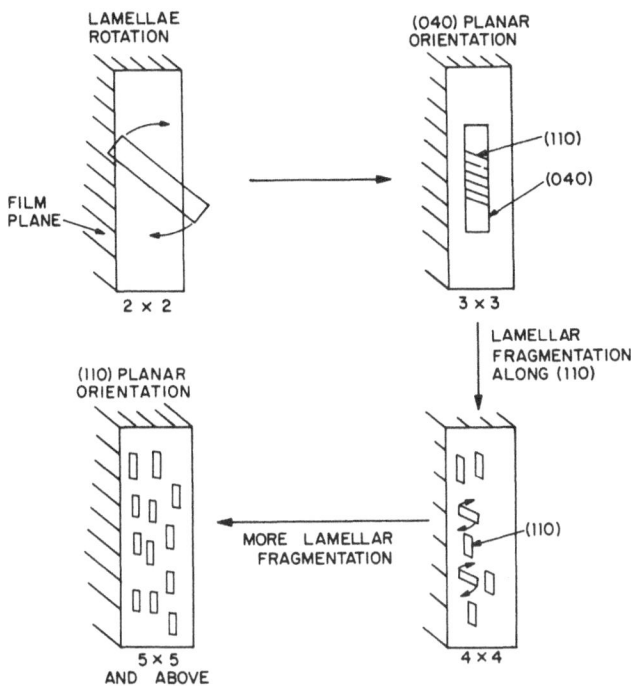

FIG. 17. Model showing lamellae rotation, fragmentation and
subsequent reorientation for biaxially stretched polypropylene
film

on biaxially stretched polyolefin film. A dual crystal orientation
has been reported on drawn and rolled linear polyethylene [12].
As with biaxially stretched polyethylene, a (200) planar orientation and an orientation resulting from (310) twinning on (200)
planar oriented crystals occurs in drawn and rolled film. The
orientation resulting from (310) twinning results in crystals with
(110) at an angle of 3.0° from being planar with the rolling plane
or plane of the film [12]. However, the orientation of one set of
crystals occurring on biaxially stretched linear polyethylene results in exactly (110) planar orientation, so that the structural
changes experienced by rolled film are different from those occurring in biaxially stretched film. Illusion was made in the biaxially stretched polyolefins. The present study on biaxially
stretched polypropylene adds credence to that suggestion. The
polypropylene work is further enlightening since (110) planar
orientation emerges only above a 4X4 stretch ratio. This conforms
to an origin via lamellar fragmentation. It is proposed that the
fragmentation occurs only after sufficient tension is established
in the tie molecules. A c-axis orientation into the biaxial plane
of stretching is expected to be complete well below a 4X4 stretch

ratio and to be essentially unchanged during fragmentation. This is shown to be the case as seen in Figure 10. The second crystal orientation is a manifestation of crystal rotation about the molecular axis.

The low angle photographs clearly show a peak maximum at an angle wider than that in unstretched cast sheet. This is further evidence for lamellar fragmentation. The equatorial location of this peak maximum enables interpretation as diffraction from the lateral sides of the crystal which the wide angle work indicates are the (110) crystal faces.

Calorimetry studies on biaxially stretched polypropylene show similar profile changes to that for the polyethylene work. The presence of species melting at lower than the main melting peak supports lamellar fragmentation. As with the biaxially oriented polyethylene films no assignment can be made of calorimetry melting peaks with each of the two crystal populations. More evidence for species of two different sizes is obtained from line broadening results determined in the 110 normal and b-axis directions.

This study of orientation effects on biaxially stretched polypropylene film and the earlier work on polyethylene film are considered of some significance because of the diffraction effects due to biaxial stretching of polyolefins. The results can be interpreted as a process of crystal disruption not heretofore evident on uniaxially stretched specimens. The disadvantage of uniaxial stretching is that the two crystal populations discussed above cannot be detected. A planar constraint on crystal rotation about the chain axis is provided by biaxial stretching. In the absence of such a constraint, as with uniaxial stretching, diffraction effects from (hko) planes are intermixed so that crystal species of different lateral dimensions (along a direction perpendicular to the chain axis) remain undetected. It is hoped that this study will stimulate a greater effort on structural changes as occur during biaxial stretching.

REFERENCES

[1]   R. J. Samuels, J. Pol. Sci., Part A, $\underline{3}$, 1741 (1965).
[2]   A. Peterlin, J. Pol. Sci., C, $\underline{15}$, 427 (1966).
[3]   a) S. Okajima et al., J. Appl. Pol. Sci., $\underline{11}$, 1703 (1967);
       b) S. Okajima et al., J. Appl. Pol. Sci., $\underline{12}$, 411 (1968);
       c) H. Tanaka et al., J. Pol. Sci., Part A-1, $\underline{7}$, 1997 (1969);
       d) H. Tanaka et al., J. Pol. Sci., Part A-1, $\underline{7}$, 3351 (1969).
[4]   H. Uejo and Sadao Hoshino, J. Appl. Pol. Sci., $\underline{14}$, 317 (1970).
[5]   Z. W. Wilchinsky, J. Appl. Phys., $\underline{31}$, 1969 (1960).
[6]   S. Hoshino et al., J. Pol. Sci., $\underline{58}$, 185 (1962).
[6a]   D. R. Morrow, Polymer Preprints, $\underline{9}$, No. 2, 1192 (1968).

[7]   G. Natta and P. Corrandini, Nuovo cimento, Suppl. to Vol. 15, <u>1</u>, 40 (1960).
[8]   W. Titow, Chemistry and Industry, p. 401, March 29, 1969.
[9]   R. J. Samuels, J. Pol. Sci., Part A-2, <u>6</u>, 2021 (1968).
[10] G. C. Adams, J. Pol. Sci., A-2, <u>9</u>, 1235 (1971).
[11] D. R. Morrow, Trans. N. Y. Academy of Sci., <u>30</u>, 1130 (1968).
[12] F. C. Frank, A. Keller and A. O'Connor, Phil. Mag., <u>3</u>, 64 (1958).

QUANTITATIVE STRUCTURAL CHARACTERIZATION OF THE MECHANICAL

PROPERTIES OF POLY(ETHYLENE TEREPHTHALATE)*

Robert J. Samuels

Hercules Incorporated, Research Center

Wilmington, Delaware   19899

INTRODUCTION

Poly(ethylene terephthalate) (PET) is a polycrystalline polymer, with structural order on the molecular and interlamellar levels. The character of this structure will vary with the conditions of fabrication. Certainly the mechanical properties of a sample of PET will depend on its particular structural arrangement.

The structural arrangement present in a fabricated sample of fiber or film can be characterized quantitatively on all structural levels using a modus operandi developed by the author [1]. The procedure has been successful in characterizing the structural changes occurring during the fabrication of isotactic polypropylene fibers [2,3] and films [4,5], and the deformation of PET fibers [6] and of hydroxypropylcellulose film [7].

Realizing that the observed mechanical properties of a polycrystalline polymer are intimately related to the internal morphological structure of the polymer, the author [8] examined the character of these relationships in isotactic polypropylene. That study demonstrated that large amounts of mechanical data could be simply and quantitatively correlated through the use of structural criteria. Are the intimate relationships between structure and properties observed for polypropylene specific for that polymer, or are they, as would seem more reasonable, characteristic of poly-crystalline polymers in general? The purpose of the present study

---

*Published with the permission of John Wiley and Sons, Inc.

is to demonstrate that the morphological model developed for polypropylene is a general one. This is accomplished by examining the
structure-property relations of PET (a crystalline polymer very
different from isotactic polypropylene). All of the data used for
this study were obtained from the literature [9,10,11]. Application
of the data in terms of a general structural model leads directly
to simple, quantitative relationships between the structural state
of PET and such properties as tenacity, shrinkage, dynamic loss
modulus, tensile modulus, long spacing, and the intensity from
small-angle x-ray scattering.

EXPERIMENTAL

All of the data used in this paper were obtained from a series
of studies on PET published by J. H. Dumbleton et al. [9,10,11].
The shrinkage, tenacity, modulus, and small-angle x-ray data
reported in Table 1 were obtained from samples treated as follows
[10,11]:

Spun PET of low crystallinity (less than 2% by x-ray measurements) and of birefringence 0.002 was drawn to draw ratios up to
5X over a hot pin at 80°C. Yarns of 3X and 5X draw were used in
the annealing experiments. Two types of annealing procedures were
used:

(1) Five meter lengths of both the 3X and 5X drawn yarn, in a
wire mesh basket, were immersed in hot silicone oil for one minute,
and (2) two and one half meter lengths of the 5X drawn yarn were
immersed in a test tube of heated air for ten minutes. These
annealing times were chosen to obtain maximum shrinkage. All samples
were quenched into carbon tetrachloride at room temperature after
annealing was completed. The shrinkage ratio S was calculated from
the formula

$$S = \frac{\text{Initial length} - \text{Final length}}{\text{Initial length}} \qquad (1)$$

Crystallinity measurements were made with a Norelco x-ray diffractometer, which was also used to obtain the crystallite orientation
from scans on the (105) reflection. The orientation parameter
employed was f_c given by Equation (2), [1].

$$f_c = 1/2 \ (3 \ \overline{\cos^2\theta} - 1) \qquad (2)$$

where $\overline{\cos^2\theta}$ is the mean-square cosine of the angle, θ, between a
(105) plane normal and the fiber axis.

Table 1 – Mechanical and Structural Data for PET Yarns (10,11)

Sample	Crystallinity (%)	Birefringence	Crystallite Orientation Function f_c	f_{AM}	Shrinkage (%)	Modulus (g./d.)	Tenacity (g./d.)	Long Period (Å)	Intensity in Small-Angle Maximum
3X PET in oil at 20°C	12	0.140	0.915	0.505	-	-	2.9	-	-
100°C	16	0.127	0.915	0.433	11	37	3.6	132	0.5
150°C	30	0.149	0.915	0.485	19	27	3.1	120	56
Crimped 175°C	37	0.150	0.889	0.472	24	26	3.0	127	77
200°C	37	0.142	0.915	0.400	33	17	2.7	144	115
225°C	36	0.132	0.858	0.382	52	11	1.7	169	155
240°C	24	0.145	-	-	67	-	-	192	50
5X PET in oil at 20°C	35	0.206	0.943	0.790	-	-	5.4	-	-
100°C	38	0.200	0.943	0.811	8	66	6.7	124	0.8
150°C	34	0.174	0.910	0.709	23	48	5.8	122	47
175°C	38	0.164	0.910	0.668	31	35	4.8	124	76
200°C	39	0.159	0.858	0.680	43	27	3.7	147	147
Crimped 225°C	39	0.112	0.858	0.480	60	12	2.2	167	214
240°C	33	0.040	-	-	75	14	1.9	195	132
5X PET in air at 100°C	31		0.93	0.84	8	77	7.1	132	13
150°C	34		0.93	0.75	20	51	5.8	120	52
175°C	40		0.93	0.75	24	44	5.3	120	76
200°C	38		0.92	0.73	30	46	4.6	132	127
225°C	38		0.92	0.68	35	32	3.7	140	172
240°C	40		0.93	0.68	44	36	4.1	151	212

Small-angle x-ray data were obtained with a Kratky camera (slit collimation). Integrated intensities of the long period maximum were derived from the areas under the maximum after a linear background had been subtracted.

The tensile modulus and tenacity were determined on an Instron testing machine at 70°F. and 65% R.H. with single filaments.

Birefringence measurements were made on single filaments by using a compensator. Because of nonuniformity in denier, it was difficult to get consistent measurements for 3X yarns.

The intrinsic birefringences for PET ($\Delta_c^o = 0.220$, $\Delta_{AM}^o = 0.275$) were determined by Dumbleton [6] using the method of Samuels [1,4]. By combining the intrinsic birefringences, Δ_c^o and Δ_{AM}^o, the crystal orientation function, f_c, the fraction of crystals, β (from the percent crystallinity), and the measured birefringence Δ_t, through equation (3), [1]

$$\Delta_t = \beta \Delta_c^o f_c + (1-\beta) \Delta_{AM}^o f_{AM} \qquad (3)$$

it was then possible to calculate f_{AM}, the orientation function for the noncrystalline chains.

The modulus and tenacity values listed in Table 1 for the 3X drawn yarns were not listed in the references but were taken from plotted values in reference [10].

The loss modulus, E", data listed in Table 2 were obtained from samples treated as follows [9]:

PET yarns drawn 3X and 4.25X over a hot pin at 80°C. were employed. Samples were heated for 6 hrs. at temperatures up to 240°C. in vacuo under conditions in which the samples were free to shrink. All samples were subsequently boiled in water, under relaxed conditions, for 1 hour.

Dynamic measurements were made with a Vibron direct reading viscoelastometer operated at a frequency of 11 c/s. Samples were heated at 1°C./min. in a nitrogen atmosphere under relaxed conditions and measurements of the tensile modulus, E', and the loss factor, tan δ, were made at increments of 5°C. except near the transition region when smaller increments were used. The loss modulus, E", was calculated from the relation E" = E' tan δ.

The crystallinity, crystallite orientation and the birefringence of these samples were obtained using the same procedures as those described above.

Table 2 - Vibron and X-Ray Data for 3X and 4.25X Drawn PET Yarns Annealed for 6 hrs. Followed by Treatment in Boiling Water(9)

Sample	Crystallinity (%)	f_c	f_{AM}	Temperature of E''_{MAX} (°C.)
3X				
No heat	33	0.91	0.53	126
100°C.	33	0.915	0.54	128
125°C.	31.5	0.920	0.54	130
150°C.	38	0.91	0.57	128
175°C.	43	0.92	0.57	123
200°C.	56	0.92	0.56	110
220°C.	57	0.91	0.56	110
240°C.	59	0.91	0.54	100
4.25X				
No heat	38	0.92	0.68	147
100°C.	40	0.93	0.68	150
125°C.	42	0.92	0.64	153
150°C.	45	0.92	0.67	142
175°C.	47	0.93	0.69	140
200°C.	51	0.94	0.72	140
220°C.	53	0.93	0.73	124
240°C.	57	0.93	0.70	120

DISCUSSION

A polycrystalline polymer is composed of both crystalline and noncrystalline regions. The amount and structural arrangment of these regions within a given sample will depend on how the polymer was fabricated. Since the properties of both the crystalline and noncrystalline regions are anisotropic and different from each other, their amount and arrangement will govern the bulk properties of the polymer.

In order to understand and predict the mechanical behavior of the polycrystalline polymer, it is logical that the structural state of the polymer must be known. By concentrating on the structural state of the polymer the confusing dependence on fabrication process variables can be avoided, and a simple, structurally sound structure-property theory can be developed. In this way interest concentrates on the structural state needed to yield certain predictable properties, and not on the multitude of fabrication processes that can lead to that structural state.

The simple, direct and logical manner in which structure relates to mechanical properties was demonstrated in a recent study of isotactic polypropylene fibers and films by the author [8]. In that study the properties of thirty different structural states produced by four different fabrication processes were examined. Here the different fabrication processes were shown to produce equivalent structural states. The mechanical properties of the samples were then shown to depend not on the fabrication process, but on the structural state of the polymer at the time of the test. In fact, the fabrication process and the test procedure could be considered as simply different deformation processes leading to some given structural state of the polymer.

The structural state of the polymer is defined by three parameters; (1) the fraction of crystals present in the sample, β, and its complement, the fraction of noncrystalline polymer present in the sample, $(1-\beta)$; (2) the orientation of the crystal chain axis, f_c; and (3) the orientation of the noncrystalline chains, f_{AM}. Here $f_{AM} = (3 \cos^2\theta_{AM}-1)/2$, where θ_{AM} is the average angle the noncrystalline polymer helical chain axis makes with the reference direction in the sample.

The deformation state model is very simple. The molecules in both the crystalline and noncrystalline regions will each have some angle of orientation with respect to the reference direction in the sample. The value of f can vary anywhere from -0.5 for chains oriented perpendicular to the reference direction, to zero for randomly oriented chains, to +1.0 for chains that are fully oriented in the reference direction. Thus the orientation states for each region can vary only within the limited orientation range -0.5 to 0 to +1.0 irrespective of the fabrication process used.

The orientation state model of a material should be independent
of the polymer under examination. The same structural definitions
will apply to PET as apply to isotactic polypropylene and thus
structure-property relations of a generally similar form should be
expected for both polymers. The structural relations examined for
polypropylene [8], fell into two main catagories: (1) those
directly related to the average orientation of the sample, f_{AV} [Here
$f_{AV} = \beta f_c + (1-\beta)f_{AM}$], and (2) those directly related to the
noncrystalline orientation, f_{AM}. The modulus, the failure envelope,
the yield behavior, and the resilience were all found to be directly
related to the average orientation state of the polymer. On the
other hand, the true fabrication strain (which is defined as $\ln\lambda_F$,
where λ_F is the fabrication draw ratio) was found to be a linear
function of the noncrystalline orientation function, f_{AM}, and the
tenacity of those samples which were drawn at a slow enough rate to
allow maximum orientation to occur before failure was also directly
proportional to the noncrystalline orientation function.

Tables 1 and 2 contain both the structural data required to
define the structural state of the PET samples and the following
mechanical data: tenacity, shrinkage, fabrication draw ratio,
tensile modulus, and the dynamic loss modulus, E". Also included
are the crystal long spacing and small-angle x-ray intensity of the
fibers. With the structure-property relations determined for
isotactic polypropylene as a guide, the data from each of these
measurements are examined structurally below. In this way the
general character of the observed structure-property relationship
will be demonstrated, and any deviation of the PET data from the
general form of the structural relationships developed for polypropylene will become obvious.

Tenacity

When a cast film or unoriented fiber is deformed, the randomly
oriented molecules tend to become oriented in the deformation
direction. If the sample is deformed slowly enough and at a
sufficiently high temperature, it will follow the stress-strain
path shown in Figure 1. Under these conditions the molecules have
enough mobility and time available to relieve the applied stress
by rearranging the internal structure of the sample. Opposed to
this plastic deformation of the sample is the presence within the
sample of a distribution of flaws (stress concentration regions)
which, under the proper deformation conditions, can cause crack
formation and ultimate sample failure. As the deformation process
proceeds through the yield point and into the strain hardening region
of the stress-strain curve, stress builds up in the highly oriented

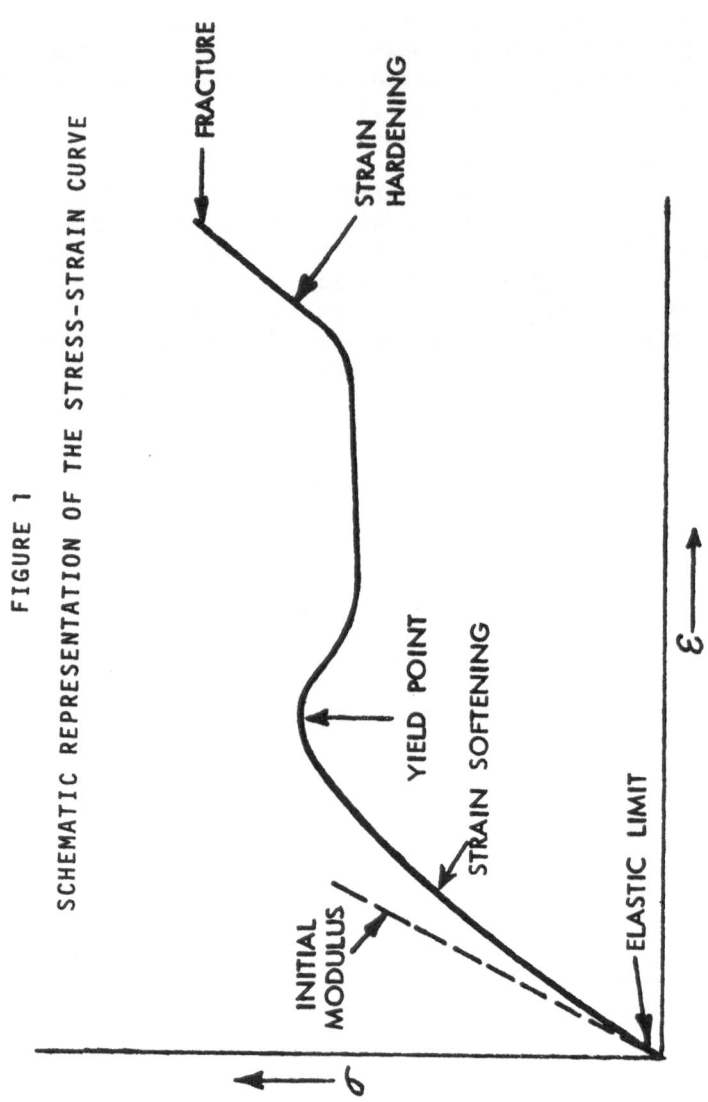

FIGURE 1

SCHEMATIC REPRESENTATION OF THE STRESS-STRAIN CURVE

structure until no further plastic deformation mechanisms are available to the molecules. At this point the flaw mechanism predominates and sample failure occurs.

Tenacity is a measure of the failure strength of a fiber. It is defined as the tensile stress required to break the sample (grams) divided by a measure of mass per unit length (denier is the weight in grams of 9,000 meters), and thus tenacity has the units of length. Tenacity represents the tensile stress measured at break divided by the original dimensions of the fiber before deformation.

Studies of the failure envelope of isotactic polypropylene demonstrated that, when the rate of deformation at a given draw temperature is slow enough for the molecules to reach the highest orientation possible before the flaw mechanism predominates, the force required for failure of a given breaking cross-sectional area of the sample will be a constant. That is, the true stress at break (the strength per breaking cross-sectional area of the sample) is independent of the starting orientation of the sample. This is true because the molecules in all of the samples are equally oriented at the time of break irrespective of the sample's starting orientation state. Under these conditions of constant true breaking stress, the tenacity (which measures the breaking force in terms of the unstretched dimensions of the sample and not in terms of the dimensions at the time of break) is simply a measure of the dimensional change available to the sample before failure occurs. Thus, under these special conditions, tenacity is essentially a reciprocal measure of the extensibility (since the smaller the extensibility range the greater the tenacity) normalized to the dimensions of the starting sample.

The tenacity of the drawn and annealed PET samples are plotted as a function of annealing temperature in Figure 2. The tenacities of both the 3X and the 5X yarns decrease with annealing temperature after an initial increase over that of the original draw yarn. Except for the general shape similarity between the curves for the differently treated yarns, and the fact that the tenacities of the 3X yarns are lower than those of the 5X yarns, there seems to be little that can be correlated.

Considering these same tenacity data from a structural point of view, we see a different picture emerging. Free shrinkage occurs primarily as a consequence of the relaxation of the strained, oriented, noncrystalline chains. This will become obvious when shrinkage is examined in the next section. The greater the annealing temperature the greater the molecular relaxation possible. Since the tenacity value depends on the distance the starting orientation state is from the maximum value

FIGURE 2

RELATION BETWEEN THE ANNEALING TEMPERATURE
AND THE TENACITY OF PET FIBERS

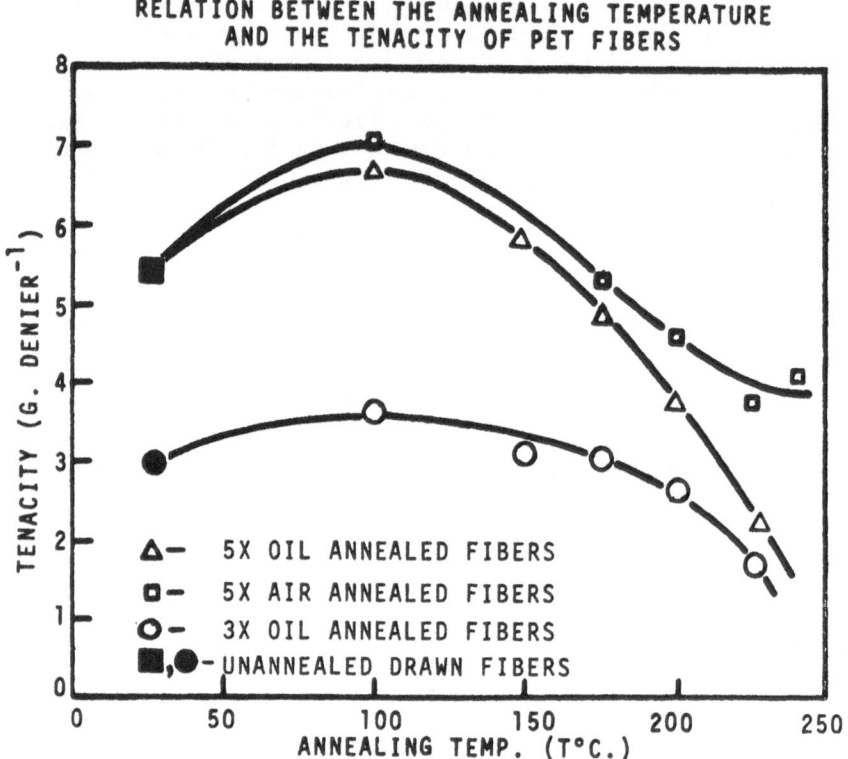

allowable at the temperature of the tenacity test, lower tenacities will be expected from samples relaxed to lower orientation states.

The tenacity studies on isotactic polypropylene showed that the controlling structural feature for tenacity relations was the noncrystalline orientation. When log tenacity was plotted against f_{AM}, the tenacities obtained from thirty structural states, fabricated by four different processes, all fell on a single line. The primary experimental requirement here was that the tensile draw rate be slow enough for all of the samples to reach the same limiting structure before failure. An extension rate of 50%/min. or less was required for the polypropylene samples to satisfy this criterion.

A strain rate of 50%/min. is typical of the slow rates of deformation generally used in fiber laboratories for standard tenacity determinations. The rate of deformation used to determine the tenacity of the PET yarns was not reported [10,11]. The extension ratio at break was also missing [10,11].

Without the extension ratio at break the true stress cannot be calculated and thus the independence of the failure true stress with sample structure cannot be checked. However, since the standard rate of deformation of tenacity samples is slow and the limiting rate of strain for PET is not known, it is reasonable to assume that the experimental PET measurements were made in the desired constant true stress region. If this is true then the isotactic polypropylene study would predict that a semilog plot of tenacity against f_{AM} would be linear with all of the PET samples fitting the same line.

Figure 3 shows a semilog plot of tenacity against f_{AM}. All of the points fall around a single line as predicted. These points include the original 3X and 5X drawn samples (solid filled points), each of the original drawn samples after free shrinkage in oil at different temperatures, and the 5X drawn sample after free shrinkage in air at different temperatures. All of these samples were then further drawn during the tenacity measurement until they failed. The correlations obtained in Figure 3 between the orientation state of the noncrystalline chains before the tenacity test and the final measured tenacity, show that the rate of deformation during the tenacity test was slow enough for all of the samples to fail under the same structural conditions irrespective of their starting structure. It also shows that under these conditions the measured tenacity of the fibers depends on the noncrystalline orientation state of the fiber at the time of the test and not on the particular fabrication history required to reach that strucutural state. In this sense the structural dependence of tenacity for PET is the same as that previously observed for isotactic polypropylene.

FIGURE 3

RELATION BETWEEN THE TENACITY AND f_{AM} FOR ANNEALED PET FIBERS

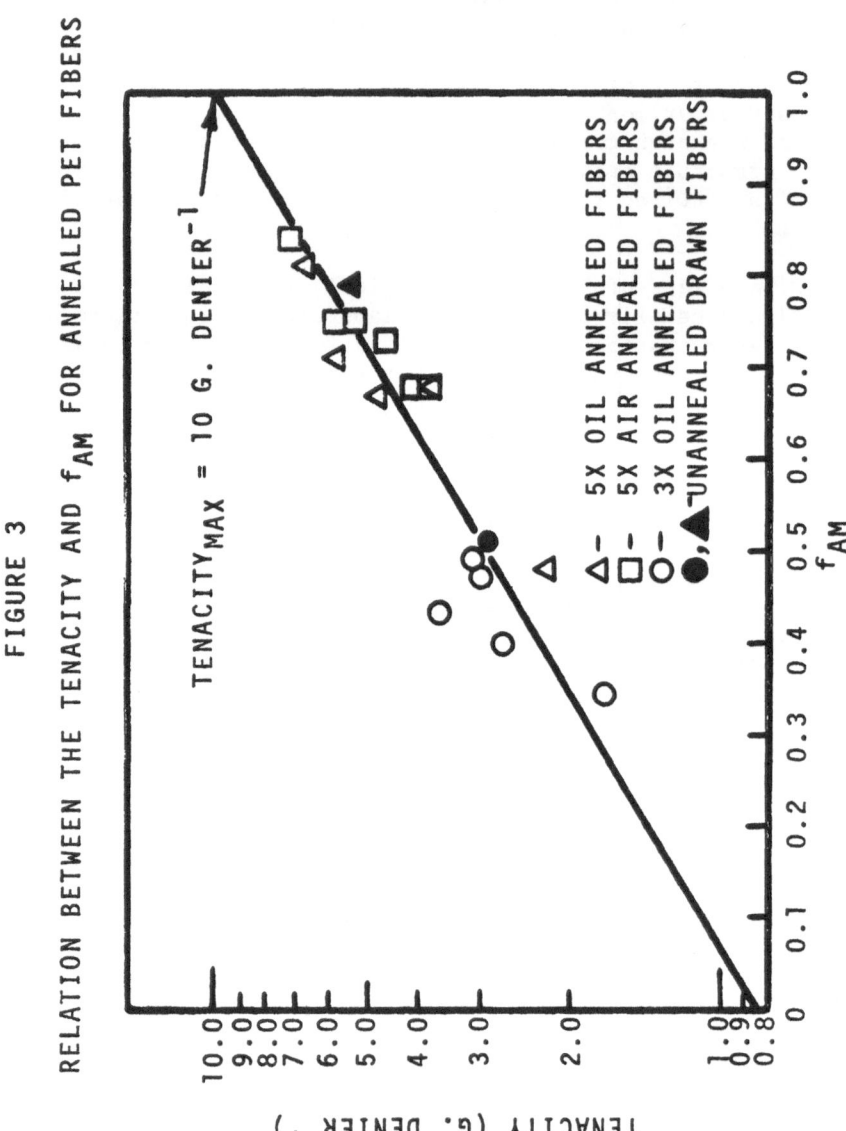

Once a quantitative correlation between tenacity and structure has been established it is possible to predict the maximum tenacity to be expected from a fully oriented PET fiber. This is done by extrapolating the tenacity – f_{AM} line to a value of $f_{AM} = 1.0$. Extrapolation of the line in Figure 3 leads to a maximum tenacity of 10 grams per denier. An experimental tenacity as high as 9.6 grams per denier has been reported for PET [12].

Thermal Annealing

(a) Shrinkage. When the essentially unoriented noncrystalline PET spun fiber is drawn to different extensions at 80°C., two processes occur; (1) the noncrystalline chains become oriented, and (2) crystallization occurs. If these drawn fibers are then subjected to increased temperature and are allowed to shrink freely for a fixed time, two further processes occur; (1) the orientation of the noncrystalline chains decreases, and (2) the crystallites thicken.

What is the relation between the structures developed during the drawing processes and the subsequent shrinkage process? Is the observed shrinkage simply related to orientation relaxation processes? Is there a fundamental structural difference between the shrinkage mechanism that occurs in oil at different temperatures and the shrinkage mechanism that occurs in air? Can characterization of the shrinkage phenomena be placed on a quantitative structural foundation?

The effect of annealing temperature and medium on the shrinkage of the drawn PET fibers is shown in Figure 4. The shrinkage is seen here to increase with increasing temperature in both oil and air, with the fibers annealed in oil showing a significantly greater shrinkage at higher temperatures than those annealed in air.

The fibers in question are highly amorphous (ca. 60–70%) and the amorphous chains are oriented. Though the glass transition temperature of unoriented amorphous PET is 67°C., that of the oriented PET can vary from 100°C. up to at least 160°C., depending on the amount of orientation (see Table 2 and the section on Dynamic Loss Modulus). Thus the annealing temperatures used to shrink the fibers are in the correct range for imparting mobility to the oriented noncrystalline chains. The amount of disorientation of the noncrystalline chains will depend on the annealing temperature and the time. The greater the annealing temperature for a given time in a given medium, the greater would be the expected disorientation and, hence, the greater would be the shrinkage.

The shrinkage data are plotted in Figure 5 as a function of the noncrystalline orientation present in the sample after annealing [here the term (1-β) corrects for the fraction of noncrystalline

FIGURE 4

RELATION BETWEEN ANNEALING TEMPERATURE
AND SHRINKAGE OF PET FIBERS

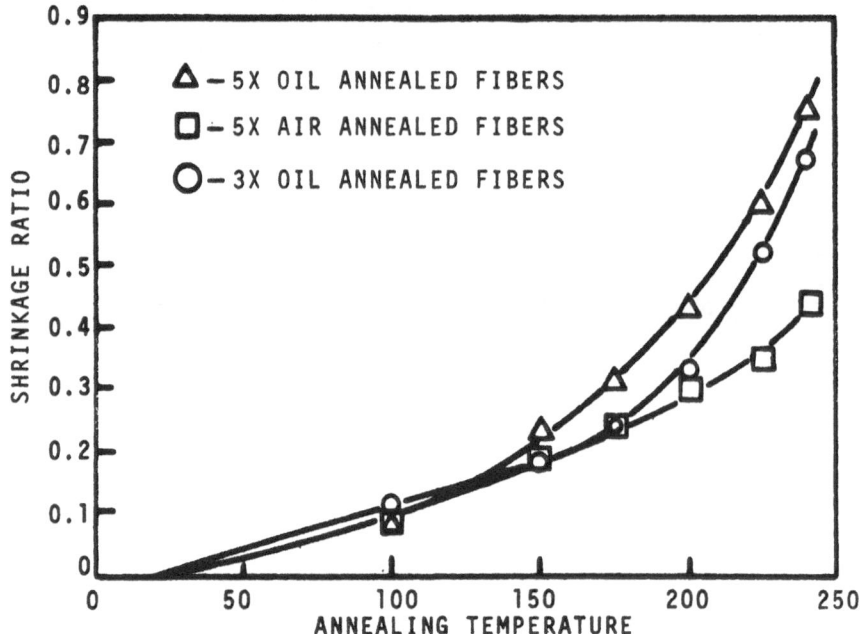

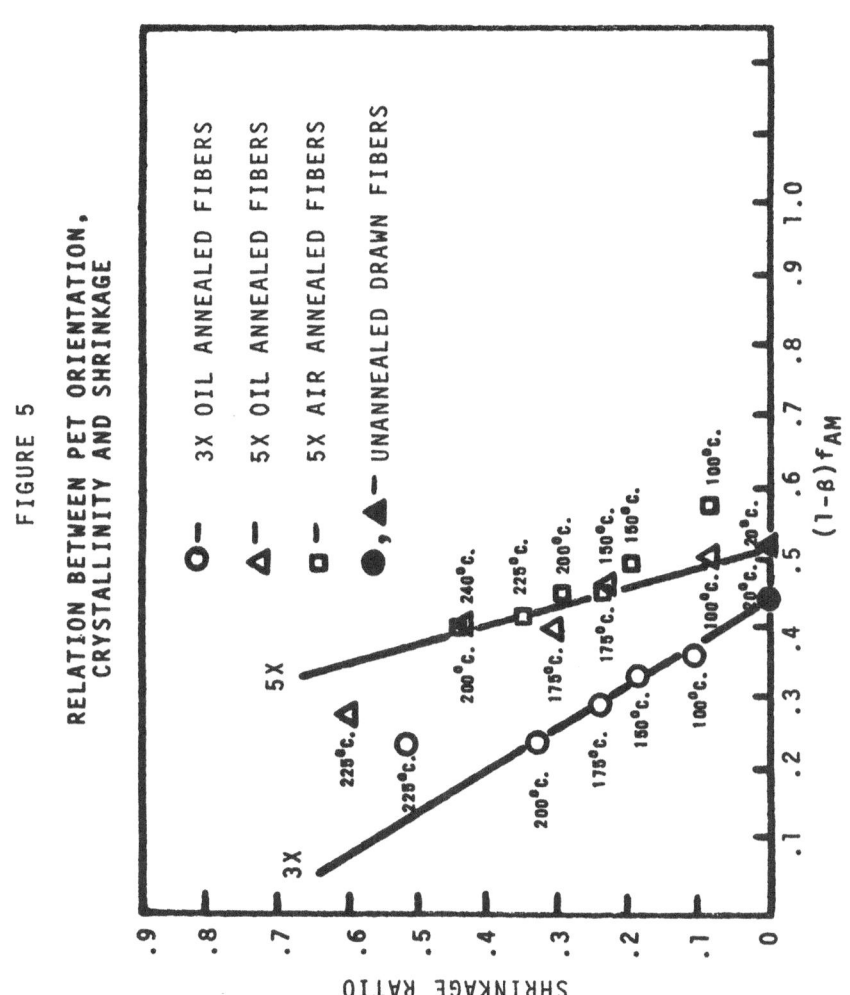

FIGURE 5

RELATION BETWEEN PET ORIENTATION,
CRYSTALLINITY AND SHRINKAGE

chains present in the sample]. The noncrystalline chains are seen
to decrease in orientation with increasing temperature by an amount
proportional to the shrinkage that occurred in the sample. The
shrinkage mechanism in oil and air media is seen to be the same,
since the shrinkage-structure relation is the same for the 5X drawn
fibers in the two media, the lower shrinkage in air occurring solely
as a consequence of poor heat transfer in this medium. The amount
of shrinkage is also seen to be directly related to the initial
starting structure present in the original drawn fibers before
shrinkage, as the shrinkage-structure relations for the 3X yarn and
the 5X yarn fall on two distinct lines with intercepts corresponding
to the structure of the original drawn fiber in each case.

The structure produced during the initial drawing process and
the subsequent structural change that occurs on annealing are thus
intimately related. Is the character of the shrinkage process in
fact determined by the fabrication conditions used to produce the
original drawn fibers? If the fabrication process begins with an
unoriented amorphous fiber, and it is assumed that the extension
of the fiber proceeds primarily by orientation of the noncrystalline
polymer, then the amount and degree of orientation of the noncrystalline polymer should be directly proportional to the extension of the
fiber. If the thermally shrunk fiber is considered as the final
product of a fabrication process in which the fiber was first drawn
to some extension (characterized by a measurable noncrystalline
polymer content and degree of orientation), and then relaxed to a
lesser extension (characterized by a new noncrystalline content and
degree of orientation), then there should be a direct relation
between the true strain (logarithm of the extension ratio) of the
final fiber and the amount and degree of orientation $[(1-\beta)f_{AM}]$ of
the noncrystalline polymer in the final fiber.

The true strain is used here as the measure of extension, as
this was shown in the earlier study on polypropylene [8] to be the
correct strain term to use when describing plastic deformation of
a polymer. The true strain, in the polypropylene study, was found
to be a linear function of the noncrystalline orientation over the
whole draw range for all of the fabrication processes examined.

The original extension ratio, λ, describing the deformation
of the unoriented, spun fiber, is given by the expression:

$$\lambda = L/L_0 \qquad\qquad (4)$$

where L_0 is the length of the original spun fiber, and L is the
length of the fiber after it has been drawn. Consider a two-stage
process in which the fiber is first drawn an extension ratio λ,
and subsequently relaxed (through shrinkage in air or oil). The
extension ratio for the two-stage process, λ' (the residual

extension ratio), can be defined as:

$$\lambda' = L'/L_o \tag{5}$$

where L' is the length of the fiber at the end of the two-stage
process. The residual extension ratio describes the total deformation process, and thus has structural significance. The shrinkage
ratio, S, describes only a portion of the deformation process, and
is therefore incomplete.

To interpret the shrinkage data from the PET fibers meaning-
fully, the shrinkage ratio must be converted to a residual extension
ratio. The initial sample length in the shrinkage experiment
(equation 1) is simply the final length, L, of the original drawing
process. The final length of the sample after shrinkage is L', the
length of the fiber at the end of the two-stage process. Thus the
shrinkage ratio, S, can be defined as:

$$S = \frac{\text{Initial Length} - \text{Final Length}}{\text{Initial Length}} = \frac{L-L'}{L} \tag{6}$$

and

$$L' = L \ (1-S) \tag{7}$$

The residual extension ratio of a fiber which has undergone a two-
stage process of extension and subsequent shrinkage is thus given
by the expression:

$$\lambda' = L'/L_o = L(1-S)/L_o = \lambda \ (1-S) \tag{8}$$

where λ is the original drawn ratio (extension ratio) of the PET
fibers before shrinkage (in this case either 3X or 5X), and S is
the measured shrinkage ratio of the sample (see Table 1).

The residual extension ratio of the PET fibers is plotted as
true residual strain on the ordinate of Figure 6. The abscissa is
$(1-\beta)f_{AM}$, which describes the state of noncrystalline orientation
at the end of the fabrication process. Included in the figure are
the original single-stage draw fibers (3X and 5X), the fibers drawn
3X and 5X and subsequently shrunk in oil at different temperatures,
and the fibers drawn 5X and subsequently shrunk in air at different
temperatures; a total of eighteen structural states and deformation
processes. All of these data fall along a single straight line with
a zero intercept, suggesting that the sample orientation returns
along the original extension path during shrinkage. The figure also
demonstrates that all of the shrinkage behavior can be quantitatively
expressed by the structural state equation,

$$\log \lambda' = K \ [(1-\beta)f_{AM}] \tag{9}$$

FIGURE 6

EFFECT OF PET ORIENTATION AND CRYSTALLINITY
ON THE RESIDUAL EXTENSION RATIO AFTER SHRINKAGE

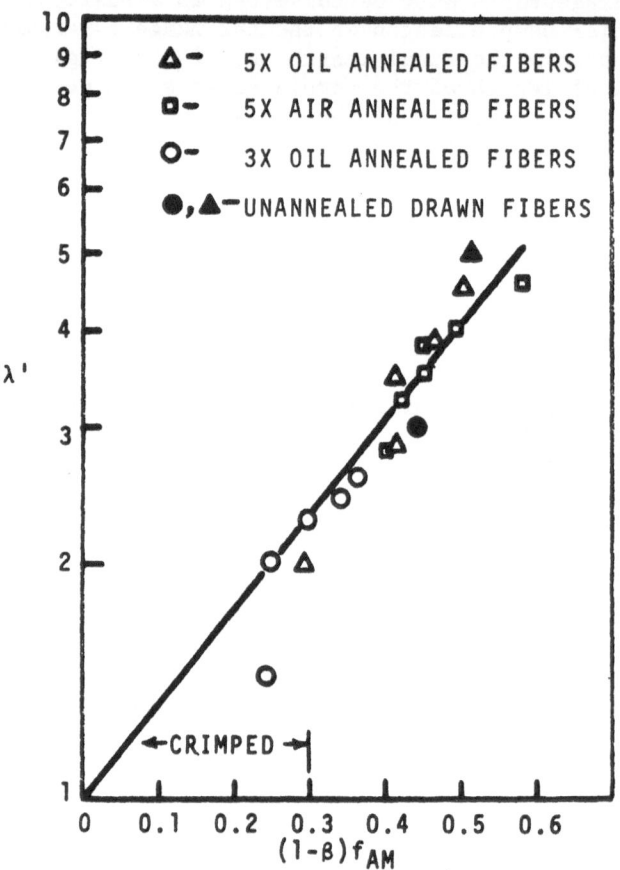

The fact that all eighteen samples fit this simple quantitative
expression shows that the shrinkage mechanism is primarily
controlled by the behavior of the noncrystalline chains.

(b) Crimp. Another observation made in this study was that
during oil annealing the 3X yarns developed a crimp at annealing
temperatures above 175°C., while the 5X yarns did not develop crimp
until an annealing temperature of 225°C. The 5X yarn did not
develop any crimp during air annealing. No explanation was given
for this phenomenon.

Once one realizes that the shrinkage is controlled by the noncrystalline chains, it is reasonable to examine the question of
crimp in the fiber. Crimp is characterized by a bending, curling,
or crumpling of the fiber and can be considered as occurring either
from a rapid "snap-back" of the fiber or from some structural relaxation resulting at some minimum orientation. When the 3X drawn
yarn is annealed in oil and allowed to shrink freely, crimp first
occurs at 175°C. At that temperature the noncrystalline orientation
has been reduced to a $(1-\beta)f_{AM}$ value of 0.297. When the 5X drawn
fiber is annealed in oil and allowed to shrink freely, crimp does
not occur until the annealing temperature of 225°C. is reached. At
that temperature the noncrystalline orientation of the 5X sample
has been reduced to a $(1-\beta)f_{AM}$ value of 0.293. Thus in both the
3X and 5X drawn oil shrunk samples, crimp does not appear until the
noncrystalline orientation has been reduced to the same value (0.29).
This seems to be a critical structural value for crimp formation in
the PET fibers, a structural condition that occurs at different
temperatures in the two fibers as a consequence of their initial
difference in noncrystalline orientation. All fibers with $(1-\beta)f_{AM}$
values of 0.29 or less were crimped (see Figure 6).

None of the thermally annealed 5X drawn samples crimped in
air. The lack of crimp in these fibers is not due to some unusual
air directed shrinkage mechanism, but is simply a consequence of the
poorer heat transfer in air. The poorer heat transfer led to less
relaxation of the noncrystalline chains and, as a consequence,
$(1-\beta)f_{AM}$ never reached the critical crimp value of 0.29. Instead,
the lowest $(1-\beta)f_{AM}$ value the air annealed sample reached was 0.408
at 240°C.

(c) Long Spacing. When the drawn PET fiber is annealed under
unrestrained conditions two processes occur: (1) the length of the
fiber decreases (the fiber shrinks) and (2) the small-angle long
spacing increases. The samll-angle long spacing is a measure of the
average repeat distance of the crystal lamellae as measured parallel
to the helical chain axis of the molecules. Thus the average
distance between crystal centers increases as the length of the
fiber decreases.

How can this seemingly anomolous result be explained? Can a structural relationship be found between the relaxation of the noncrystalline chains and the growth of the crystallites? How closely are the shrinkage and the crystallite growth related? The effect of annealing temperature and medium on the long spacing of the PET fibers is shown in Figure 7. After an initial slight decrease in long spacing, the long spacing increases with increasing temperature for all of the fibers. The 3X and 5X drawn yarns annealed in oil follow the same long spacing curve, increasing quite rapidly with increasing temperature. The 5X drawn fiber annealed in air also increases in long spacing with increasing temperature, but at a much slower rate than the oil annealed samples.

Except for the higher long spacing values at an annealing temperature of 100°C., the shape of the long spacing annealing-temperature curves in Figure 7 is very similar to the shape of the shrinkage ratio-annealing temperature curves in Figure 4. This suggests that there is a direct relationship between the shrinkage ratio and the long spacing developed in the PET fibers. The shrinkage ratio is plotted against the long spacing in Figure 8. A linear relation is obtained between the shrinkage ratio and the long spacing for each of the PET fiber shrinkage sets, except for those samples annealed at 100°C., suggesting that the decrease in sample length and the increase in long spacing are intimately related.

The shrinkage of the fiber should be considered as the second-stage of a two-stage fabrication process which includes extension as well as shrinkage. Is the size of the annealed crystal a direct function of the relaxation of the noncrystalline chains? Can the change in long spacing be directly related quantitatively to the residual extension of the two-stage fabrication process?

If the size of the crystal is controlled by the relaxation of the noncrystalline chains, then the maximum orientation controlled long spacing would occur when the noncrystalline orientation of the particular drawn sample is allowed to relax to the random state ($f_{AM} = 0$). This condition will be achieved when the sample shrinks to its original length before extension (i.e., a residual extension ratio, λ', of 1.0). For the 3X drawn sample this occurs at a shrinkage ratio, S, of 0.67, while for the 5X drawn sample this occurs at a shrinkage ratio of 0.80. If the linear relation between the shrinkage ratio S, and the long spacing, L, for each shrinkage set in Figure 8 is extrapolated to the value of S corresponding to the original length of the sample, the maximum relaxation controlled long spacing for that set, L_{MAX}, can be determined. For the 3X drawn sample, maximum shrinkage was achieved at 240°C., and L_{MAX} = 192 Å (see Figure 8). The 5X drawn sample almost reached L_{MAX} at 240°C. in oil; L_{MAX} for the oil samples is 200 Å. The 5X samples annealed in air were some distance from L_{MAX} = 211 Å, even at 240°C., as a consequence of the poor heat transfer in this medium.

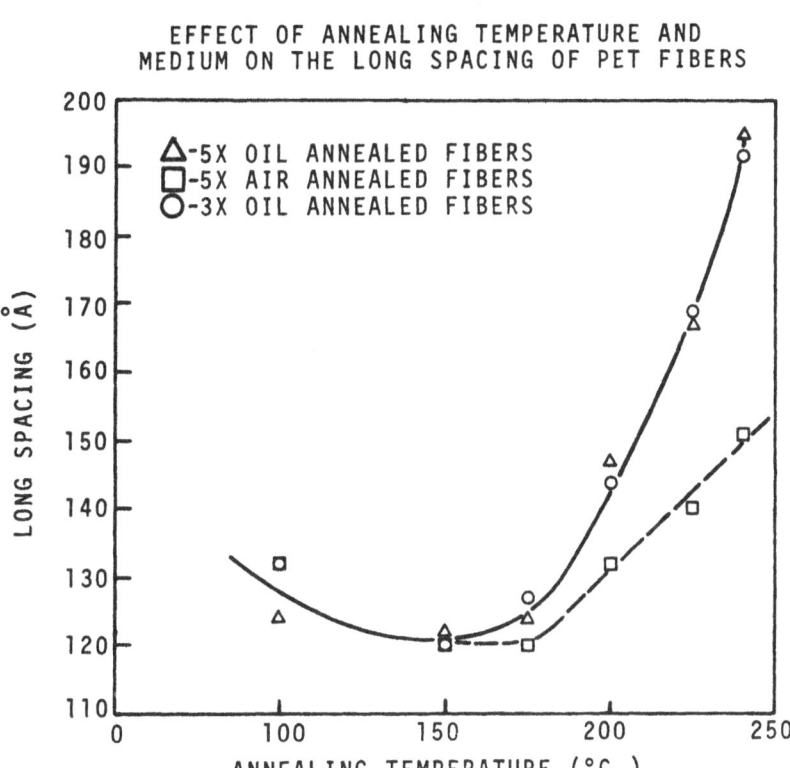

FIGURE 7

EFFECT OF ANNEALING TEMPERATURE AND
MEDIUM ON THE LONG SPACING OF PET FIBERS

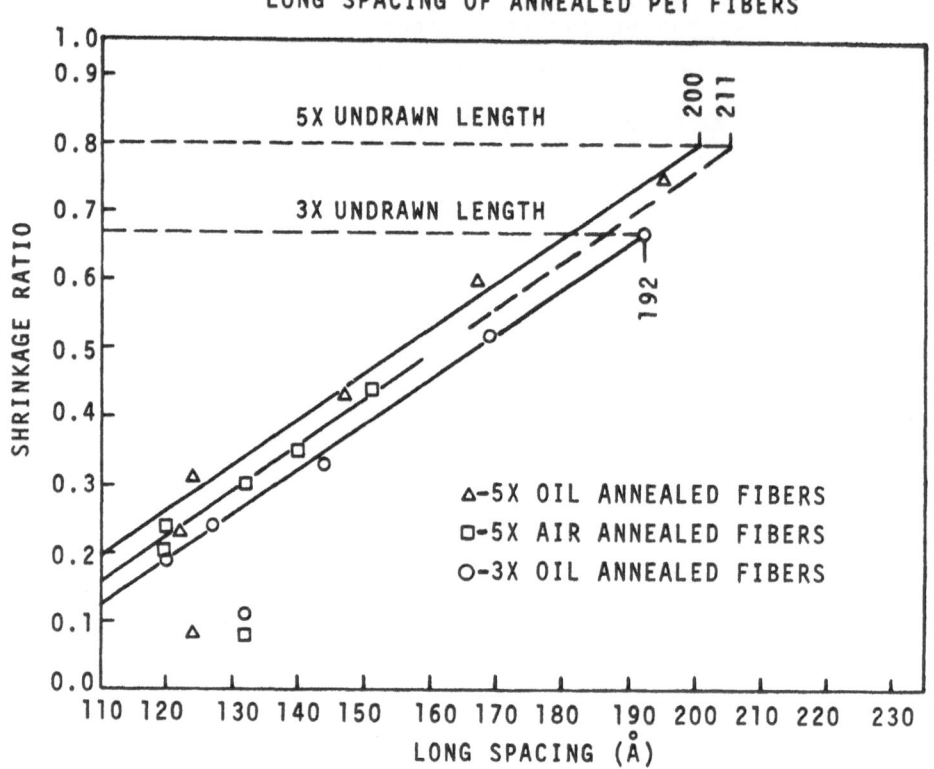

FIGURE 8

RELATION BETWEEN THE SHRINKAGE RATIO AND THE
LONG SPACING OF ANNEALED PET FIBERS

The distance the measured long spacing, L, is from the maximum long spacing it could have if it was allowed to shrink completely, L_{MAX}, is ($L_{MAX}-L$). The proportion of the original extension of the fiber which this change in spacing represents is simply the draw ratio, λ_F, times ($L_{MAX}-L$). Thus the term, $[\lambda_F(L_{MAX}-L)]$, represents the crystallite growth analog of the residual extension, and normalizes all of the long spacing data to the fabrication conditions of both drawing and shrinkage (i.e., the complete two-stage fabrication process).

The general long spacing term, $[\lambda_F(L_{MAX}-L)]$, is plotted against the residual extension ratio, λ', in Figure 9. All of the data, except for the samples annealed at 100°C., fit a straight line with a positive slope and a zero intercept, having the form:

$$[\lambda_F(L_{MAX}-L)] = K \lambda' \tag{10}$$

This includes the 3X and 5X drawn samples which were annealed in oil, and the 5X drawn samples that were annealed in air. Thus, it seems, the change in long spacing with annealing of drawn fibers can be directly related quantitatively to the residual extension of the two-stage fabrication process.

The fact that the general long spacing term is directly and quantitatively related to the residual extension suggests that the crystallization process is controlled by the relaxation of the non-crystalline chains. This conclusion is reached as a consequence of the fact that the true residual strain is directly proportional to the amount and degree of orientation of the noncrystalline chains (Figure 6) and therefore $[\lambda F(L_{MAX}-L)]$ will have a similar proportionality when plotted in a similar manner (Figure 10). Thus it seems that shrinkage proceeds by a two-step process in which first the noncrystalline chains relax along their original deformation path causing a decrease in the length of the fibers, and then the relaxed chains are utilized by the crystals for growth.

(d) Small-Angle X-Ray (SAXS) Intensity. According to small-angle x-ray theory [13] the intensity of the small-angle x-ray diffraction is proportional to the square of the difference in electron density between the crystalline, ρ_c, and noncrystalline ρ_{AM}, regions, thus,

$$\sqrt{I} \underset{\sim}{\sim} (\rho_c-\rho_{AM}) \tag{11}$$

For an oriented polymer the intensity is also proportional to the degree of orientation of the regions, since the discrete scattering intensity is anisotropic and only occurs parallel to the molecular helical chain axis direction.

FIGURE 9

RELATION BETWEEN [DRAW RATIO (L_MAX-L)]
AND THE RESIDUAL EXTENSION RATIO, λ',
OF ANNEALED PET FIBERS

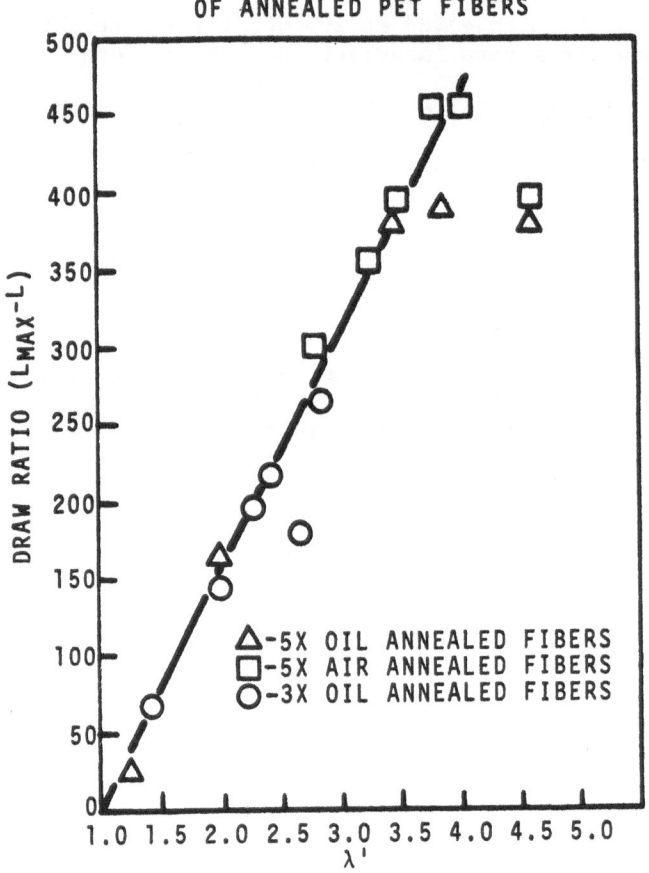

FIGURE 10

RELATION BETWEEN [DRAW RATIO (L_{MAX}-L)]
AND [$(1-\beta)f_{AM}$] FOR ANNEALED PET FIBERS

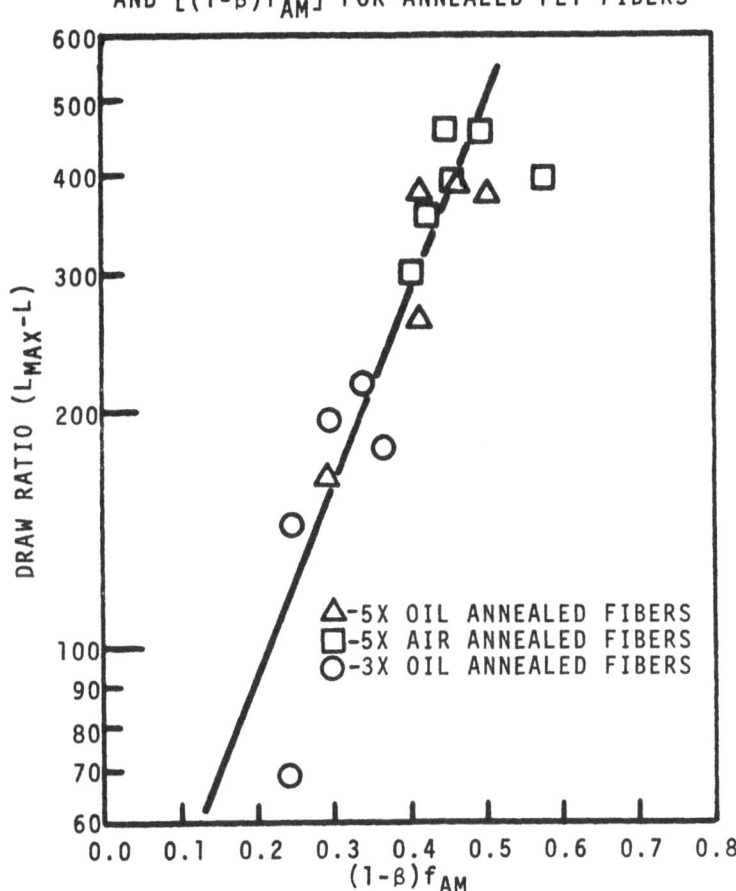

The greatest difference in $(\rho_c-\rho_{AM})$ will occur when the non-crystalline chains are fully relaxed and $f_{AM} = 0$. Under this special condition the molecules in the crystalline regions are aligned and well ordered and those in the noncrystalline region are disordered. This large difference in structural character will lead to the highest SAXS intensity for a given fraction of crystals. The other intensity extreme occurs when the noncrystalline region is fully oriented and all of the noncrystalline chains are aligned. For this special case the noncrystalline polymer has an ordered character more nearly approximating the ordered character of the molecules in the crystals, and this structural similarity between the crystalline and noncrystalline regions leads to the lowest SAXS intensity for a given fraction of crystals. Of course, intermediate degrees of orientation of the noncrystalline chains will lead to intermediate SAXS intensity values. Thus, the small-angle x-ray diffraction intensity will be some function of the amount and degree of orientation of the polymer chains.

Since the SAXS intensity is proportional to the square of the difference in the electron density of the two regions; the square root of the SAXS intensity, $\sqrt{I}$, is the proper variable to use for structural analysis. Figure 11 shows the effect of annealing temperature and medium on the observed SAXS intensity. Except for the PET fibers annealed in oil at 240°C., for which there are no structural data, the SAXS intensity increases with increasing annealing temperature. This character of the annealing curve is similar to that previously observed for the effect of annealing temperature on the shrinkage ratio (Figure 4), and the long spacing (Figure 7).

In view of the similarity of the SAXS behavior to that of the shrinkage ratio, S, and the long spacing, L, and the fact that SAXS intensity like S, and L, will depend on the orientation of the polymer, the SAXS intensity was plotted against the amount and degree of noncrystalline orientation, $(1-\beta)f_{AM}$ (Figure 12). The dependence of the SAXS intensity on the noncrystalline orientation is obvious, and almost identical with that previously observed for the shrinkage ratio (Figure 5). Thus the SAXS intensity is seen to increase with increasing temperature as the noncrystalline orientation decreases and the sample shrinks. The character of the SAXS intensity increase is the same whether it occurs in an oil or an air medium since the $\sqrt{I}$-structure relation is the same for the 5X drawn fibers annealed in both media. This suggests that the same noncrystalline chain disorientation mechanism is responsible for the intensity increase in both media, a conclusion identical with that previously drawn for the shrinkage process. The increase in $\sqrt{I}$ is also seen to be directly related to the initial starting structure present in the drawn fibers before shrinkage, as the $\sqrt{I}$-structure relation for the 3X yarn and the 5X yarn falls on two

FIGURE 11

RELATION BETWEEN THE SAXS INTENSITY AND
THE ANNEALING TEMPERATURE OF PET FIBERS

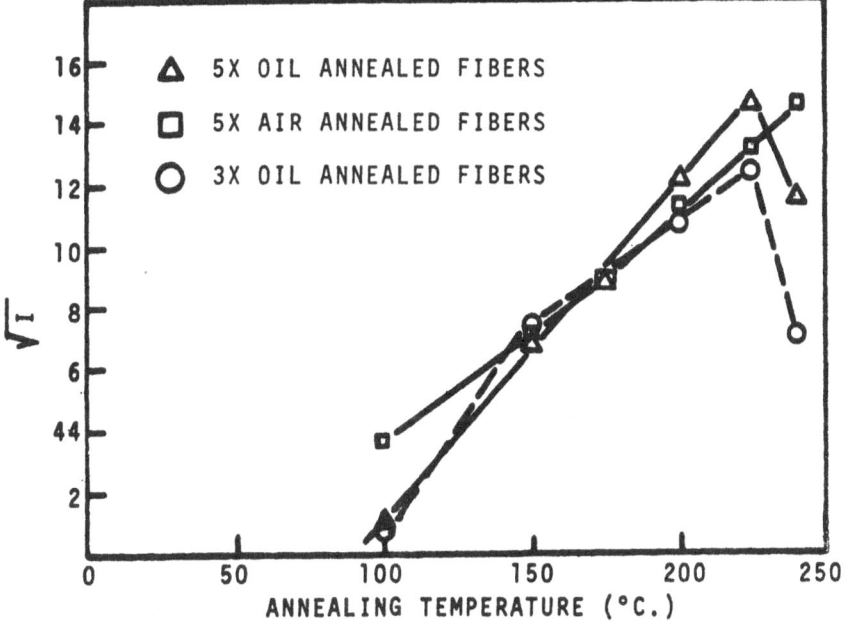

FIGURE 12

RELATION BETWEEN THE SMALL-ANGLE X-RAY INTENSITY,
$\sqrt{I}$, AND THE NONCRYSTALLINE ORIENTATION,
$(1-\beta)f_{AM}$, IN ANNEALED PET FIBERS

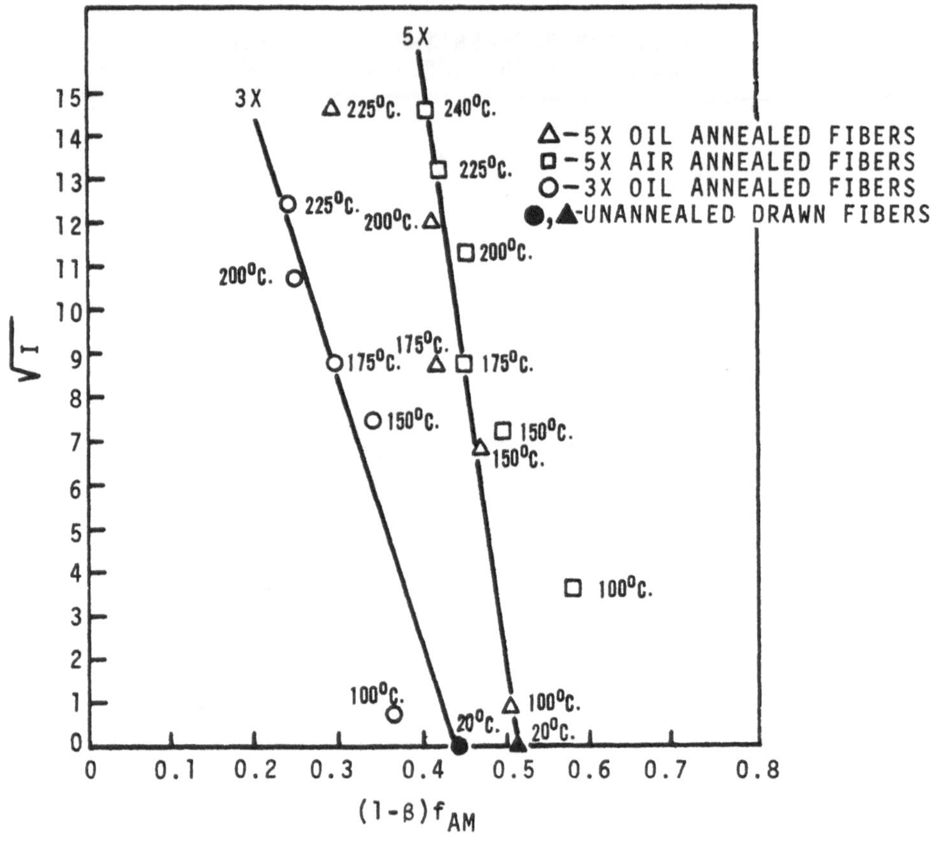

distinct lines with intercepts corresponding to the structure of the original drawn fibers in each case.

The similarity between the SAXS intensity behavior and the behavior of the shrinkage ratio suggests that there is a direct relationship between these two parameters. The shrinkage ratio is plotted against the SAXS intensity, $\sqrt{I}$, in Figure 13. All of the data, except for the PET fibers annealed in oil at 240°C., for which there are no structural data and which act anomalously throughout (see Figure 11), follow a single straight line with zero intercept. The experimental magnitude of $\sqrt{I}$ for the unannealed drawn fibers is taken as zero, as the integrated intensity of these samples was reported as "so low as to be hardly measurable for the 3X and 5X samples as drawn, and there appeared to be little intensity difference between the samples" [10]. The fact that within the scatter of the data there is a direct relation between the shrinkage ratio and the $\sqrt{I}$, shows that the decrease in sample length and the increase in SAXS intensity are intimately related.

The shrinkage of the fiber must be considered as the second stage of a two-stage process. The SAXS intensities from both the drawing process and from the subsequent shrinkage process are intimately related (Figure 12). The SAXS data must be treated in terms of the total two-stage process before a single quantitative structural correlation can be achieved. The maximum anisotropic structure controlled SAXS intensity will occur when the noncrystal-line chains are allowed to relax to the random state ($f_{AM} = 0$) under those thermal conditions that achieve the largest most perfect crystals. This will occur when the drawn sample is allowed to free shrink back to the undrawn length of the original spun fiber (a residual extension ratio, λ', of 1.0). As pointed out in the dis-cussion of the long spacing, this occurs at a shrinkage ratio of 0.67 for the 3X drawn fiber, and at a shrinkage ratio of 0.80 for the 5X drawn fiber. By extrapolating the linear relation between the shrinkage ratio and $\sqrt{I}$ to the corresponding shrinkage ratio for complete shrinkage of that fiber ($\lambda' = 1.0$), the value of the maximum anisotropic structure directed SAXS intensity, $\sqrt{I}_{MAX}$, can be determined. The maximum anisotropic structure directed SAXS intensity is $\sqrt{I}_{MAX} = 19.7$ for the 3X drawn fiber, and $\sqrt{I}_{MAX} = 23.5$ for the 5X drawn fiber (Figure 13).

The amount the measure $\sqrt{I}$ value is from the value it could have if full shrinkage was allowed to occur, $\sqrt{I}_{MAX}$, is ($\sqrt{I}_{MAX} - \sqrt{I}$). The proportion of the original extension of the fiber which this change in spacing represents is simply the draw ratio times ($\sqrt{I}_{MAX} - \sqrt{I}$). Thus the term, $[\lambda_F (\sqrt{I}_{MAX} - \sqrt{I})]$, represents the SAXS intensity analog of the residual extension, and normalizes all of the SAXS intensity data to the conditions of the two-stage fabrication process.

FIGURE 13

RELATION BETWEEN THE SHRINKAGE RATIO AND THE SMALL-ANGLE
X-RAY INTENSITY, $\sqrt{I}$, OF ANNEALED PET FIBERS

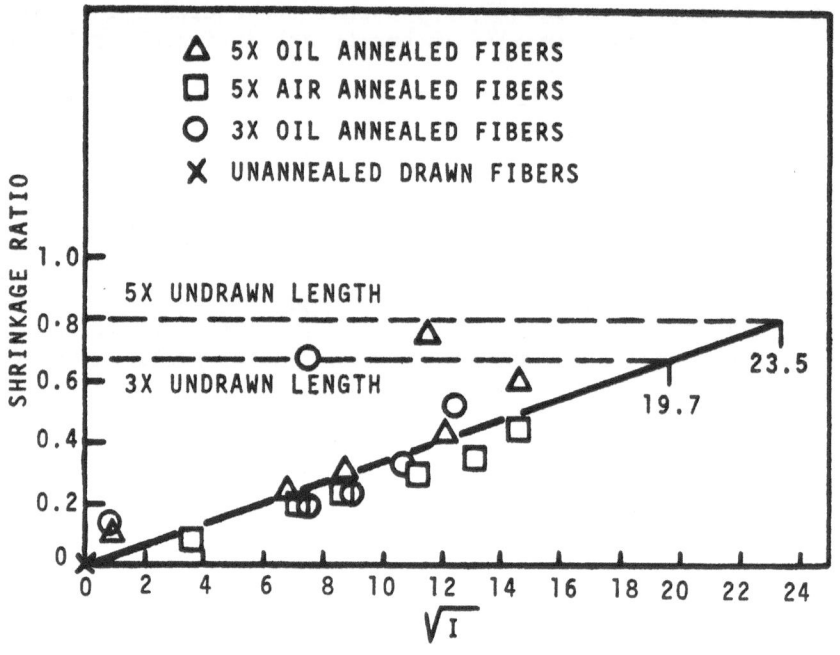

The general SAXS intensity term, $[\lambda_F(\sqrt{I_{MAX}}-\sqrt{I})]$, is plotted against the residual extension ration, λ', in Figure 14. All of the data, except for the samples annealed in oil at 240°C., fit a straight line with a positive slope and a zero intercept, having the form

$$[\lambda_F(\sqrt{I_{MAX}}-\sqrt{I})] = A\lambda' \qquad (12)$$

Thus it seems the change in SAXS intensity that occurs when drawn PET fibers are thermally annealed and allowed to shrink freely is directly related quantitatively to the residual extension of the two-stage fabrication process.

The SAXS intensity will increase as the oriented noncrystalline chains relax, disorient, and become less crystal-like. Since the term $[\lambda_F(\sqrt{I_{MAX}}-\sqrt{I})]$ is directly proportional to the residual true strain, and the residual true strain is directly proportional to the amount and degree of orientation of the noncrystalline chains (Figure 6), the correlation between the SAXS intensity and the non-crystalline orientation can be placed on a quantitative basis. A plot of log $[\lambda_F(\sqrt{I_{MAX}}-\sqrt{I})]$ against $(1-\beta)f_{AM}$ is linear (Figure 15), and can be expressed as:

$$\log [\lambda_F(\sqrt{I_{MAX}}-\sqrt{I})] = A + B[(1-\beta)f_{AM}] \qquad (13)$$

where A and B are constants. This expression fits data from the drawn and unannealed fibers, as well as the data from the drawn fibers shrunk in different media at different annealing temperatures, demonstrating that SAXS intensity data can be quantitatively correlated with the structural state of the polymer.

The thermal annealing of the PET fibers is part of a total fabrication process whereby the fiber is drawn from one structural state (that of the spun fiber), to a second structural state (that of the drawn fiber), and then to a final structural state (that of the fiber after shrinkage). The change in the properties of the sample represents the change that has occurred in the fiber as it goes from its initial state to the final structural state, and are intimately related to the state functions. In order to show how the structural state change has produced the observed changed pro-perties of the polymer it is necessary to describe the total change of the property in terms of the initial and final stage of the fabrication process. Thus shrinkage ratio must be converted to the residual extension, long spacing must be related to the draw ratio and the final recovered spacing, and SAXS intensity must be treated similarly. Once this is done the intimate relation between

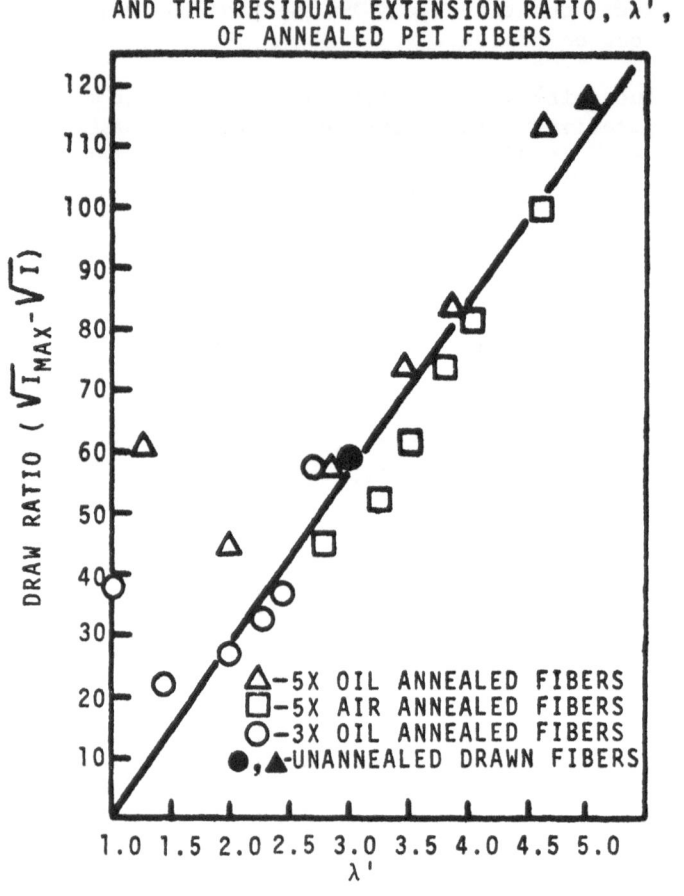

FIGURE 14

RELATION BETWEEN [DRAW RATIO ($\sqrt{I}_{MAX} - \sqrt{I}$)]
AND THE RESIDUAL EXTENSION RATIO, λ',
OF ANNEALED PET FIBERS

FIGURE 15

RALATION BETWEEN [DRAW RATIO ($\sqrt{I_{MAX}} - \sqrt{I}$)]
AND $(1-\beta)f_{AM}$ FOR ANNEALED PET FIBERS

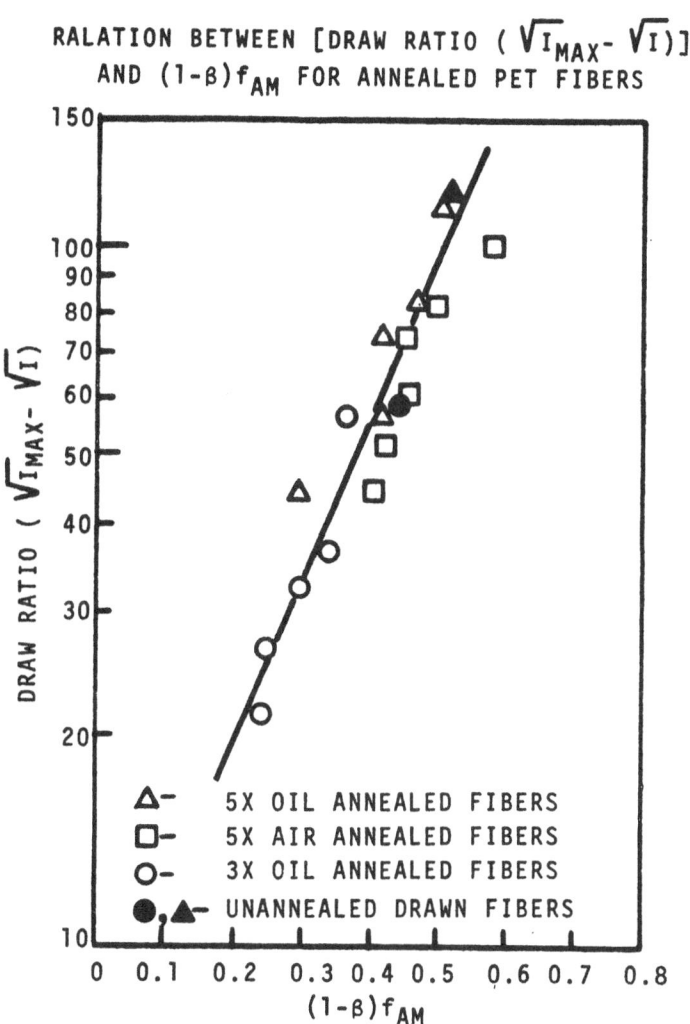

shrinkage, crimp, long spacing, and SAXS intensity and their
quantitative identification with the change in the noncrystalline
orientation states become obvious.

In this way thermal annealing can be understood mechanistically
as a process by which a sample which has been drawn to some
orientation state, gets enough thermal energy from its environment
to allow the noncrystalline chains to disorient. The decrease in
the chain alignment causes a decrease in the length of the fiber
(shrinkage) and makes less strained chains available for the
crystals to use for growth (long spacing increases). The increasing
difference between the ordered molecular arrangement within the
annealed crystals, and the disordered arrangement of the noncrystalline chains as the temperature is raised, leads to an increase in
the SAXS intensity with increasing annealing temperature. If the
disordering of the noncrystalline chains goes beyond a critical
value, the sample crimps. Above that critical value no crimp will
appear. Thus the property changes that occur on thermal annealing
are all intimately and quantitatively correlated to the changing
structural state of the polymer.

                            Dynamic Loss Modulus

The dynamic mechanical loss modulus is a measure of the energy
dissipated by the polymer as a consequence of molecular relaxation
processes. When the loss modulus is measured at a fixed frequency
at different temperatures a dynamic spectrum will be produced. At
specified temperatures in the spectra the loss modulus will have
maxima, which represent specific molecular relaxation processes.
In PET the α loss peak, E_{MAX}'', represents energy dissipated by
molecular relaxation processes in the noncrystalline region of the
polymer [14]. It is, in effect, a measure of the glass transition
temperature, T_g, of the polymer.

The molecular mechanism that occurs in the thermal relaxation
process involves a transition from a thermally frozen molecular
(glassy) state to one in which the molecules are allowed to relieve
the frozen in stresses by movement in the noncrystalline region.
Though the ends of the sample are restrained during the dynamic
loss modulus measurement, the molecular motions that occur at the
α transition are similar in kind to that which occurs from thermal
relaxation of the built-in stresses during free shrink. The loss
modulus maximum, E_{MAX}'', will be reached only when the temperature
has imparted enough energy to the molecules to overcome the thermodynamic stability of the particular energy state represented by the
orientation state of the molecules in the noncrystalline region of
the polymer. Shrinkage is controlled by similar forces. Thus, just

as with shrinkage, the observed temperature at which the α process loss maximum, E''_{MAX}, occurs in the PET fibers should be some function of the amount and degree of orientation of the noncrystalline chains in the starting sample.

The starting fibers used for the loss modulus measurements had a complicated fabrication history [9]. PET yarns initially drawn 3X and 4.25X over a hot pin at 80°C. were subsequently heated for six hours in vacuo at temperatures up to 240°C., and allowed to free shrink. These shrunk fibers were then given a further treatment in which they were boiled in water, under relaxed conditions, for one hour. The structural state present in the fibers after this complicated fabrication procedure [Table 2] is the one that determines the subsequent loss modulus results.

The relation between the measured temperature of E''_{MAX} and the annealing temperature used during the initial shrinkage in vacuo step of the fabrication process is shown in Figure 16. The essentially flat region of the curves in the annealing temperature region of room temperature (actually the point for these drawn fibers might best be placed at 80°C., [the hot pin temperature during drawing]), to 125°C. occurs as a consequence of the subsequent boiling water fabrication step, in which the lower temperature annealed fibers are all further annealed to the same higher temperature structural state. As the other fibers were initially relaxed during free shrinkage at temperatures higher than that of boiling water, their final structure was predominately developed during the initial shrinkage in vacuo. The earlier shrinkage study showed that the greater the annealing temperature the greater will be the relaxation of the oriented noncrystalline chains. Both the 3X and 4.25X drawn and annealed fibers show a continual decrease in $T_{E''_{MAX}}$ in Figure 16 with increasing temperature of thermal annealing after 125°C. This decrease in $T_{E''_{MAX}}$ corresponds to the decreasing degree of orientation and decreasing amount of noncrystalline chains present in the starting test fibers.

How the PET α loss peak maximum temperature, $T_{E''_{MAX}}$, depends on the initial structural state of the test fibers is shown in Figure 17. Here log $T_{E''_{MAX}}$ is plotted against $(1-\beta)f_{AM}$ for the test fibers (Table 2). Log $T_{E''_{MAX}}$ is directly proportional to $(1-\beta)f_{AM}$ (the structural state of the noncrystalline region) for all of the 16 structural states examined. When the best line through the points is extrapolated to the fully amorphous, unoriented state [$(1-\beta)f_{AM} = 0$], a $T_{E''_{MAX}}$ of 67°C. is predicted. The known glass transition temperature, T_g, of unoriented fully amorphous PET is 67°C. [16].

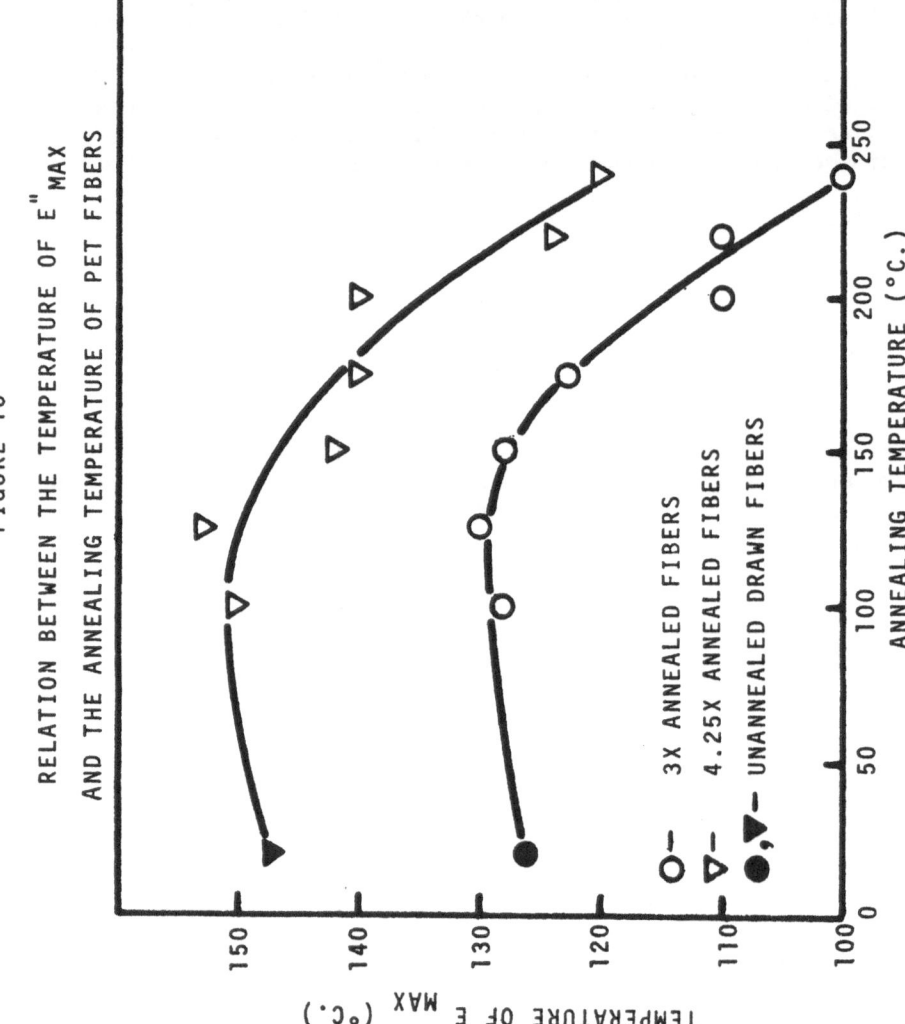

FIGURE 16

RELATION BETWEEN THE TEMPERATURE OF E"MAX

AND THE ANNEALING TEMPERATURE OF PET FIBERS

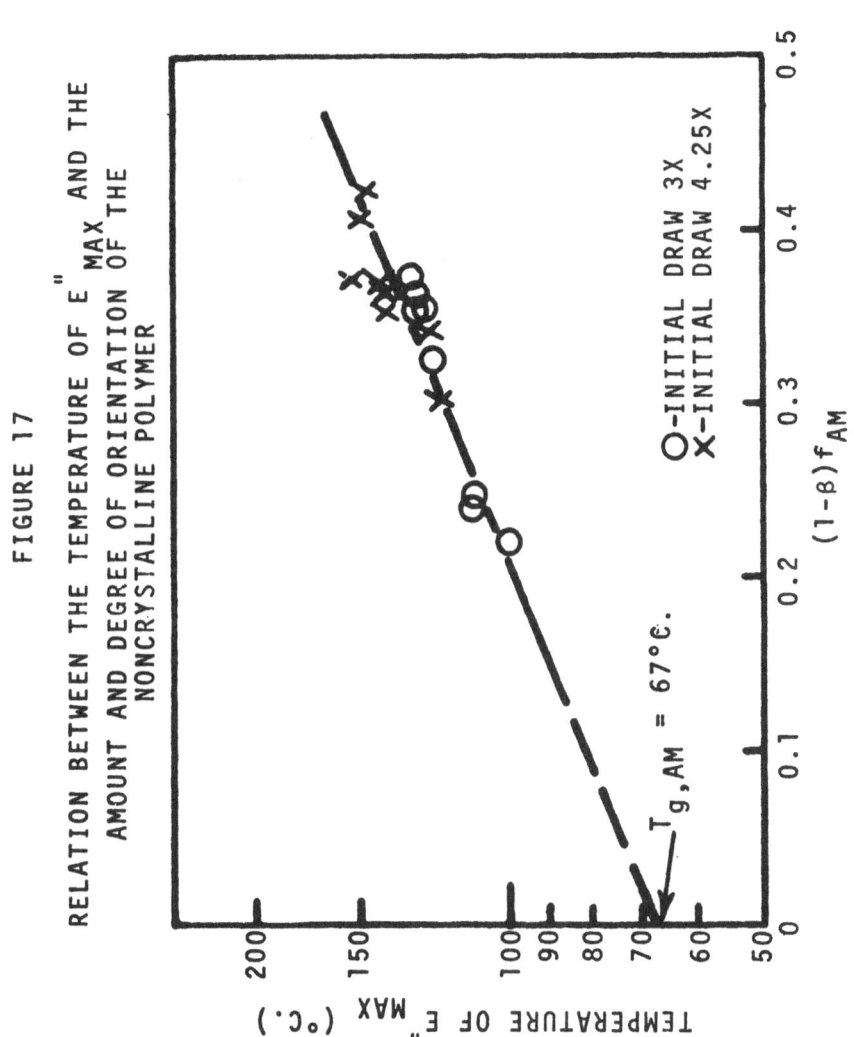

FIGURE 17

RELATION BETWEEN THE TEMPERATURE OF E'' MAX AND THE
AMOUNT AND DEGREE OF ORIENTATION OF THE
NONCRYSTALLINE POLYMER

The fact that the temperature of E''_{MAX} depends on the state of orientation of the noncrystalline polymer is not unexpected. The effect of orientation on the molecular mobility of crystallizable polystyrene [16], nylon 66 [17,18,19], and PET [20,21] has been examined quantitatively by nuclear magnetic resonance, dynamic mechanical, and thermal distortion techniques. These studies all showed that there is an increase in resistance to thermal mobility of the molecules with increasing orientation, and the higher the molecular orientation, the higher the observed glass transition temperature, T_g. What these studies did not show, as they had not quantitatively defined the orientation state of the test samples, was any quantitative correlation between the orientation state and the thermal mobility of the molecules. Figure 17 shows such a quantitative correlation. Here,

$$\log(T_{E''_{MAX}}) = \log T_{g,AM} + K\ (1-\beta)f_{AM} \qquad (14)$$

where K is the slope constant. Thus $T_{E''_{MAX}}$, just as the tenacity, shrinkage, long spacing, and SAXS intensity, can be characterized in terms of structural mechanisms, provided the structural state of the polymer is quantitatively defined.

Tensile Modulus

The tensile modulus of the fiber should be a measure of the slope of the stress-strain curve in the region of the elastic limit (Figure 1). Since for polycrystalline polymers the elastic limit occurs at very low extensions (<0.5%), the observed stress-strain curve, even in its observed lower limit, only approximates elastic behavior. For this reason the measured tensile modulus usually includes some viscoelastic contributions, especially at the slow rates of strain generally used for tensile measurements.

When a high frequency sound pulse is sent along a fiber, the sound wave propagates by molecular compression and extension (i.e., a stress-strain curve of molecular dimensions). The combination of high frequency (in the KHz range) and small displacement of the polymer by the sound wave, makes the sample respond elastically to its propagation. As no structural changes occur from pulse propagation, the observed sonic velocity depends on the structure of the polymer investigated. Since the sonic modulus, E_s, is simply the density times the square of the sonic velocity, the sonic modulus also depends directly on the average structure of the polymer investigated.

The relation between the sonic modulus and the internal structure of a uniaxially oriented polymer is given by the expression [1,4],

$$(3/2) \; \Delta E^{-1} = \beta f_c / E^{\circ}_{t,c} + (1-\beta) f_{AM} / E^{\circ}_{t,AM} \tag{15}$$

$$\text{where } \Delta E^{-1} = E_u^{-1} - E_{or}^{-1} \tag{16}$$

$$\text{and } E_u^{-1} = (2/3) [\beta / E^{\circ}_{t,c} + (1-\beta) / E^{\circ}_{t,AM}] \tag{17}$$

Here E_{or} is the sonic modulus of an oriented sample, E_u is the sonic modulus of an unoriented sample of the same crystallinity as the oriented sample being measured, $E^{\circ}_{t,c}$ and $E^{\circ}_{t,AM}$ are the intrinsic lateral Young's modulus of the crystalline and noncrystalline regions, respectively, and β, f_c, and f_{AM} have the same definitions as given previously.

The intrinsic lateral moduli of the crystalline and noncrystalline regions of PET are 3.68×10^{10} (dyne/sq.cm.) and 1.82×10^{10} (dyne/sq.cm.), respectively [6]. Converting this information to grams per denier units through the equation:

$$E_s \, (g./denier) = 0.1133 \times 10^{-8} E_s \, (dyne/sq.cm.)/density \, (g./cu.cm.) \tag{18}$$

and applying this, together with the structure data in Table 1, to equations (15) and (16) yields E_s (grams/denier) for the oriented PET fibers.

The measured tensile moduli, E_{TM}, and structure calculated sonic moduli, E_s, of the thermally annealed PET fibers are not expected to be identical for the same sample, as a consequence of the extreme strain rate difference between the two measurements, and the fact that some viscoelastic mechanisms influence the tensile modulus. Their trends should be similar, however, if the tensile modulus is structure dominated.

The similar structure dependence of the measured tensile modulus and the predicted sonic modulus (calculated from structure data) can be readily demonstrated for the PET fibers. Equation (15) can be rearranged to yield:

$$1/E_{or} = [(1/E_u) - (2\beta f_c / 3E^{\circ}_{t,c})] - (2/3E^{\circ}_{t,AM})[(1-\beta) f_{AM}] \tag{19}$$

Since the crystallinity and crystallite orientation data in Table 1 are fairly constant, the bracketed term $[(1/E_u) - 2\beta f_c / 3E^{\circ}_{t,c}]$ can be treated as a constant (except for four samples the bracketed term has values between 0.027 and 0.029 with an average value of 0.0277). Thus equation (19) predicts, specifically for the data in Table 1,

that a straight line relation exists between the calculated sonic
compliance $(1/E_{or})$ of the oriented fibers [the compliance is the
recriprocal of the modulus] and $(1-\beta)f_{AM}$. Figure 18 shows a plot
of the sonic compliance against $(1-\beta)f_{AM}$ for all of the samples in
Table 1. The solid line was drawn using the average intercept value
(0.0277) calculated from the structure data in Table 1 and the known
intrinsic sonic moduli, and a slope calculated from the known value
of $E_{t,AM}^{\circ}$. All of the data fit the predicted line except for the
3X drawn yarn shrunk at 100°C. This sample has a different crystallinity than that of the other samples and therefore does not fit
the constant βf_c criteria of equation (19).

If the measured tensile modulus has a structure dependence
similar to that of the calculated sonic modulus, it should be
revealed in a plot of the tensile compliance $(1/E_{TM})$ against $(1-\beta)f_{AM}$.
This is because the tensile modulus data are also for the samples
in Table 1, where the crystallinity and crystallite orientation are
relatively constant. Figure 19 shows a plot of the measured tensile
compliance $(1/E_{TM})$ against $(1-\beta)f_{AM}$ for the samples in Table 1.
Except for the 3X and 5X samples that crimped in oil at 225°C. all
of the data fit the predicted relationship quite satisfactorily.
Of course, the intercept and slope are no longer the structure
predicted parameters, since the tensile modulus data are taken at
a slow speed and have viscoelastic components. The similarity
between the predicted structure dependence of the sonic modulus and
the observed structure dependence of the tensile modulus is striking.
Thus, the measured tensile modulus behaves as predicted by theory,
and like the other properties of PET is dominated by the fabricated
structural state of the samples used in the measurement.

CONCLUSIONS

The purpose of this study was to see if the same quantitative
morphological criteria used to define the structural state of
isotactic polypropylene would lead to similar quantitative structure-
property correlations when applied to PET. Only data available in
the literature were used. These included tenacity, thermal shrinkage
in oil and air, tensile modulus, dynamic loss modulus, long spacing,
and SAXS intensity data, along with structural data for quantitatively describing the structural state of the sample.

When the tenacity, shrinkage, and tensile modulus were examined
in terms of defined structural criteria, the quantitative PET
structure-property correlations had the form predicted by the isotactic polypropylene study. The tenacity correlated with amorphous
orientation; shrinkage correlated directly with amorphous orientation
provided, as required by the earlier study, it was treated as a true
residual extension; and the tensile modulus followed the form
required by the structural state sonic modulus equations, all as

FIGURE 18

THEORETICAL CORRELATION BETWEEN THE SONIC COMPLIANCE

E_{or}^{-1} (CALCULATED FROM β, f_c AND f_{AM}) AND $(1-\beta)f_{AM}$

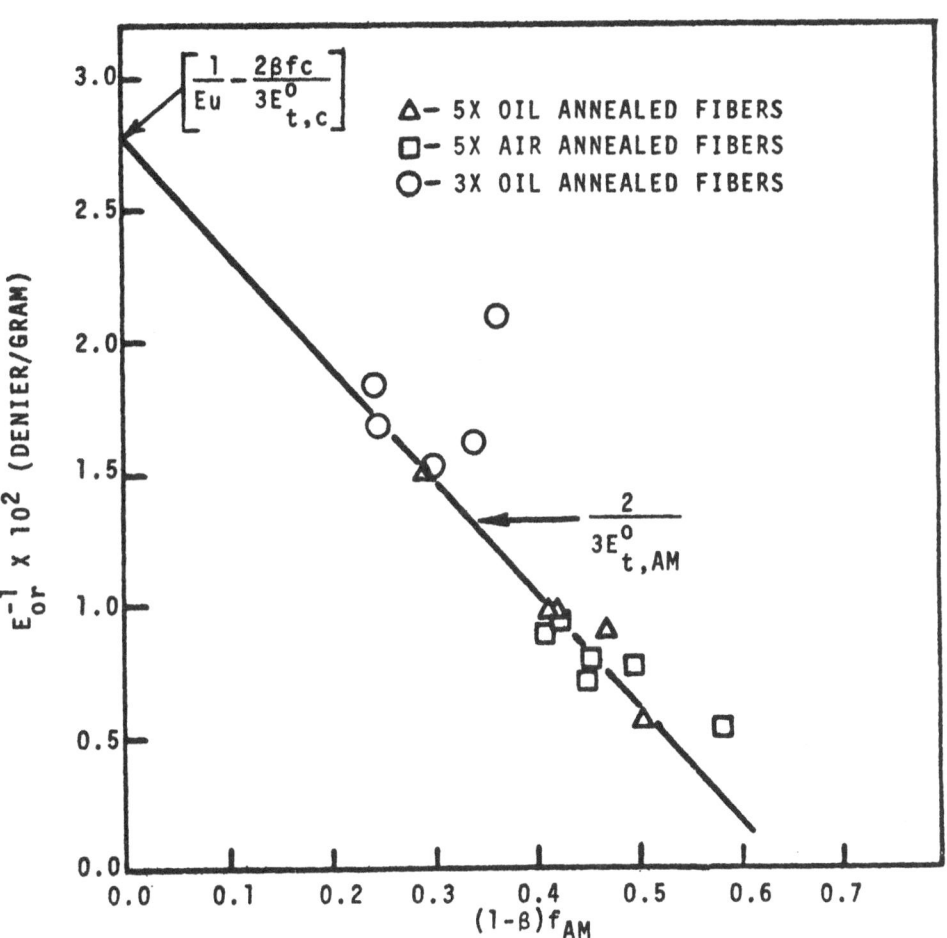

$\left[\dfrac{1}{Eu} - \dfrac{2\beta f c}{3E_{t,c}^0}\right]$

△ – 5X OIL ANNEALED FIBERS
□ – 5X AIR ANNEALED FIBERS
○ – 3X OIL ANNEALED FIBERS

$\dfrac{2}{3E_{t,AM}^0}$

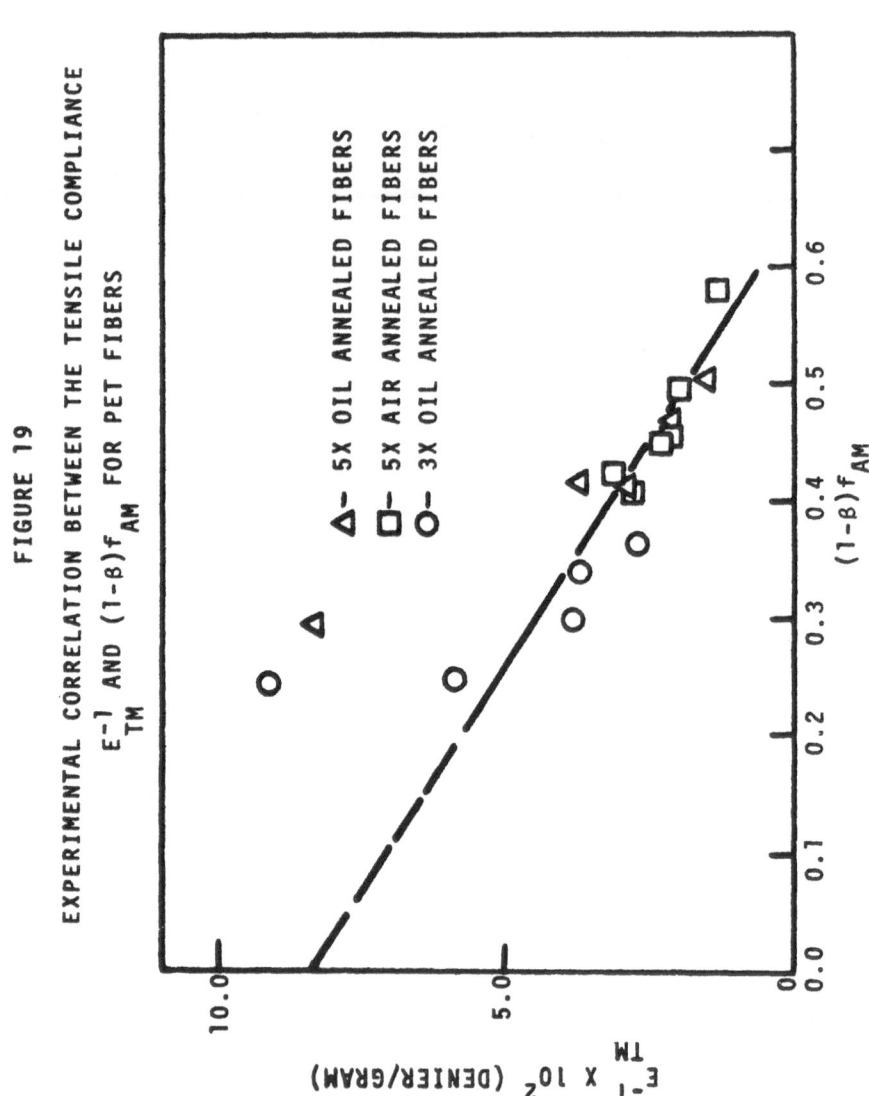

FIGURE 19

EXPERIMENTAL CORRELATION BETWEEN THE TENSILE COMPLIANCE
E_{TM}^{-1} AND $(1-\beta)f_{AM}$ FOR PET FIBERS

△— 5X OIL ANNEALED FIBERS
□— 5X AIR ANNEALED FIBERS
○— 3X OIL ANNEALED FIBERS

predicted by the isotactic polypropylene study. Since there were
no discrepancies between the predicted and observed behavior of
the PET, it can be concluded that the structural state approach to
physical property correlations is a general one and not specific
to isotactic polypropylene.

Structure-property correlations for the dynamic loss modulus,
long spacing, and SAXS intensity had not been examined in the
isotactic polypropylene study. The general validity of the
structural state approach to structure-property correlations was
therefore further reinforced, when quantitative correlations
between these properties and the structural state of PET fibers
were obtained. Thus, the temperature at which the maximum in the
α loss-modulus appears is quantitatively correlated with the amount
and degree of orientation present in the test fibers. Also,
thermal annealing with free shrinkage is shown to be a two-stage
process whereby the fiber shrinks by relaxation of the noncrystalline chains, which then become available for incorporation into the
annealing crystallite. This dependency of crystal growth on
orientation relaxation leads to quantitative correlation between
the increase in long spacing and the disorientation of the noncrystalline chains. Finally, an increase in SAXS intensity with
increased thermal annealing temperature in the free shrink experiments is shown to be due to the disorientation of the noncrystalline
chains. A quantitative correlation is obtained between the SAXS
intensity and the orientation state of the noncrystalline chains
in the PET fibers.

In all of the cases studied, examination of the properties
of the polymer in terms of the structural state of the sample has
led to simplified quantitative correlations. The internal
mechanisms controlling the observed property became obvious, and
served as a guide to further correlations. Fabrication parameters
such as draw ratio, rates of draw, and draw or annealing temperature, though interesting in their historical context, could not
be used as generalizing criteria for property analysis. Only by
considering the internal structure produced by the particular
fabrication process could this be done. Thus the major conclusion
of the study must be that only by considering the structural state
of a polymer and not its fabrication parameters, can general,
simplifying, quantitative, structure-property correlations be
achieved.

REFERENCES

[1]   R. J. Samuels, in Science and Technology of Polymer Films,
      (O. J. Sweeting, Ed), Wiley-Interscience, New York, 1968,
      Chap. 7.

[2]   R. J. Samuels, in Supramolecular Structure in Polymers
      (P. H. Lindenmeyer, Ed.), Wiley-Interscience, New York,
      1967, p. 253.

[3]   R. J. Samuels, J. Polymer Sci., (A-2)6, 2021 (1968).

[4]   R. J. Samuels, J. Polymer Sci., A3, 1741 (1965).

[5]   R. J. Samuels, J. Polymer Sci., (A-2)6, 1101 (1968).

[6]   J. H. Dumbleton, J. Polymer Sci., (A-2)6, 795 (1968).

[7]   R. J. Samuels, J. Polymer Sci., (A-2)7, 1197 (1969).

[8]   R. J. Samuels, J. Macromol. Sci. - Phys., B4, 701 (1970).

[9]   J. H. Dumbleton, T. Murayama, and J. P. Bell,
      Kolloid Z.U.Z. Polymere, 228, 54 (1968).

[10] J. H. Dumbleton, J. Polymer Sci., (A-2)7, 667 (1969).

[11] J. H. Dumbleton, Polymer, 10, 539 (1969).

[12] E. I. Du Pont, British Patent, 1,006,136 (1962).

[13] B. K. Vainshtein, Diffraction of X-Rays by Chain Molecules,
      Elsevier Pub. Co., New York, (1966), p. 383.

[14] N. G. McCrum, B. E. Read, and G. Williams,
      Anelastic and Dielectric Effects in Polymeric Solids,
      John Wiley and Sons, New York, (1967), p. 501-520.

[15] J. Brandrup and E. H. Immergut, Ed., Polymer Handbook,
      Interscience Pub., New York, (1967), p. VI-87.

[16] S. Newman and W. P. Cox. J. Polymer Sci., 46, 29 (1960).

[17] A. M. Thomas Nature, 179, 862 (1957).

[18] P. E. McMahon, Polymer Letters, 4, 43 (1966).

[19] W. O. Statton, in Supramolecular Structure in Polymers,
      (P. H. Lindenmeyer, Ed.), Wiley-Interscience, New York,
      1967, p. 117.

[20] A. R. Thompson and D. W. Woods, Trans. Faraday Soc., 52,
      1383 (1956).

[21] I. M. Ward, Textile Res. J., 31, 650 (1961).

ORIENTATION AND CRYSTAL TRANSFORMATION ASSOCIATED WITH DEFORMA-

TION AND RELAXATION PROCESSES IN POLYBUTENE-1

Tadahiro Asada and Shigeharu Onogi

Department of Polymer Chemistry

Kyoto University, Kyoto, Japan

INTRODUCTION

It is well-known that mechanical properties of a crystalline polymer change with aging time. Besides, crystalline polymers usually show change in types of crystal modification, when they are allowed to stand after preparation. One can easily imagine that the aging effect on mechanical properties is strongly related to their time dependent crystal transformation. However, observing the crystal transformation during stress-strain or stress-relaxation tests is usually very difficult, and hence little significant research on this line has hitherto been published.

Among the various rheo-optical techniques now available, the simultaneous measurement of infrared dichroism with stress and strain was first developed in our laboratory in 1964(1). Since then it has been applied to many systems, including crystalline polymers, their blends, and block copolymers, demonstrating a broad usefulness. This infrared dichroism technique enables us to follow the progress of the crystalline transformation during stress-strain and stress-relaxation measurements. The technique also affords information on molecular and crystallite orientation. Thus, the application of this technique to a crystalline polymer such as polybutene-1 may be regarded as one of the most effective ways of studying the relation between crystal transformation and mechanical properties.

The present paper will discuss the relation between crystal transformation and the variation of mechanical properties with time, taking polybutene-1 as a model substance.

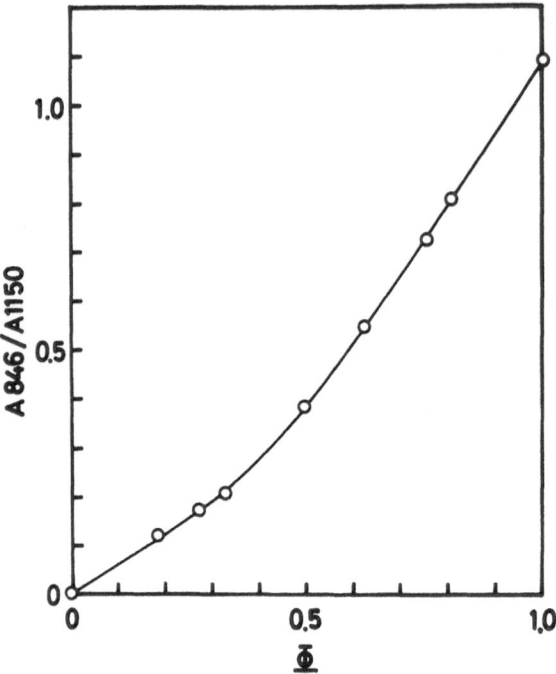

Figure 1.   The ratio of the absorbance A_{846} of the 846 cm^{-1} band to
the absorbance A_{1150} of the 1150 cm^{-1} band against Φ,
the degree of crystal transformation.   (Ref. 12).

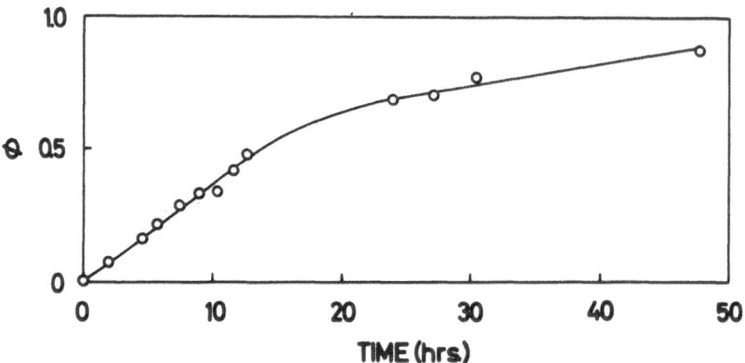

Figure 2.   The variation of Φ with elapsed time, measured from
30 min. after sample preparation.   (Ref. 12).

The tetragonal crystal modification (mod. 2) in polybutene-1 is transformed into the hexagonal crystal modification (mod. 1), when a fresh sample of this material is allowed to stand after preparation (2). Accordingly physical properties related to crystalline structure, such as birefringence, change with time (3,4). Therefore, the variation of the mechanical properties of this material with time can be fully understood only when this transformation is taken into consideration.

EXPERIMENTAL

Material

Commercial pellets of polybutene-1 placed between two aluminum plates were premelted for 3 min. and then pressed for 5 min. at a pressure of 50 kg/cm^2 in a laboratory press equipped with heating plates. The sample was then removed from the press and quenched by plunging it into an ice water bath. The films thus prepared were designated as fresh samples. The thickness of the films was about 70μ. Fresh samples kept in a desiccator for three weeks or more suffered structural change and were designated as aged samples; their infrared spectra and x-ray diffraction patterns were those characteristic to mod. 1 as reported by Natta et al. (5,6).

Measurements

The crystal transformation for mod. 2 to mod. 1 in polybutene-1 can be detected by several techniques such as x-ray diffraction (2,7), infrared absorption (8,9), differential thermal analysis (2,10), and dilatometry (2,11). Among these methods, the infrared absorption method enables us to measure the transformation most quickly.

Natta et al. (6) and Boor et al. (2) have already elucidated the infrared spectra corresponding to the two modifications of polybutene-1. From these spectra it can be seen that the bands at 10.81, 11.78, and 12.25μ (925, 849, and 816 cm^{-1}) are associated with mod. 1, while another band at 11.03μ (906 cm^{-1}) is associated with mod. 2.

In our measurements at no longer than 30 min. after film preparation, the fresh samples exhibited the x-ray diffraction intensity curve and infrared spectra for mod. 2. The aged samples, on the other hand, gave those for mod. 1. From our experimental results, the band at 1025 cm^{-1} was also shown to be associated with mod. 1.

In order to obtain the relation between the content of mod. 1 crystals and the absorbance A_{846} of the 846 cm^{-1} band, which is associated with mod. 1, the ratio of A_{846} to the absorbance A_{1150} of the 1150 cm^{-1} band, used as a measure of thickness, was determined for laminated films of fresh and completely aged samples of different thicknesses. The ratio A_{846}/A_{1150} is plotted against Φ, the degree of crystal transformation, in Fig. 1. The degree of crystal transformation Φ is here defined as the weight fraction of crystals changed from mod. 2 to 1, and was assumed to be equal to the weight fraction multiplied by the degree of crystallinity of the aged film in the laminates. Such a curve of Fig. 1 provided the standard for infrared measurements of Φ.

The variation of the infrared dichroic ratio for some key bands characteristic of polybutene-1 was measured on the film during stress-strain tests at a constant rate of elongation, 1.64%/min., at 20°C. The apparatus employed is the combination of an infrared spectrometer and a self-recording tensile tester, similar in principle to the Instron Tensile Tester. The details of the apparatus and measuring techniques have been described elsewhere (1). The key bands employed were those at 760, 905, and 1025 cm^{-1}.

The instrument used for measurement of birefringence and stress relaxation was a combination of an Instron-type tensile tester (Tensilon UTM-III), and an optical system for measuring the birefringence continuously by the intensity method. The details of the instrument were described in a previous paper (13). In order to carry out the measurements at constant temperature, an oven equipped with a thermostat was devised.

The stress relaxation measurements were carried out with film specimens 5 cm in length and 1.5 cm in width under an initial strain of about 2% resulting from fast stretching at a rate of 400%/min. in the temperature range from 14 to 80°C.

RESULTS AND DISCUSSION

Crystal Transformation in Unstrained Samples

The crystal transformation in a fresh sample at 20°C begins at about 30 min. after film preparation. The variation of Φ with elapsed time is shown in Fig. 2. The elapsed time in this figure is measured from 30 min. after the preparation of the film. Φ increases with increasing time and reaches about 0.9 at 50 hours and 1.0 at about 3 weeks after sample preparation.

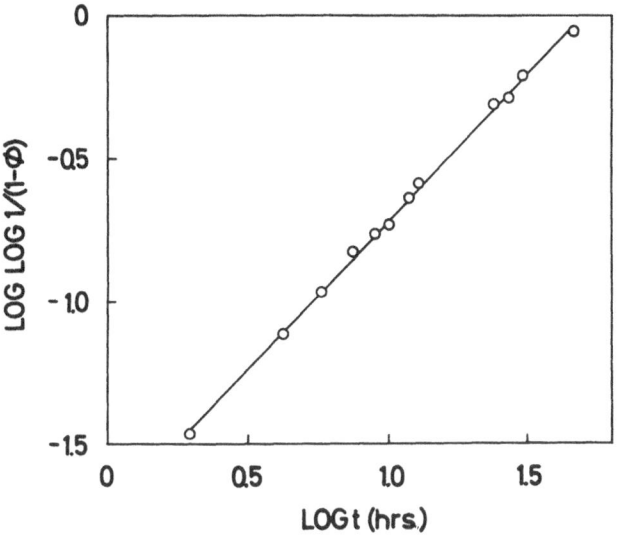

Figure 3. Log (log 1/1-Φ) vs. log (elapsed time). (Ref. 12).

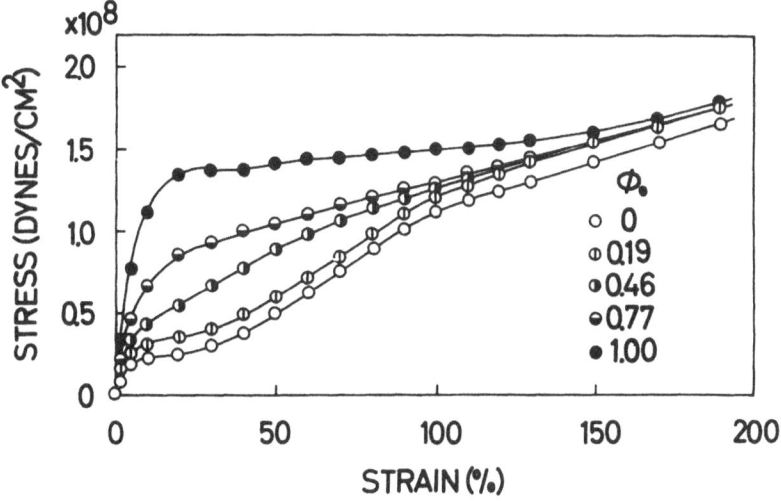

Figure 4. Stress-strain curves for film specimens having different
           initial degrees of crystal transformation. (Ref. 12).

To clarify the time effect upon the crystal transformation, log (log 1/1-ϕ) is plotted against the logarithm of elapsed time t in Fig. 3. The plot gives a straight line having a slope of unity. We therefore obtain the following relation between ϕ and t:

$$\log \frac{1}{1-\phi} = \frac{1}{2.303} kt \tag{1}$$

It is noteworthy that this equation is quite similar to that for one-dimensional crystallization rate (13). The rate of crystal-crystal transformation is thus described by an equation of the same form as that for crystallization itself. The same relation between ϕ and t is also applicable to the relation between the refractive index and t, though the details are omitted here (12).

Stress-Strain Behavior and Crystal Transformation

As demonstrated above, a fresh sample of polybutene-1 suffers crystal transformation even when it is not stretched. It has also been known that the crystal transformation is accelerated by stretching the sample. However, the relation between the crystal transformation and the mechanical behavior has not been fully investigated hitherto. We therefore measured the mechanical and orientation behavior at the same time by the use of rheo-optical techniques (12).

Stress-strain curves for film specimens having different initial degrees of crystal transformation ϕ_0 were obtained under a constant rate of elongation (1.64%/min.), and are shown in Fig. 4. Except for the case of $\phi_0 = 1$, no macroscopic necking was observed during the tests. For the sample with $\phi_0 = 1.0$, necking was observed at the strain of 30%. The curves in this figure differ very much at low strains, but they draw closer at higher strains. The lower is ϕ_0, the lower is the yield stress. In other words, aged samples need much energy to deform up to the yield point than do fresh ones. This difference below the yield point is due to the difference between fresh and aged samples in resistance to the disintegration of spherulites. On the other hand, the rate of increase in the stress beyond the yield point is lower for the aged samples than for the fresh ones.

It was surmised that the great differences in stress and rate of increase in stress beyond the yield point were associated with the crystal transformation. Therefore the variation of the degree of crystal transformation with strain was measured simultaneously with the stress-strain behavior. The results are shown in Fig. 5. As is evident in this figure, ϕ for the fresh and incompletely aged samples increases with increasing strain. The lower the initial degree of transformation ϕ_0, the greater the variation of ϕ

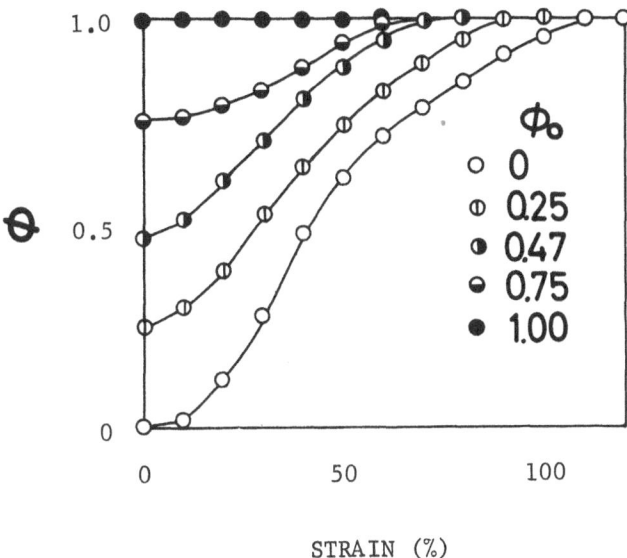

STRAIN (%)

Figure 5.   The variation of the degree of crystal transformation
            with strain during stress-strain measurements.   (Ref. 12).

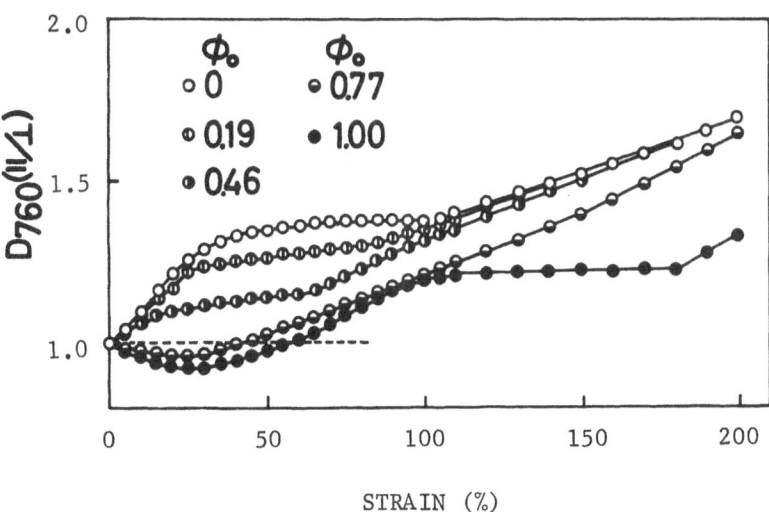

STRAIN (%)

Figure 6.   The variation of the dichroic ratio D_{760} ($\parallel/\perp$) of the
            760 cm^{-1} band with strain, determined during stress-
            strain measurements.   (Ref. 12).

with strain. It follows from these results that one must pay at-
tention to the crystal transformation phenomenon when interpreting
the stress-strain curves of polybutene-1. Φ for the fresh sample
($\Phi_0 = 0$) increases slowly up to about 10% elongation, but then
very rapidly up to about 50% elongation, and reaches its ceiling
value, i.e. 1, at about 105% elongation. Similar behavior is observed for the other samples, excepting the fully aged one. The
higher the Φ_0 value, the smaller the strain at which Φ reaches the
ceiling value.

Next, the dichroic ratio D_{760} ($||/\perp$) of the 760 cm^{-1} band was
determined during stress-strain measurements; it is plotted against
strain in Fig. 6. Since the band at 760 cm^{-1} is associated with
the crystallite orientation of both modifications 1 and 2, D_{760}
can serve as a measure of total crystallite orientation. As is
seen from the figure, D_{760} for the fresh sample ($\Phi = 0$) increases
with increasing strain up to about 40% strain, remains almost constant from 40% to 105% strain, and then increases rapidly again at
larger strains. It is noteworthy that the strain of 105% coincides
with the strain at which Φ reaches its ceiling value (cf. Fig. 5).
For the sample with $\Phi_0 = 0.46$, D_{760} increases with increasing
strain up to about 20% and exhibits a plateau extending to about
60% strain. The height of this plateau is lower than that for the
fresh sample with $\Phi_0 = 0$.

On the other hand, for the sample having $\Phi_0 = 1.0$ and 0.77,
D_{760} decreases at first and then increases after passing through a
minimum at about 30% strain. The plateau observed for the aged
sample between 110% and 180% strain is only an apparent plateau,
due to partial necking. It is noteworthy that the strain at which
D_{760} takes its minimum coincides with the strain at which macroscopic necking starts (compare Fig. 4 with Fig. 6).

The following two propositions may be considered as reasons
why a plateau is observed in the D_{760} vs. strain curves for partially aged samples: (i) the orientation of crystallites, both in
mod. 1 and mod. 2, does not change during deformation in this region; or (ii) the orientation of mod. 1 crystals increases while
that of mod. 2 crystals decreases, or vice versa, so that the total
crystallite orientation does not change. In order to determine
which is the true explanation, the dichroic ratio D_{1025} ($||/\perp$) for
the 1025 cm^{-1} band was measured as a function of strain and is
shown in Fig. 7. Since the band at 1025 cm^{-1} is a parallel band
and is associated only with mod. 1 crystals, the dichroic ratio
D_{1025} ($||/\perp$) is directly related to the degree of orientation of
mod. 1 crystals. The variations of D_{1025} with strain for the
samples having $\Phi_0 = 1.0$ and 0.7 are similar to those of D_{760}, with
D_{1025} taking a minimum at about 30% strain. Similarly, the variation of D_{1025} with strain for the fresh and partially aged samples

is very similar to that of D_{760}, though D_{1025} could not be measured accurately at the smaller strains. The conclusion is that the plateau observed in the D vs. strain curves for the fresh and other samples is due to the fact that the orientation of crystals in both mod. 1 and mod. 2 does not change with strain in the inter-mediate region.

These results suggest that the orientation of mod. 2 crystals predominates at small strains, and that the oriented crystals show a higher degree of orientation or higher D_{1025} even after they have changed into mod. 1. This idea is supported also by the measurement of the dichroic ratio D_{905}, the measure of the orien-tation of mod. 2 crystals (12).

To summarize the above discussion, the stress-strain behavior of the fresh sample, for example, can be divided into the following three processes: (i) $0 \leq$ strain $\gamma \leq$ 30%: Both orientation and transformation of crystals take place. The orientation of mod. 2 crystals proceeds, and crystals change into mod. 1.

(ii) $30 \lesssim \gamma \lesssim 105$%: Crystal transformation occurs predominantly, and the orientation remains almost constant. The crystal trans-formation results in more extended chain conformation, since the molecular chains are more extended in mod. 1 than in mod. 2 crys-tals. Therefore, the orientation due to external straining com-pensates for the disorientation due to self-induced extension during the crystal transformation from mod. 2 to mod. 1.

(iii) $\gamma \geq 105$%: Crystal transformation is complete, and orienta-tion increases again.

On the other hand, the deformation process in the aged sample is very different from that in the fresh sample. That is to say, the crystallite orientation initially decreases with strain. The strain at which the dichroic ratio takes its minimum coincides with the strain at the yield point. Such behavior of aged poly-butene-1 is very similar to that of polypropylene described else-where (14). Below the yield point, spherulites or superstructure consisting of lamellae show elastic deformation, and the lamella axes orient in the direction of elongation. At the yield point, the spherulites disintegrate but show great resistance, or a high yield value. Beyond the yield point, the disintegrated spherulites or superstructure show irrecoverable deformation due to the un-folding of folded chains, which orient in the direction of stretch-ing as the strain increases.

On the other hand, the spherulites or superstructures of the fresh sample consist of crystallites in mod. 2 and are easily dis-integrated; the yield value is very low. Moreover, unfolding of folded chains also occurs very easily.

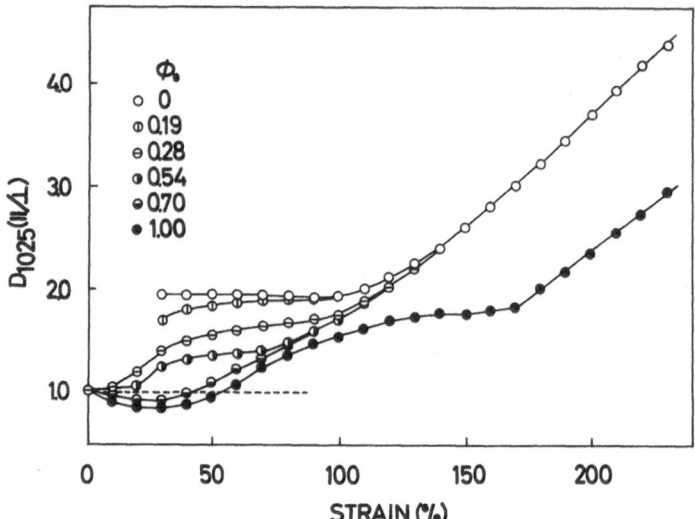

Figure 7. The variation of the dichroic ratio D_{1025} ($\parallel/\perp$) for the 1025 cm^{-1} band with strain during stress-strain measurements. (Ref. 12).

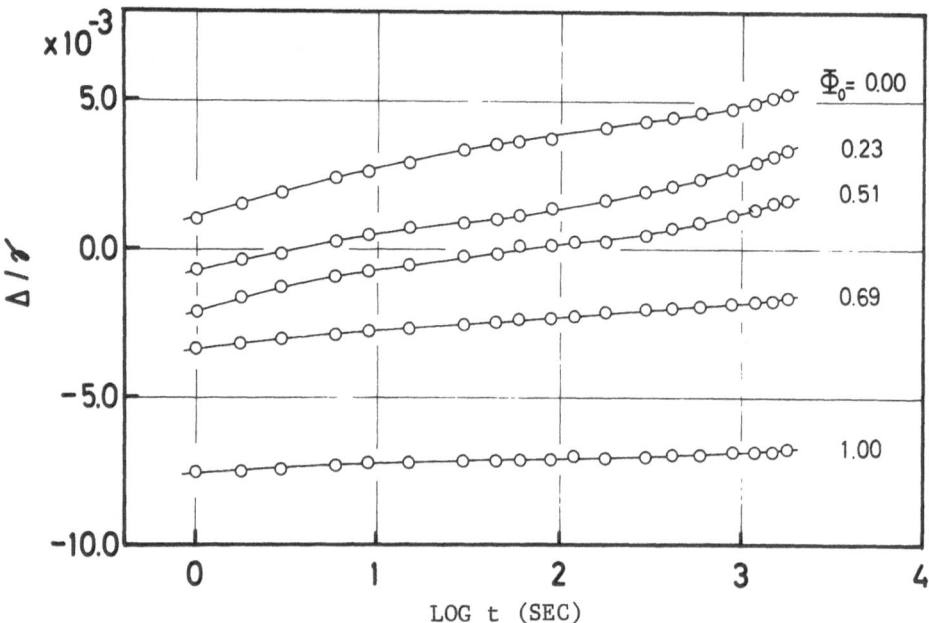

Figure 8. The variation of strain-optical coefficient with time for film specimens having different initial degree of crystal transformation Φ_0. (Ref. 18).

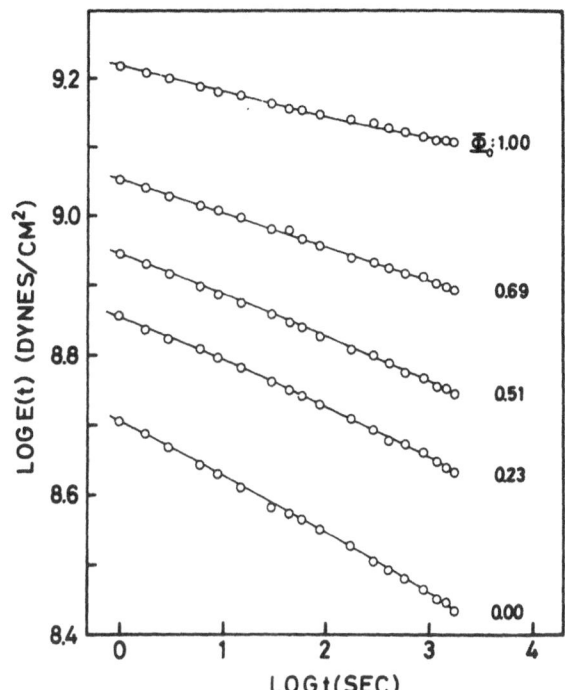

Figure 9.   The variation of relaxation modulus with time for film
            specimens having different initial degree of crystal
            transformation Φ_0.   (Ref. 18).

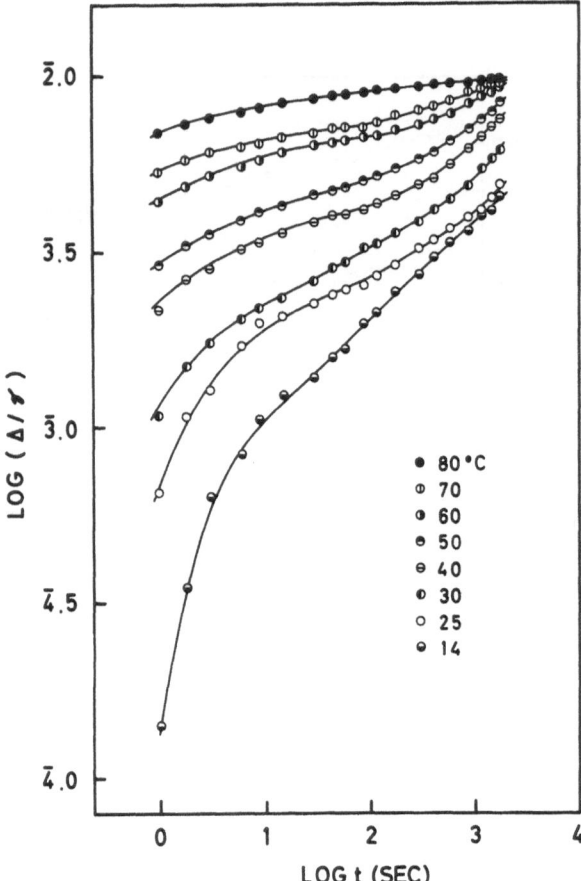

Figure 10.   The variation of strain-optical coefficient with time
             for fresh sample at various temperatures.   (Ref. 18).

The deformation process in samples having Φ_0 between 1 and 0 is intermediate between these two extreme cases.

Stress and Optical Relaxation and Crystal Transformation

The change in the birefringence with time was recorded simultaneously with stress relaxation. The results for the strain-optical coefficient Δ/γ (birefringence/strain) and the relaxation modulus for the samples of various degrees of Φ_0 are shown in Fig. 8 and Fig. 9. The values of the strain-optical coefficient are changing from plus value to minus value with increasing Φ_0. The Δ/γ are increasing with time and the rate of increase is greater for smaller Φ_0 samples. And also it can be seen on the curve for lower Φ_0 samples that there is an inflection point, in striking contrast to the case of other crystalline polymers such as polyethylene. The absolute value of the relaxation modulus for the fresh sample is smaller than that for aged one, but the rate of decrease of the modulus for the fresh sample is much greater than that for the aged one. These results may be related to the crystal transformation.

For the fresh sample, simultaneous measurements of birefringence and stress during stress relaxation at various temperatures ranging from 14∿80°C have been carried out. The observation was started after the sample was maintained for 20 min. at the definite temperature (it was mounted in the oven 10 min. after the preparation). The results are shown in Fig. 10. Δ/γ is larger, the higher the temperature is. The time dependency is quite unique, showing no equilibrium value in striking contrast to the case of other crystalline polymers. Especially below 70°C, these curves show inflection at a time between 20∿100 sec. and after that time the curves are rapidly increasing. It seems that the time-temperature superposition is applicable to the curves before the inflection points. The composite curve at 25°C is shown in Fig. 11. As is seen from the figure, the superposition is good in the shorter time region from the inflection point, and the composite curve is similar to that for polyethylene (16). Not only horizontal but also vertical shift is necessary to obtain the composite curve. There are parts of the curves which deviate from the composite curve. The fact that there are parts which deviate from the composite curve appears to originate mainly from the crystal transformation during the relaxation process. The unsuperposable portion $(\Delta/\gamma)_s$ of the strain-optical coefficient for each curve is plotted against time and is shown in Fig. 12. The subtracted strain optical coefficient $(\Delta/\gamma)_s$ at each temperature changes with time almost linearly, except in the case of 14 and 70°C.

The degree of crystal transformation (Φ) under the same

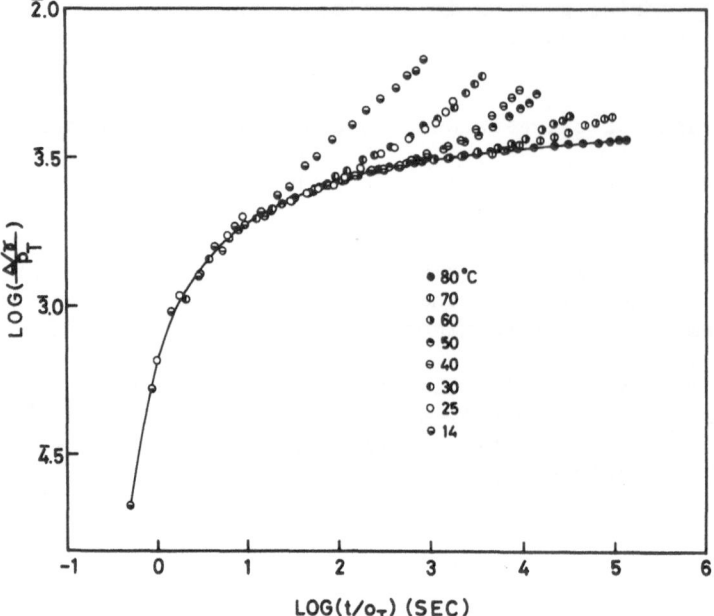

Figure 11.   Composite curve of strain-optical coefficient for
             fresh sample at 25°C.   (Ref. 18).

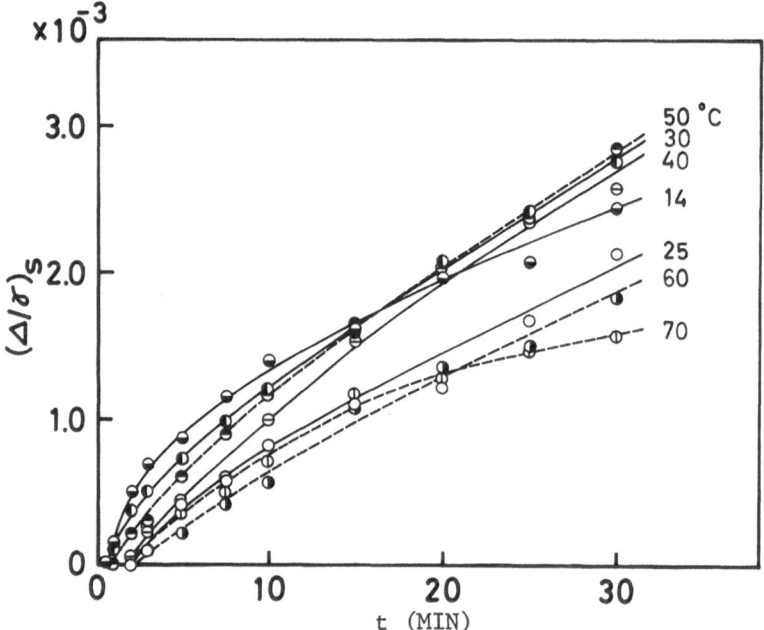

Figure 12.   Unsuperposable part of the strain-optical coefficient
             $(\Delta/\gamma)_s$ against time.   (Ref. 18).

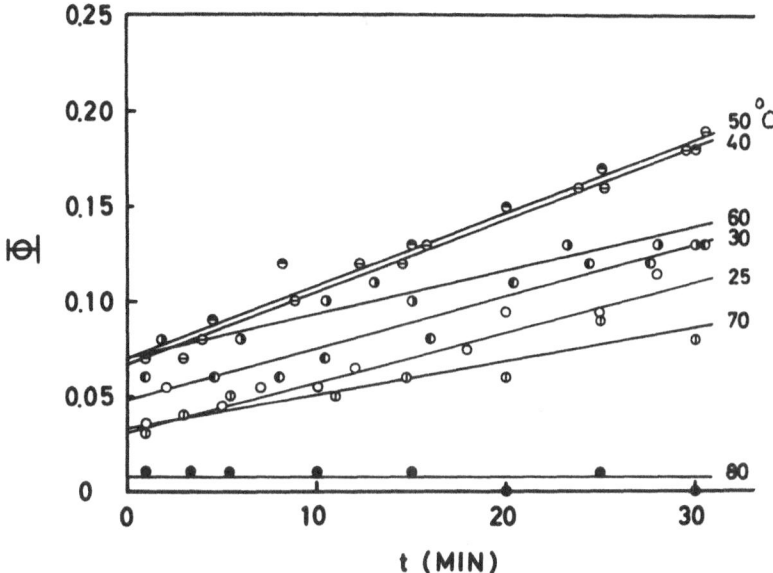

Figure 13.   The variation of the degree of crystal transformation
with time during stress relaxation at various temperature.   (Ref. 18).

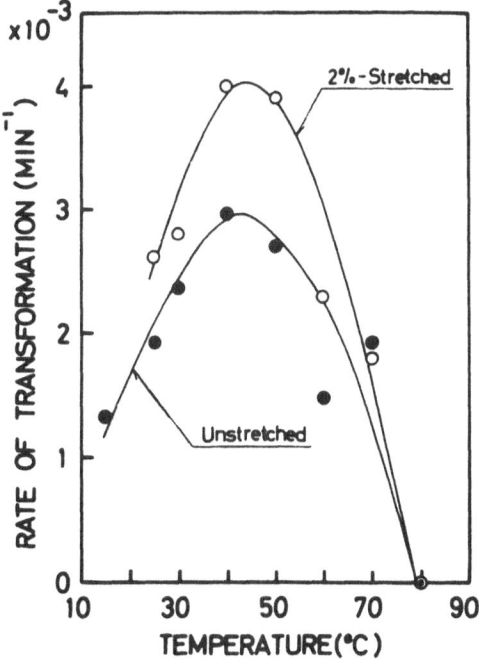

Figure 14.   The variation of rate of crystal transformation with
temperature.   (Ref. 18).

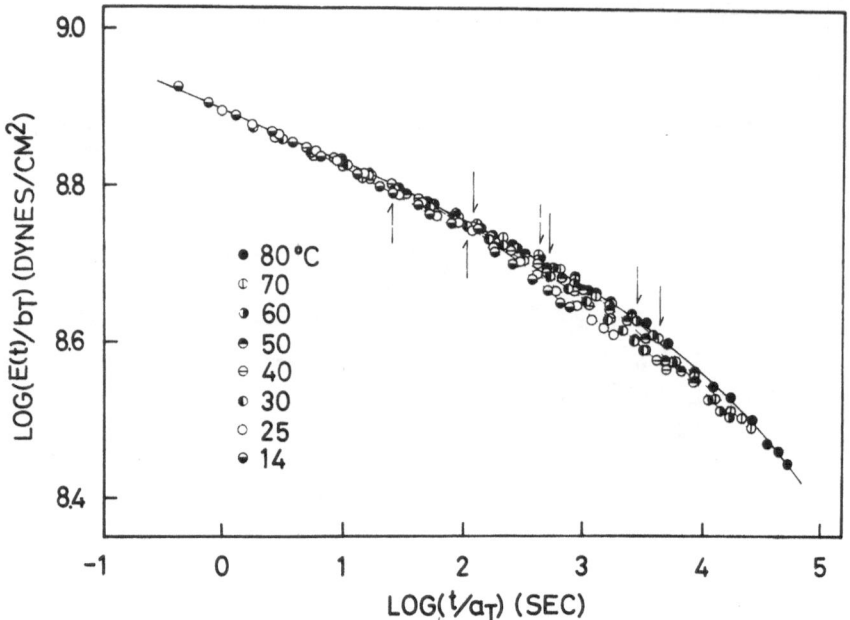

Figure 15.   Composite curve of relaxation modulus for fresh sample
              at 25°C.   (Ref. 18).

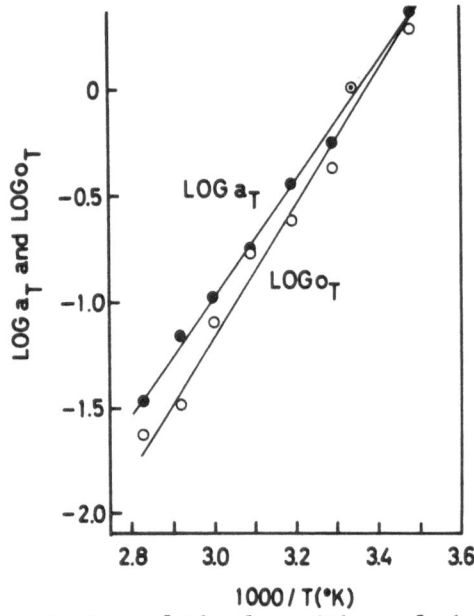

Figure 16.   The variation of the logarithms of the horizontal shift
              factors (log o_T, log a_T) against reciprocal absolute
              temperature.   (Ref. 18).

condition was obtained by the infrared method, and the results are
shown in Fig. 13. Φ changes with time linearly. The rate of crystal transformation obtained from the slope is plotted against temperature and shown in Fig. 14. It is seen from the figure that
the rate of transformation takes it maximum between 40 and 50°C.
These facts tell us that the subtracted strain-optical coefficient
$(\Delta/\gamma)_s$ corresponds to the degree of crystal transformation Φ. It
is obvious from Fig. 13 that Φ at 80°C does not change with time,
corresponding well to the fact that the Δ/γ vs. t curve at 80°C
shows no deviation from the composite curve. This also implies
that the composite curve in Fig. 11 is the one for the polybutene-1
sample which contains mod. 2 crystals only.

The trial of the time-temperature superposition for the relaxation modulus is shown in Fig. 15. In this case also, both
horizontal and vertical shifts are necessary. The arrows in the
figure show the points where the deviations of the strain-optical
coefficient from the composite curve take place. As is seen from
the figure, in the case of relaxation modulus, the deviations
from the composite curve also occur at the same time that the
strain-optical coefficient deviations occur. And it can be concluded that the crystal transformations accelerate the stress relaxation. (The deviations are downward.)

When the logarithms of the horizontal shift factors (log o_T,
log a_T) needed for obtaining the composite curve in Fig. 11 and
Fig. 15 are plotted against reciprocal absolute temperature (1/T),
straight line relations are obtained between them (Fig. 16). The
apparent activation energies obtained from the slopes are 14.3
and 12.7 kcal/mole, respectively. These values are considered to
be the activation energy which is required to change the orientation of crystals in the polybutene-1 film containing only mod. 2
crystals.

According to A. T. Jones (17), the transformation from mod. 2
to mod. 1 causes the extension of the monomer unit in the direction
parallel to c-axis and the shrinking in the transverse direction.
So, the increase of birefringence given by the crystal transformation should be due to the increase of the intrinsic birefringence
which is given by the chain extension of the oriented crystals.
And the stress-relaxation given by the crystal transformation
should be due to the crystal transformation of those oriented crystals whose c-axes are parallel to the stretching direction.

The crystal orientation during the relaxation is also observed
by x-ray method, though the details are omitted here (18). The
observations also support the conclusion that the crystal transformation under stretching occurs predominantly within the crystals
whose c-axis is parallel to the stretching direction. And while

the orientation of the transformed crystal (mod. 1) does not depend so much on temperature, still the rate of crystal transformation is much dependent on temperature.

REFERENCES

1. S. Onogi, H. Kawai, and T. Asada, Kobunshi Kagaku, $\underline{21}$, 746 (1964).
2. J. Boor, Jr. and J. C. Mitchell, J. Polym. Sci., Part A. $\underline{1}$, 59 (1963).
3. R. Yang and R. S. Stein, J. Polym. Sci., Part A-2, $\underline{5}$, 939 (1967).
4. R. Y. Yee and R. S. Stein, J. Polym. Sci., Part A-2, $\underline{8}$, 1661 (1970).
5. G. Natta, P. Corradini, and I. W. Bassi, Makromol. Chem., $\underline{21}$, 240 (1956).
6. G. Natta, P. Corradini, and I. W. Bassi, Nuovo Cimento Suppl., $\underline{15}$, 52 (1960).
7. F. Sakaguchi, Sen-i to Kogyo, $\underline{1}$, 402 (1968).
8. J. P. Luongo and R. Salovey, J. Polym. Sci., Part A-2, $\underline{4}$, 997 (1966).
9. G. Goldbach and G. Peitsher, J. Polym. Sci., Part B, $\underline{6}$, 783 (1968).
10. I. D. Rubin, J. Polym. Sci., Part A, $\underline{3}$, 3803 (1965).
11. J. Powers, J. D. Hoffmann, J. J. Weeks, and F. A. Quinn, Jr., J. Res. NBS, $\underline{69A}$, 335 (1965).
12. T. Asada, J. Sasada and S. Onogi, Polymer J., Japan, in press.
13. S. Onogi, T. Asada, Y. Fukui, and T. Fujisawa, J. Soc. Materials Sci. Japan, $\underline{15}$, 389 (1966).
14. I. D. Lauritzen and J. D. Hoffmann, J. Res. NBS, $\underline{64A}$, 73 (1960).
15. S. Onogi, T. Asada, and T. Takaki, Zairyo, $\underline{16}$, 746 (1967).
16. Y. Fukui, T. Sato, M. Ushirokawa, T. Asada, and S. Onogi, J. Polym. Sci., Part A-2, $\underline{8}$, 1195 (1970).
17. A. T. Jones, Polymer, $\underline{7}$, 23 (1966).
18. N. Sugimoto, T. Asada, and S. Onogi, unpublished data.

MICROFIBRILLAR STRUCTURE, RADICAL FORMATION, AND FRACTURE OF HIGHLY DRAWN CRYSTALLINE POLYMERS

A. Peterlin

Research Triangle Institute
Post Office Box 12194
Research Triangle Park, N. C.   27709

SUMMARY

The highly drawn crystalline polymer with the almost perfectly oriented fibrous structure differs from the unoriented more or less spherulitic material not only in the orientation of crystal lattice and crystal lamellae but also in the basic structural elements: stacks of parallel lamellae in the latter and well-aligned microfibrils in the former case. This difference explains very well not only the anisotropy of mechanical properties but also the superior elastic modulus and strength of fibrous material. The extremely long and thin macrofibril consists of folded chain crystal blocks connected axially with a great many tie molecules. At the ends of microfibrils the axial molecular connection in the fiber is interrupted. Under tensile load such point vacancies yield microcracks -- about 10^{15} per cm^3 -- thus concentrating the stress on adjacent microfibrils. Depending on the ratio of microfibril strength and autoadhesive forces between adjacent microfibrils the coalescence and growth of microcracks occurs by axial (longitudinal) cracks along the boundary between the microfibrils or by radial (transverse) cracks through adjacent microfibrils. In the former case, very few and in the latter case a great many tie molecules are ruptured producing radicals detectable by ESR. Polyethylene and polypropylene with weak Van der Waals forces between adjacent microfibrils yield few and nylon 6 and 66 with stronger hydrogen bridges a great many radicals in excellent suppport of the microfibrillar model of fibrous structure.

INTRODUCTION

Drawing of crystalline polymers transforms the more or less spherulitic structure of the unoriented material obtained by crystallization from unstrained melt into the highly oriented fibrous structure. The transformation is either highly localized in the neck propagating through the sample (cold drawing) or occurs more or less uniformly all over the sample (hot drawing without necking). The general morphology of the drawn material seems to be little affected by this difference. In both cases the chains in the crystal lattice are oriented parallel and the amorphous layers between subsequent crystals are more or less perpendicular to the draw direction as amply demonstrated by wide- and small-angle X-ray scattering. The crystallinity and the long period are of the same magnitude as in the spherulitic state.

As a consequence of the high degree of orientation, the fibrous material exhibits a conspicuous anisotropy of mechanical properties.[1,2] The elastic modulus and the tensile strength in the draw direction may be ten times larger or even more than the corresponding values of the unoriented spherulitic material. A smaller increase is also observable in the perpendicular direction. These changes are very often considered as caused by the chain orientation[3] although it is a little difficult by this effect to explain an increase of elastic modulus in all directions above that of the unoriented material. The best argument against the chain orientation as the source of mechanical properties of fibrous structure can be found in the properties of well-oriented poly-ethylene single crystal mats obtained by precipitation of crystals grown from dilute solutions. Such a mat has a fairly good orientation of crystal lattice and crystal lamellae well comparable with a moderately drawn material. The mechanical properties, however, exhibit the maximum possible contrast. The oriented mat has a low elastic modulus and strength. It is a brittle, mechanically useless material, much worse than spherulitic polyethylene. If one anneals a drawn sample and an oriented single crystal mat at gradually increasing temperature, one finds a deterioration of mechanical properties in the former and an improvement in the latter case although the orientation remains practically unchanged.[4] One must conclude that although chain orientation is the most conspicuous molecular attribute of drawn material and easily detectable by X-ray scattering, birefringence, or dichroism it is not the cause of the observed increase of elastic modulus and strength produced by drawing.

The small change of crystallinity[5] and crystal dimensions[6] as reflected in density and long period, respectively, excludes also these two factors as explanation of the drastic change of mechanical properties as a result of transformation from spherulitic to

fibrous structure. One has indeed to consider in more detail the
supercrystalline morphology of both structures and the molecular
connections of the basic elements, folded chain lamellae in the
former and microfibrils in the latter case, if one wishes to understand and explain the observed strain hardening.

The unoriented material obtained by crystallization of relaxed
melt contains spherulites. They are large and well developed if
only few primary nuclei were formed in the melt. With increasing
number of such nuclei the spherulites become smaller and may be
soon confined to the embryonic regions, i.e., to stacks of parallel
lamellae, each grown at one primary nucleus (Fig. 1). Similar
stacks of parallel lamellae are also the basic element of the well-
developed spherulites. The lamellae of such a stack are connected
by some tie molecules originating from molecules which during
solidification started to be included into the crystal lattice of
two different lamellae so that the intervening section of the chain
remained prevented from crystallization. The number of such tie
molecules increases with molecular weight and crystal growth rate.

But for the tie molecules there is no molecular connection
between adjacent lamellae through the "amorphous" surface layer
mainly containing chain folds, free chain ends, and some rejected
impurities. Therefore, above the glass transition temperature of
the amorphous component the amorphous layer between the lamellae
acts as a layer of lubricant permitting shear gliding of lamellae
at relatively low stresses. A substantial blocking of such deformation is caused by the helical twist of lamellae in polyethylene
spherulites. On the other hand, the radial structure of spherulites and the parallel packing of lamellae favors a force transmission along the crystals without a high stress load on the

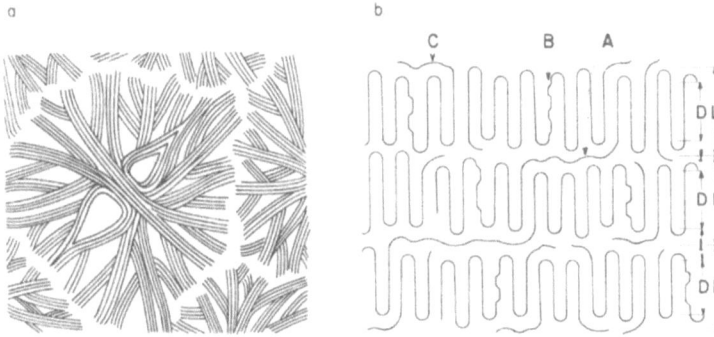

Fig. 1 - Model of (a) spherulitic structure and (b) stack of par-
        allel lamellae of the spherulite showing (A) tie molecules,
        (B) crystal defects, and (C) free chain ends. The long
        period is L, the crystal core thickness is D, and the
        thickness of the amorphous layer is ℓ.

amorphous layers with exception of those perpendicular to the principal tensile or compression stress. As a rule, under a tensile
load the mechanical breakdown of such layers through the center of
the spherulites or along spherulite radii produces well observable
microcracks. The outer boundaries of stacks of parallel lamellae
and the boundaries between adjacent spherulites are particularly
weak areas of the unoriented polymer solid containing impurities
rejected during crystallization and exhibiting even a smaller amount
of molecular connection by tie molecules between close by lamellae
than between adjacent lamellae of the same stack. Hence a well-
developed spherulitic structure as obtained by slow crystallization
at small supercooling is rather brittle with early crack formation
and poor drawing ability.

The fibrous structure obtained by drawing seems to consists
of parallel lamellae oriented more or less perpendicular to the
draw direction. Such a structure explains well the small- and
wide-angle X-ray scattering data. The electron microscopy study
of the morphological changes during deformation and of the final
structure, however, has very convincingly demonstrated that the
lamellae are more an artifact than a true basic element of highly
drawn material. They are formed by good lateral fit of crystal
blocks of adjacent microfibrils. But almost all the material
connection by tie molecules among blocks is in axial direction
within the microfibril and practically none in perpendicular
direction between adjacent blocks of the same lamella.

MICROFIBRILLAR MODEL OF FIBROUS STRUCTURE

Such a model, a combination of folded chain and fringed
micelle concept of microfibrils as the basic element of fiber
structure (Fig. 2) was developed by Peterlin[7-9] on the basis of
deformational studies of single crystals,[10-13] thin layers,[14] and
bulk samples[15-17] of PE. The model was corroborated by electron
microscopy[18] and X-ray diffraction studies[19] of drawn polypropylene,
polyoxymethylene,[20] and nylon 6.[21] It provides a straightforward
explanation of the enormous amount of experimental morphological
and mechanical data on highly drawn films and fibers of crystalline
polymers.[22] In that which follows, a short sketch of the experimental background of the model will be given in order to make
understable its claims and details. A more extensive description
is given in Ref. 9.

During plastic deformation the stacks of parallel lamellae of
the unoriented more or less spherulitic starting material obtained
by crystallization from the melt rotate and slide into the direction of maximum plastic compliance of lamellae. Concurrently the
chains in the lamellae tilt and slip until the lamellae crack.
With favorable crack orientation one obtains a great many micro-

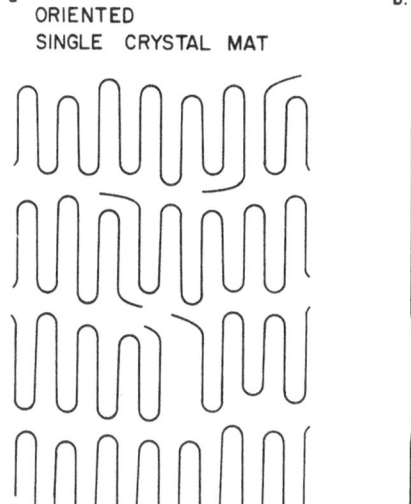
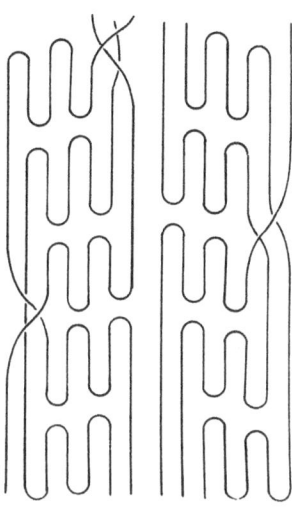

Fig. 2 - Model of (a) highly oriented single crystal mat and (b)
         fibrous structure. In both cases the crystal lattice is
         perfectly oriented.

fibrils bridging the crack. Their lateral dimensions are between
100 and 300Å. Two micronecks at the opposite sides of the crack
gradually transform a narrow section of the lamella into one microfibril which consists of fully oriented crystal blocks connected
by a great many highly stretched intrafibrillar tie molecules.
The long period, depending almost exclusively on the temperature
of drawing, is of the same magnitude as that of a spherulitic
sample crystallized from a relaxed melt at the same temperature as
that of drawing.

     The higher the fraction of tie molecules and the more stretched
they are the higher will be the longitudinal elastic modulus and
elastic strength and the smaller the strain to break of the microfibrils. The finite dimensions of the primary lamella yield a
finite length of the microfibril. The finite width of the section
of the lamella feeding one microfibril results in the microfibril
in a spacial separation of blocks which were originally adjacent
in the lamella. As a consequence, the partially unfolded chains
at the original block boundary traverse more than one amorphous
layer between consecutive blocks of the microfibril. Hence they
are counted and act more than once as tie molecules thus increasing
the longitudinal strength of the microfibril. One expects proportionality between the average fraction of tie molecules

per amorphous layer and the draw ratio of the microfibril which
in turn is proportional to the width of the section feeding one
microfibril.

A certain variation of the average draw ratio of microfibril
bundles is expected in a bulk sample as a consequence of the variation of the growth plane orientation of the stacks of parallel
lamellae at the moment when the cracking of the whole stack starts.
The micronecks in the destruction zones at both sides of the crack
transform the stacked lamellae into a fibril which differs in draw
ratio from the adjacent fibril as a consequence of the difference
of crack to lamella orientation. But the orientation for all
lamellae of a stack is practically the same thus yielding practically
the same draw ratio for the whole bundle of microfibrils resulting
from the same stack and forming the same fibril. The variation of
draw ratio among the microfibrils of such a fibril is mainly the
consequence of the variation of microneck density at the single
lamella crack. This variation is expected to be smaller than that
between stacks of different orientation.

The length of the microfibril is proportional to the volume
of the section of the lamella feeding one microneck, i.e., to the
product of section width and lamella extension in the direction
perpendicular to the crack. Hence the whole bundle of microfibrils
from a single stack of parallel lamellae will have practically the
same length proportional to the draw ratio. If the draw ratio is
10 and the lamella extension perpendicular to the crack is 2μ, the
microfibrils are expected to have a length of 20μ. The ends will
be located on the outer boundary of the fibril with some concentration at the ends of the fibril.

Each microfibril consists of alternating crystal blocks and
amorphous layers consisting of chain folds, free chains ends, and
tie molecules. The boundary between the crystalline and amorphous
component in the microfibril must be very diffuse, the differences
in density less than between ideal crystal and supercooled melt
and the fluctuation of long period quite substantial as one can
derive from the small intensity of meridional SAXS and the absence
of higher order maxima.

The drawn material consists of fully oriented microfibrils
and some remains of partially deformed original spherulitic structure. The mechanical properties are mainly the consequence of the
presence of microfibrils and hence in first approximation proportional to the fraction of microfibrils. The same applies to
the orientation. This close correlation between mechanical properties and orientation is very often taken as a causal connection
between them. Actually, both effects are only the manifestation
of the presence of fibrous structure. That orientation does not

mean increased strength and elastic modulus is best demonstrated
by an oriented single crystal mat which is a brittle, mechanically
useless material in spite of high orientation (Fig. 2).

The strength and elastic modulus of the fibrous structure
equals the strength and elastic modulus of microfibrils modified
by some sliding possibilities of microfibrils upon each other and
by the defects of the microfibrillar superlattice caused by the
interruption of molecular connection at the ends of the microfibrils.
The same applies to the strain to break. According to this model
one expects the elastic modulus and tensile strength of the fibrous
structure and the fraction of tie molecules to increase almost
linearly with the draw ratio[23-25] up to the complete transformation
of the spherulitic into fibrous structure and very little change
with draw ratio after that in excellent agreement with experimental
data (Fig. 3).

The tendency of lateral fit of crystalline blocks of adjacent
microfibrils produces the lamella structure documented by electron
microscopy[15] and SAXS. But the lamellae are not the primary structural element. They are indeed an artifact masking the underlying

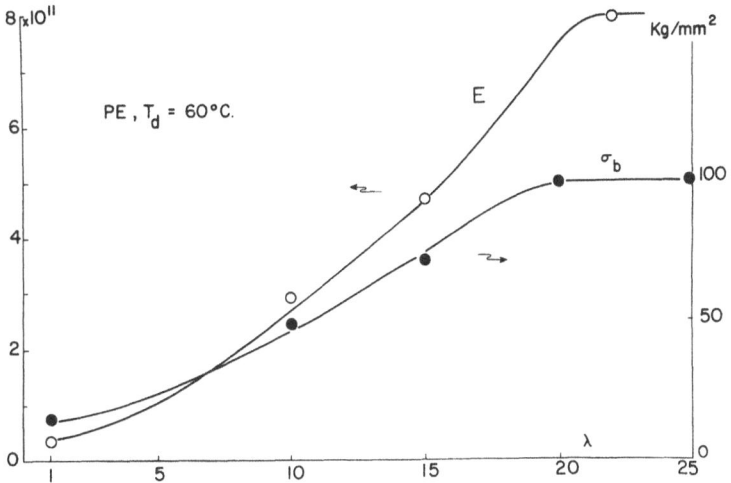

Fig. 3 - Elastic modulus E and stress to break σ_b, measured at
        -180°C, of linear polyethylene drawn at 60°C to different
        draw ratios λ.[24] E in dynes/cm^2 σ_b in kg/mm^2.

microfibrillar structure. There is no or very little lateral
material connection by macromolecules between the crystal blocks
of the lamella but a great many axial connections by tie molecules
between the blocks of the same microfibril and belonging to
different lamellae. Therefore, one can rather easily separate the

microfibrils from each other thus breaking the lamellae.[18] Such
a separation involves merely work done against the van der Waals
cohesion forces between adjacent microfibrils but need not rupture
any chain. But it is practically impossible or at least very
difficult to separate the lamellae since that demands beyond the
work against the van der Waals cohesion forces in the amorphous
surface layer the rupture of all the great many tie molecules
connecting the blocks in all the microfibrils passing through the
lamellae.

MOLECULAR CHAIN RUPTURE

The fact that any fracture plane through the polymer solid
cuts a great many macromolecules leads to the conclusion that one
must expect chain rupture to occur if the chains have no or not
sufficient chance to be pulled out of the solid. Early experiments
by Bressler et al.[26] and Butyagin et al.[27] on low temperature
grinding supplemented by cutting experiments by DeVries, Backman,[28]
and Panzoni et al.[29] have indeed demonstrated the existence of
radicals generated by the unsaturated free ends of broken polymer
chains. As a rule, one observes secondary radicals formed by
hydrogen abstraction from the unbroken chains. The primary radicals
at the broken chain ends are too unstable for observation. One
assumes that each broken chain results in two primary and after
excitation transfer to two secondary radicals. Their number
between 10^{12} and 10^{13} per cm^2 of new surface corresponds to a
fraction of the maximum possible number $1/2 A = 2.5 \times 10^{14}$ of chains
in amorphous and $1/A = 5 \times 10^{14}$ in crystalline conformation if the
cut is perpendicular to the chain direction. Here $A = 20 Å^2$ is
the cross section of the nylon or polyethylene chain.

The experiments of DeVries and Backman on nylon 6,6, polypropylene, and polyethylene show a remarkable decrease with increasing temperature of the number of secondary radicals per cm^2
of new surface and the nearly identical limiting number below the
glass temperature 1.5, 0.7, 0.7×10^{13} cm^2, respectively, for
all three materials investigated. Since it is safe to assume that
the cut primarily proceeds through the weakest areas of the sample,
i.e., through the "amorphous" boundaries between adjacent spherulites,
stacks of lamellae and lamellae, one can conclude that about 1 in
50 chains in the crystal lattice is acting as tie molecule broken
by the cutting. Moreover, there is no difference between the chain
rupture frequency in nylon 66, polypropylene, and polyethylene which
would request the conclusion that the chains in the former material
are better anchored in the crystal lattice than in the latter one.
If a crack proceeds through the amorphous layer, the tie molecules
bridging the crack are either ruptured or pulled out from one
crystal lamella. With the assumption that the number of tie molecules does not vary by an order of magnitude between nylon 66 and

polyethylene the nearly identical number of broken chains per unit crack area means that the probability ratio between chain rupture and pulling out of the crystal lattice cannot be appreciably different in both materials.

In the tensile experiment of highly oriented nylon 6 and nylon 66 samples one observes radicals in the whole strained specimen well before the fracture occurs, at about 40% of the strain to break ε_b, and their number per cm^3 is about 10^3 times larger than expected per cm^2 of the fracture plane.[30] This means that long before the sample fails one breaks all over the sample a great many polymer chains which at the applied strain ε are strained so much that their lifetime is reduced to or below the time range of the tensile experiment. The number of detectable radicals, extremely high in strained rubber (10^{18}) and nylon 6 and 66 (10^{17} – 10^{18}), is much smaller in polyethylene (10^{16}), polypropylene (10^{16}) and poly(ethylene terephthalate) (10^{15}).[31-33] It is still undecided whether this difference is the consequence of a smaller number of chains broken or of a shorter lifetime of radicals formed.

The straining of chains in stressed PP was followed up by the displacement of the IR absorption band at 975 cm^{-1}. The maximum stress on C-C bonds surpassed at least ten times the average stress in the specimen.[34] In rubbery material such highly strained chains which eventually break in the tensile experiment are connecting two consecutive chemical cross-links. In crystalline polymers they are tie molecules connecting two crystals which act as strong mechanical cross-links. But the number of tie molecules per cm^3 is about 10^3 times higher than the number of broken chains. That means that in spite of so many ruptured chains the strength of the material is hardly affected and any attempt to correlate the deviation from the linear stress-strain relationship, the creep and relaxation behavior with chain rupture is bound to fail. This conclusion is further corroborated by the fact that in a second tensile loading experiment no new radicals are formed up to the maximum strain of the first loading in spite of the practically identical stress-strain curve in both runs.

The small fraction of ruptured tie molecule is most likely the consequence of the fact that the rupture only occurs at the vacancy defects of the microfibrillar lattice of the fibrous structure, i.e., at the ends of individual microfibrils where the material connection in the axial direction is interrupted[35] (Fig. 4). The stress concentration on the adjacent microfibrils seems to be sufficient for the rupture of the most affected microfibrils. According to SAXS investigation by Zhurkov et al.[36] one indeed has in nylon 6 fibers strained almost to fracture between 10^{15} and 10^{16} cm^{-3} flat disk-like microcracks of about 250Å diameter in extremely good agreement with the estimate derived from the

a.   b.

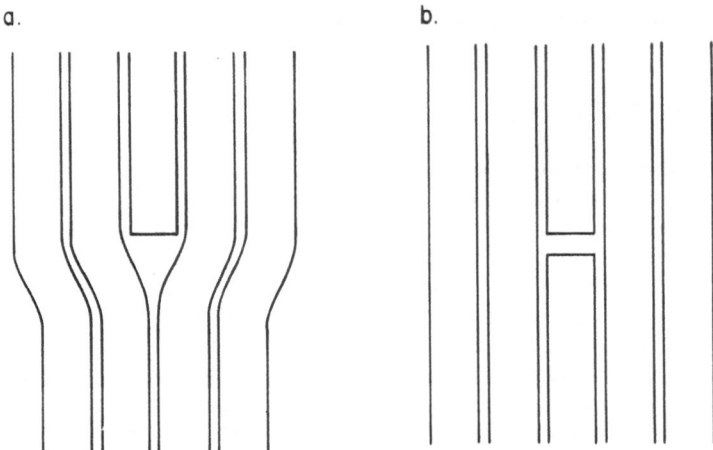

Fig. 4 - Point vacancies at the end of a microfibril: (a) the
          microfibril ending at the vacancy acts as an interstitial
          defect of the microfibrillar lattice, (b) the end of one
          microfibril and the start of another one act merely as an
          interruption of molecular connection.

microfibrillar morphology and the number of observed radicals[35]
assuming that at least in one cross section of the microfibrils
adjacent to the superlattice vacancy all the tie molecules are
broken。 With increasing strain the microcracks tend to coalesce
either by cracks through microfibrils perpendicular to the fiber
axis or between microfibrils parallel to the fiber axis (Fig. 5).
In the former case (radial crack propagation), new chains are
broken and radicals formed and in the latter case (axial crack
propagation) very few such ruptures occur。 As soon as by such
growth the critical size of crack is reached the sample fails
catastrophically。 But the number of ruptured chains at this last
step -- $\beta/A = 5\times10^{13}$ cm^{-2} where $\beta \sim 10\%$ is the fraction of molecules in the crystal lattice acting as tie molecules -- is unde-
tectably small compared to the total number, $10^{17} - 10^{18}$ cm^{-3},
ruptured in the great many microcracks before the catastrophic
crack propagation begins.

     Such a difference of microcrack propagation along the boundary
between microfibrils and through the microfibrils is very likely
the main reason for the difference in observed radicals between
polyethylene and nylon 6 or 66。 The strong hydrogen bridges in
nylon very efficiently enhance the interfibrillar autoadhesion
forces thus making the crack propagation along the boundary
between adjacent microfibrils relatively difficult so that instead
the microcracks prefer to grow by rupture of adjacent microfibrils。

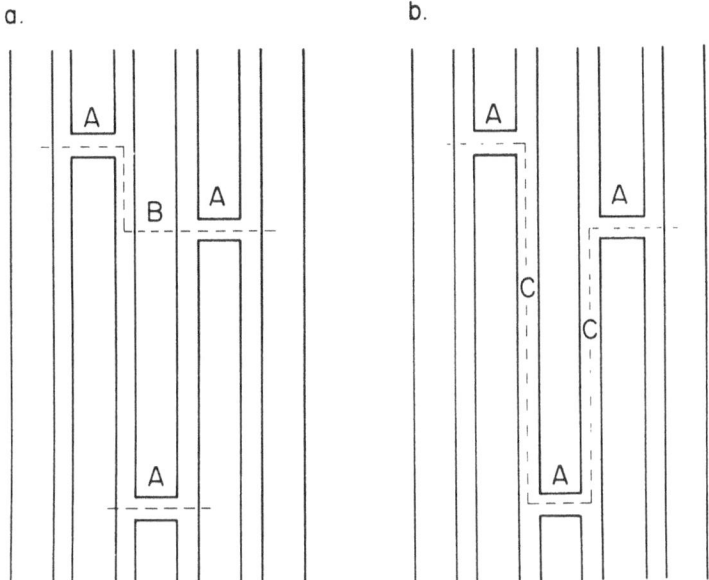

Fig. 5 - Crack coalescence by (a) radial and (b) axial crack pro-
         pagation.

Such a rupture breaks the great many intrafibrillar tie molecules
and hence yields a high radical population. In polyethylene,
however, the autoadhesion forces between adjacent microfibrils
seem to be smaller as demonstrated by the extremely small shear
resistance of low density polyethylene drawn to a draw ratio 4.5.
Linear, unbranched polyethylene with the usual draw ratio between
10 and 20 does not exhibit a similarly high shear compliance.
This difference one suspects may be more the consequence of the
higher draw ratio with a proportionately longer microfibrils than
a consequence of a better autoadhesion per unit area. If that is
indeed so, then the smaller interfibrillar forces favor crack propagation along the boundary between adjacent microfibrils instead
of through the microfibrils thus breaking substantially less tie
molecules.

      The effect may be enhanced by the fact that the low draw ratio
of nylon (~4) as compared with the substantially larger value
(~15) of polyethylene produces more point vacancies and less intrafibrillar tie molecules. As a consequence, the number of microcracks is larger and the strength of the individual microfibril is
smaller in nylon than in polyethylene. Both effects enhance the
rupture of tie molecules in the former case and a crack propagation
without significant chain rupture in the latter case.

ACKNOWLEDGMENT

The author would like to thank the Camille and Henry Dreyfus Foundation for the generous support of this work.

REFERENCES

1. G. Raumann and D. W. Saunders, Proc. Phys. Soc. 77, 1028 (1961); 78, 1271 (1962).
2. V. B. Gupta and I. M. Ward, J. Macromol. Sci. B2, 89 (1968).
3. R. J. Samuels, J. Polymer Sci. A-2, 6, 1101, 2021 (1968); 7, 1197 (1969); J. Macromol. Sci. B4, 701 (1970).
4. G. Meinel and A. Peterlin, J. Polymer Sci. B5, 613 (1967).
5. W. Glenz, N. Morosoff, and A. Peterlin, J. Polymer Sci. B9, 211 (1971).
6. A. Peterlin and F. J. Baltá-Calleja, Kolloid-Z. & Z. Polymere 242, 1092 (1971).
7. A. Peterlin, J. Polymer Sci. C9, 61 (1965); C15, 427 (1967); C18, 123 (1967).
8. A. Peterlin, Kolloid-Z. & Z. Polymere 216/217, 129 (1967); Man-Made Fibers, Ed. by H. F. Mark, S. M. Atlas, and E. Cernia, Interscience-J. Wiley & Sons, New York, 1967, Chapter 8, pp. 283-320; Polymer Eng. Sci. 9, 172 (1969).
9. A. Peterlin, J. Material Sci. 6, 490 (1971).
10. P. H. Geil, J. Polymer Sci. A-2, 3835 (1968); Polymer Single Crystals, J. Wiley & Sons, New York 1963.
11. A. Peterlin, P. Ingram, and H. Kiho, Makromol. Chem. 86, 294 (1965).
12. K. Sakaoku and A. Peterlin, Makromol. Chem. 108, 234 (1967).
13. P. Ingram, Makromol. Chem. 108, 281 (1967).
14. K. Sakaoku and A. Peterlin, Makromol. Chem. 108, 234 (1967).
15. A. Peterlin and K. Sakaoku, Clean Surfaces, Ed. by G. Goldfinger, M. Dekker, Inc., New York 1970, p. 1.
16. A. Peterlin and K. Sakaoku, Kolloid-Z. & Z. Polymere 212, 51 (1966); J. Appl. Phys. 38, 4152 (1967).
17. K. Sakaoku and A. Peterlin, J. Macromol. Sci. B1, 103 (1967).
18. K. Sakaoku and A. Peterlin, J. Polymer Sci. A-2, 9, 895 (1970).
19. F. Baltá-Calleja and A. Peterlin, J. Polymer Sci. A-2, 7, 1275 (1969); J. Material Sci. 4, 722 (1969); J. Appl. Phys. 40, 4238 (1969); J. Macromol. Sci. B4, 519 (1970).
20. A. Siegmann and P. H. Geil, J. Macromol. Sci. B4, 557 (1970).
21. K. Sakaoku and A. Peterlin, Makromol. Chem. (in press).
22. See for instance the September issue of J. Macromol. Sci. B4 (1970) and the June issue of J. Material Sci. 6, (1971), devoted to the plastic deformation of polymer solids and the structure and properties of oriented polymer solids.
23. W. C. Sheehan and T. B. Cole, J. Appl. Polymer Sci. 8, 2359 (1964).
24. G. Meinel and A. Peterlin, J. Polymer Sci. A-2, 6, 587 (1968).

25. J. M. Andrews and I. M. Ward, J. Material Sci. 5, 411 (1970).
26. S. E. Bressler, S. N. Zhurkov, E. N. Kasbekov, E. M. Saminsky, and E. E. Tomashevski, Zh. tekhn. fiziki 29, 358 (1959).
27. P. Y. Butyagin, A. A. Berlin, A. E. Kalmanson, and L. A. Blyumenfeld, Vysokomol. Soed. 1, 865 (1959).
28. K. L. DeVries and D. K. Backman, J. Polymer Sci. A-1, 7, 2125 (1969).
29. J. Pazonyi, F. Tudos, and M. Dimitrov, Angew. Makromol. Chem. 10, 75 (1970).
30. S. N. Zhurkov and E. E. Tomashevski, Yield and Fracture, Ed. by A. C. Strickland, Inst. Phys. & Phys. Soc. Conf. Ser. 1, Oxford 1968, p. 200.
31. See for instance the review article H. H. Kausch-Blecken von Schmeling, J. Macromol. Sci. C4, 243 (1970).
32. G. S. P. Verma and A. Peterlin, J. Macromol. Sci. B4, 589 (1970).
33. J. Becht and H. Fischer, Kolloid-Z. & Z. Polymere 240, 766 (1970).
34. S. N. Zhurkov, V. I. Vettegren, V. E. Korsukov, and I. I. Novak, Fracture 1969, Ed. by P. Pratt, Chapman and Hall, London 1969, p. 545.
35. A. Peterlin, Inst. J. Fract. Mech. 7, 496 (1971).
36. S. N. Zhurkov, V. S. Kuksenko, and A. I. Slutsker, Fracture, Ed. by P. Pratt, Chapman and Hall, London, 1969, p. 531.

A MECHANISM OF ENERGY-DRIVEN ELASTICITY IN CRYSTALLINE POLYMERS

Edward S. Clark

Plastics Department, DuPont Experimental

Station, Wilmington, Delaware 19898

ABSTRACT

It is proposed that a fundamental property of a folded-chain lamella is its ability to recover elastically from small deformations. The high degree of elastic recovery of certain extruded fibers and films of crystalline polymers is attributed to this property. Experiments on these elastic materials indicate that a common feature is a row structure of folded-chain lamellae joined together by short fibril links. Small deformations of individual lamellae are additive in the aggregate to produce a large deformation of the specimen without exceeding the elastic limit of the individual lamellae. Correspondingly, a phenomenal number of microvoids is created. These disappear with the elastic recovery of the specimen.

Experimental data show the elastic behavior is a consequence of energy-driven forces. Two alternate proposals for elastic behavior based on energy forces are offered: 1. bending of lamellae and 2. shearing of lamellae with van der Waal's forces providing the driving force for recovery.

INTRODUCTION

A series of partially crystalline fibers and films having remarkable, almost rubber-like elasticity has been developed by a number of industrial laboratories [1].

A recovery of over 80% after 50% elongation is typical and often recovery is above 95% after deformation of 5%. It is generally agreed that the forces of elasticity are energy-driven and not entropy-driven as in a classical elastomer. Polymers which have been prepared in the energy-elastic form include polypropylene, acetal homo- and co-polymer, poly-3-methyl-butene-1, polybutene-1, and polypivalolactone. In general, polymers which can be created in the energy-elastic form tend to have high crystallinity in contrast to classical elastomers which are generally amorphous.

In this paper, I present an hypothesis developed in 1962 to explain novel energy-elasticity in highly crystalline fibers of polypivalolactone [1c]. The hypothesis is updated to include recent concepts of nucleation mechanisms and additional experimental data. The basis of this hypothesis is that the source of elasticity is not reversible strain of amorphous regions but is a result of a fundamental property of the folded-chain lamella: reversible deformability.

ROW STRUCTURES

A structural feature common to these energy-elastic compositions is an ordered stacking of crystal lamellae forming an aggregate termed a "row structure". As discussed in a review paper by Clark and Garber [2], these row structures develop when polymers are crystallized under high axial stress such as in melt spinning of fibers, simple extrusion of thin films, and the blown film process. When the melt is subject to extensional-flow induced crystallization, a small fraction of molecules is forced to crystallize as very small fibrils, some tens of molecules in diameter. These fibrils, while comprising a negligibly small component in the bulk crystallized specimen, play a critical role in the crystallization process as nuclei from which highly oriented overgrowths of folded-chain lamellae develop. These row structures have been classified into two categories by Keller and Machin [3] which are designated in this paper as Type A and Type B as illustrated in Figure 1.

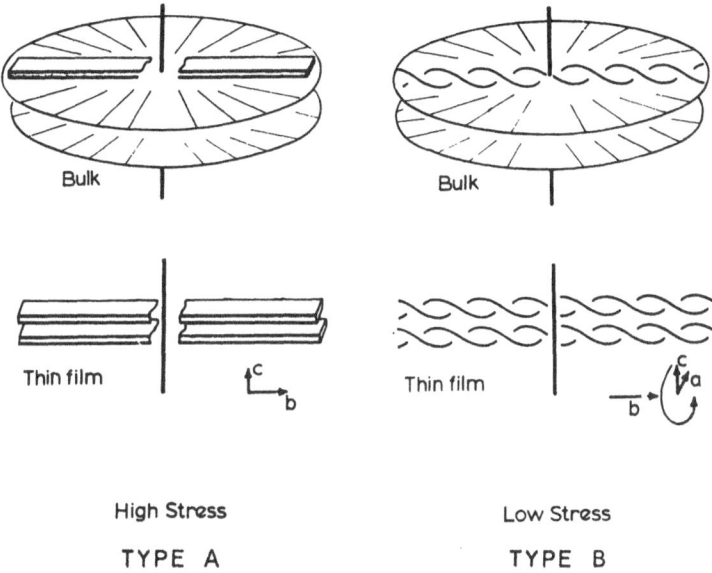

High Stress

TYPE A

Low Stress

TYPE B

FIG.1   ILLUSTRATION OF ROW STRUCTURES (REF.3)

INTERLAMELLA CONNECTIONS

An essential feature of this hypothesis is the occurrence of tie molecules of relatively short length joining adjacent lamellae together. While longer tie molecules may exist as discussed by Keith et al. [4], shorter linkages are much more important in considerations of elasticity in that they comprise the load bearing elements transferring stress from one lamella to another.

In the present hypothesis, the preferred mechanism for creation of tie molecules was devised in a study of injection molding of acetal resin [5]. Figure 2 illustrates the proposed web-like system of tie molecules which are created as large numbers of lamellae grow along a front parallel to the fibril nucleus to form a row structure. Within a localized volume, lamellae grow at a relatively uniform rate with the

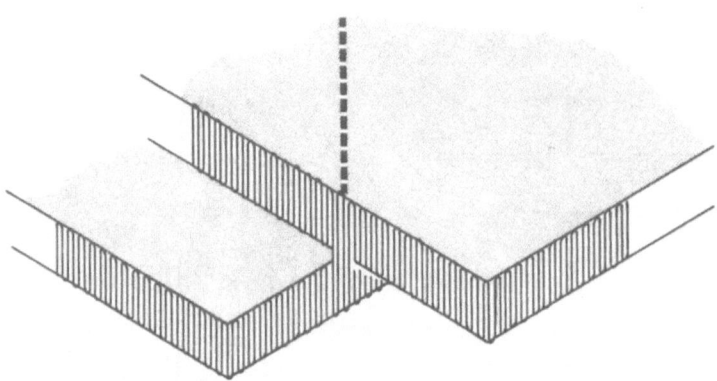

FIG.2 ILLUSTRATION OF TIE MOLECULES BETWEEN LAMELLAE

result that one growing fold plane is close to another
growing fold plane. Chains add to the growing fold
planes from the melt, usually folding to traverse a
single lamella. However, at points of overlap between
adjacent lamellae on the crystallizing front, a chain
(or more likely a group of chains) can fold across two
lamellae forming short interlamella connections. These
connections are short fibrils which by their proximity
along a locus within the lamella, create a web-like
interlamella linkage. The effect of these web-like
linkages in the growth of a row structure is the creation of a series of "tunnels" which follow tortuous
paths for substantial distances through the crystallized
specimen. The relevance of this to permeability is
discussed later.

REVERSIBLE DEFORMATION OF LAMELLAE

The growth of a Type A row structure with interlamella connections leads to an aggregate of the type
shown in Figure 3 which is highly idealized. The interconnecting fibrils, which are less regularly dispersed
than in the Figure, will distribute an applied strain
in a cooperative action to deform each lamella slightly.
This leads to a key feature of the hypothesis: small
deformations of folded-chain lamellae are *reversible*.
If an assemblage of lamellae and interconnecting fibril
links is highly oriented, as in a row structure, this
effect is additive and produces the property of elasticity in the macroscopic specimen. It should be noted
that this behavior is anisotropic. This mechanism is
applicable only for strain normal to the lamellae (parallel to the original flow direction).

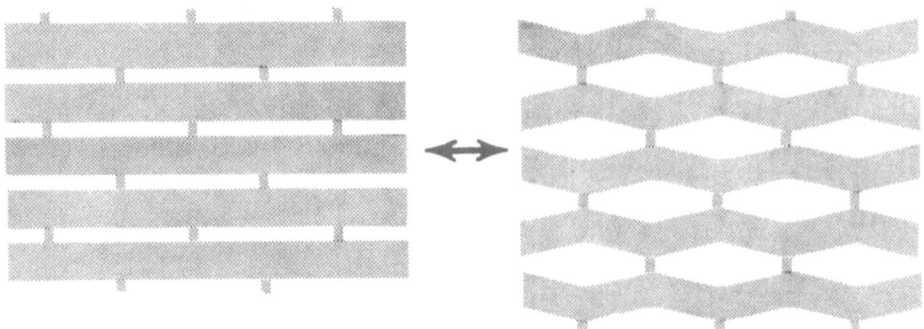

FIG.3   REVERSIBLE DEFORMATION OF ROW STRUCTURE
        WITH CREATION OF VOIDS (HIGHLY IDEALIZED)

     Mechanically, this concept can be represented by
the expandable "paper bell" party decoration shown in
Figure 4. The paper sheets correspond to the lamellae
and the glue lines to the locus of fibril interconnections. For ideal operation of this structure, the
interconnections must be distributed in a precise
pattern as in Figure 3. If some fibrils are too closely
spaced, portions of the structure may not open while
neighboring layers splay apart. Thus, in actuality, a
specimen may be expected to deform somewhat unevenly on
a microscopic scale. However, the proposed mechanism
for interlamella links will tend to give a fairly
uniform distribution of interconnections.

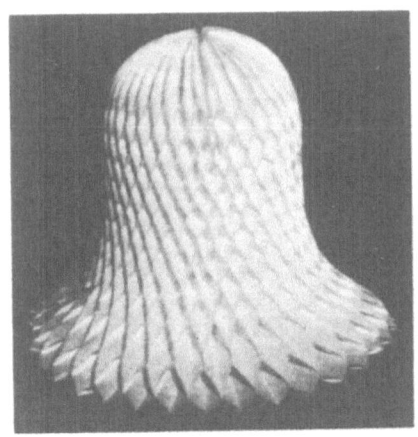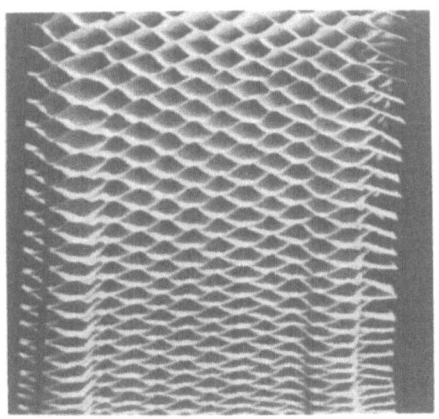

FIG.4   PAPER BELL MODEL AND ELONGATED SECTION

EXPERIMENTAL OBSERVATIONS

Experimental data in support of this "paper bell" row structure model include measurements of density, gas adsorption, high-pressure mercury porosimetry, wide and small angle x-ray diffraction, and force-temperature measurements.

Density

The proposed structure illustrated in Figure 3 shows that, with elongation, a phenomenal number of microvoids develop. In a strict interpretation of this model, the diameter of a fiber with this type of row structure should remain constant for a substantial amount of elongation. Thus the Poisson Ratio for these remarkable materials should be *zero*, whereas the value expected for a classical elastomer is 0.5. Measurements of diameter vs. elongation for a high-recovery fiber of polypivalolactone show no change in diameter for extensions up to 80%. In contrast, the diameter of a fiber of commercial nylon decreases regularly on extension with values in agreement with those calculated for constant density. Similar experiments on other elastic crystalline fibers are reported

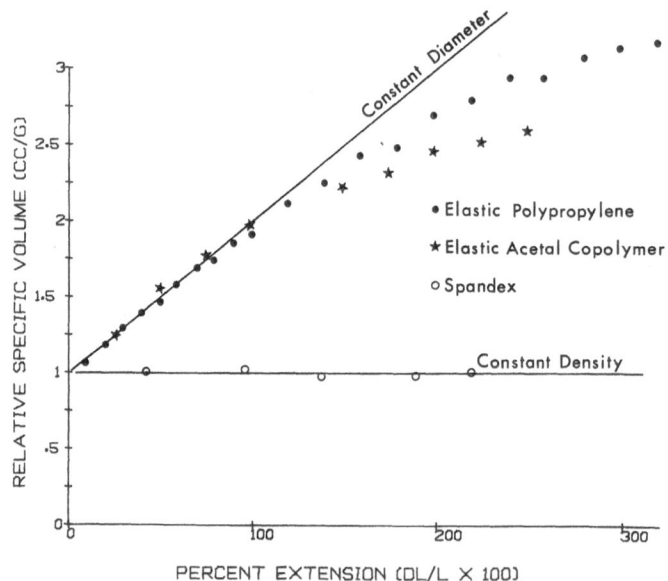

FIG.5   SPECIFIC VOLUME VS. ELONGATION (REF.6)

by Quynn and Brody [6]. Their data are replotted in
Figure 5 in terms of specific volume vs. elongation.
At higher elongations, the specific volume of these
elastic fibers becomes less than that predicted by the
model. It is in this region that the degree of recovery
from extension becomes substantially less than 100%.
This can be interpreted as the onset of other mechanisms
of deformation such as "unravelling" of lamellae with
the short fibril links being drawn irreversibly into
longer fibers. A continuation of this process leads to
a drawn fiber with many more interconnecting linkages
as described by Peterlin [7].

The development of a void filled structure is also
supported by measurements of mercury porosimetry and
gas adsorption [6]. For example, the surface area of
elastic polypropylene fibers increases linearly with
extension over the range to 50% elongation in agreement
with the predicted large increase in void content.

X-RAY DIFFRACTION

Wide angle x-ray diffraction patterns show that
these energy-elastic fibers and films are highly oriented. Elastic fibers of polypivalolactone [1c], polypropylene [1b], and poly-3-methylbutene-1 [6] show
very sharp equatorial peaks indicating that the molecular axes are aligned with a high degree of parallelism
to the fiber axis. This is consistent with a Type A
row structure in which the chains fold to form untwisted
lamellae with the chain axes normal to the lamella
surface. On the other hand, elastic films of polyoxymethylene [8] show a lower degree of molecular orientation consistent with a Type B row structure. The nature
of this material is discussed later.

Small angle diffraction data give some of the strongest experimental support for the proposed structure.
A Type A row structure under extension can be represented
by a model proposed by Schultz et al. [9]. Consider a
stack of plates of fixed thickness and density separated
by spacings of uniform thickness. That is, the plate-
void entity is the repeating unit along a line lattice
normal to the plate surface. If the dimension of the
void is small relative to that of the plate, the diffraction pattern consists of a sharp spot and its higher
orders along the meridian (direction normal to the plate
surface). Schultz et al. show that for small increases

in the dimension of the void relative to the plate, the
intensity of the diffraction spot increases by a factor
proportional to the square of the void thickness. Thus,
in the proposed Type A row structure, the initial pattern should consist of a diffraction spot on the meridian which, with extension of the specimen, increases
by a factor proportional to the square of the elongation.
Correspondingly, the Bragg angle of the meridian spot
should become smaller by the function:

$$\sin \theta = \lambda/2d$$

where λ is the wavelength and d is the repeat spacing of
the lamella-void unit. (It should be noted again that
Figure 3 is idealized; the voids are much thinner than
the lamellae and do not exhibit the "cross-hatched"
pattern of the Figure.)

Small angle x-ray diffraction data for high-recovery fibers of polypivalolactone are in good agreement
with those predicted by the simple model of Schultz et
al. As shown in Figure 6, the small angle pattern for
an unstretched fiber consists of a sharp spot on the

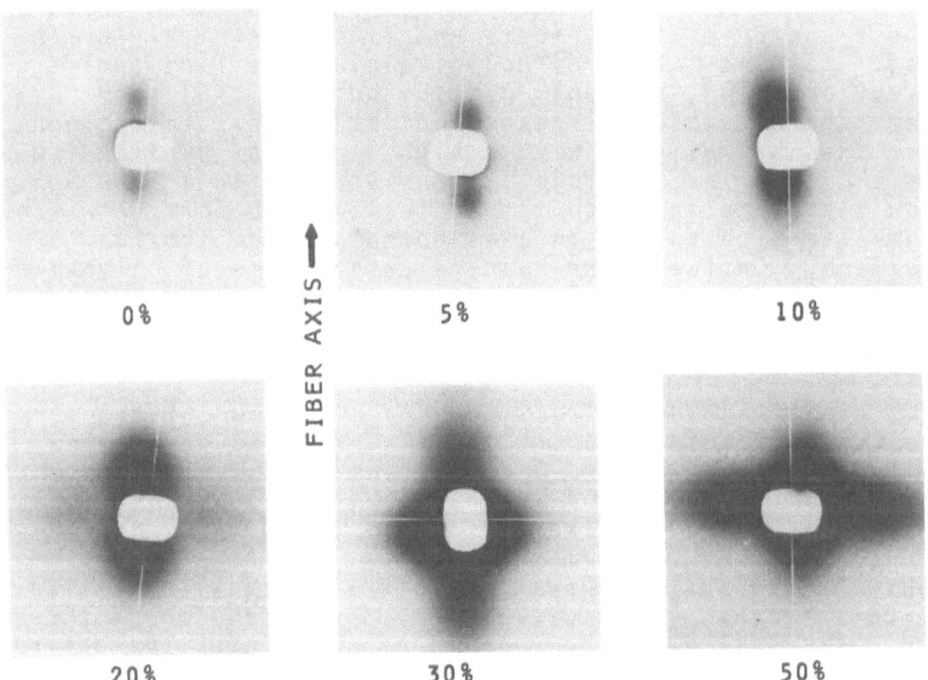

FIG.6   SMALL ANGLE X-RAY DIFFRACTION PATTERNS
OF POLYPIVALOLACTONE

meridian. With elongation of the fiber, the intensity
increases greatly and the Bragg angle of the reflection
becomes smaller. At higher elongations, a continuum
develops along the equator which may indicate the onset
of fibrillation. Figure 7 shows results obtained with
a Kratky diffractometer. The increase in intensity is
proportional to the square of the elongation. The d-
value calculated from Bragg's law shows a greater
increase than the predicted linear function. This can
be explained in terms of the intensity function. If the
structure opens somewhat unevenly, the portions of the
structure having larger voids (larger d-values) will
diffract with much higher intensity than the portions
with smaller voids. Thus, the resulting diffraction
pattern will be weighted in favor of the larger voids
giving higher d-values than predicted.

THERMOMECHANICAL BEHAVIOR

A feature which distinguishes these elastic row
structures uniquely from classical elastomers is the
effect of temperature change on the force of recovery
from extension. In a simple experiment, a rubber band
was stretched 20% and the magnitude of the stress was

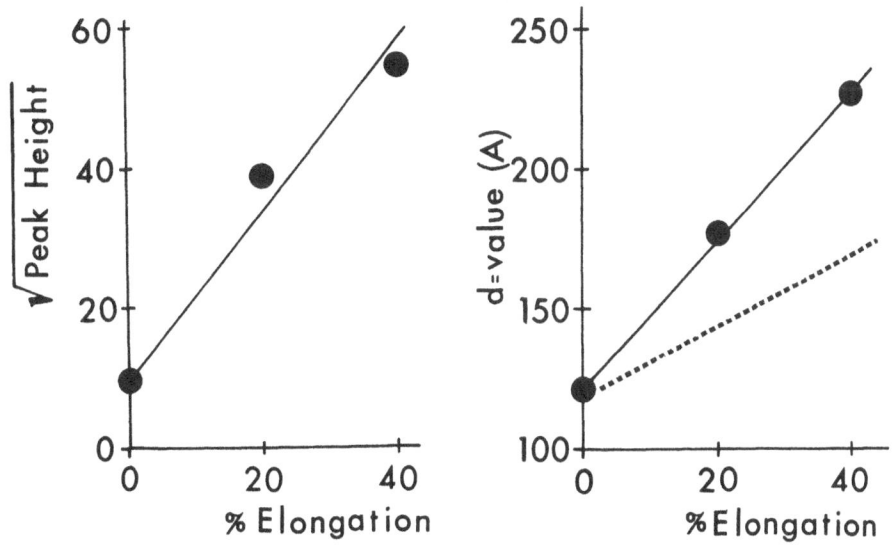

FIG.7   X-RAY DIFFRACTOMETER DATA FOR
FIBERS OF POLYPIVALOLACTONE

recorded. The temperature was then increased from 23°
to 80°C. The stress increased sharply demonstrating
the well-known entropy effect. In contrast, a similar
experiment on a high recovery fiber of polypivalolactone
showed a marked loss of stress on increase in temperature. In both experiments, the change of stress with
change of temperature at constant elongation was reversible for 10 second cycling of temperature from 25° to
80°C. However, the direction of change of stress was
opposite in the two cases, indicating that recovery
forces are energy-based in elastic fibers of polypivalolactone.

Quynn and Brody [6] give more quantitative evidence
for energy-based recovery forces in these elastic row
structures. Elastic fibers of Celcon® acetal copolymer
were extended 50% at several temperatures. Over the
range 22° to 120°C, the negative slope of the stress-
temperature curve revealed the predominance of internal
energy effects in the mechanism of elasticity. Similar
results are reported for elastic fibers of polypropylene.
These authors also report that the elastic behavior is
insensitive to the usual embrittling effects of the
glass transition.

TYPE B ROW STRUCTURE

The mechanism of deformation of a Type B row structure is more complicated than that for a Type A row
structure because of the twisting of lamellae during
crystallization. In a study of elastic film of polyoxymethylene, Garber and Clark [8] give evidence for a
Type B row structure. As shown in Figure 8A, lamellae
grow from fibril nuclei and twist about the growth
direction. On elongation, some lamellae slip past
neighboring lamellae by a shearing mechanism. This is
accompanied by rotation of these lamellae to become
nearly parallel to the surface of the film. It is difficult to explain the slippage of portions of lamellae
past their neighbors if these lamellae are joined by
short links. Therefore, the following modification of
the crystallization mechanism for a row structure is
offered for the Type B variety:

The creation of interlamella fibril links results,
as explained previously, by chains folding across two
lamellae parallel to the extensional-flow direction.
If the surfaces of the growing lamellae are perpendi-

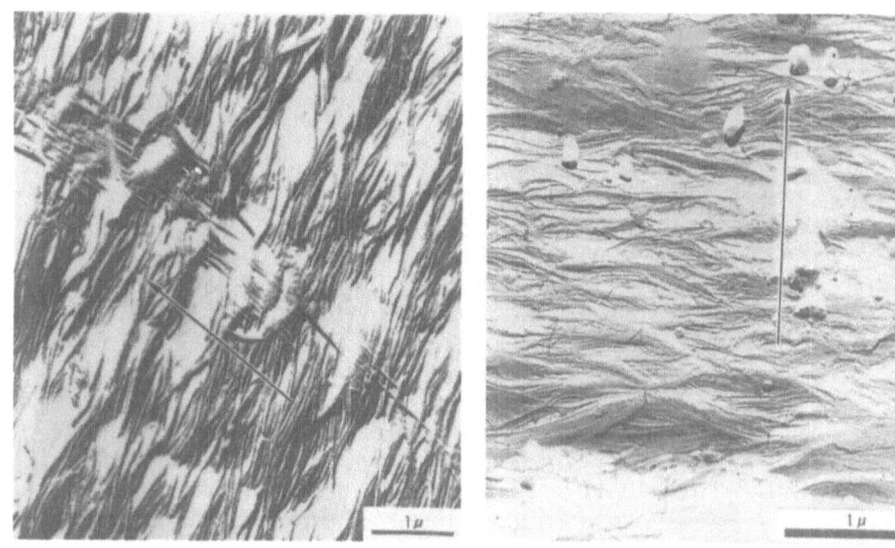

A. UNSTRETCHED   B. ELONGATED 30%

FIG.8   ELECTRON MICROGRAPHS OF ELASTIC FILM OF
        POLYOXYMETHYLENE

cular to the flow direction, such links occur frequently
since the geometry is favorable. On the other hand,
if the lamellae twist so that their surfaces are
parallel to the flow direction, the geometry for creation of interlamella fibril links is unfavorable and
few links occur. Thus, it is proposed that there is
a highly non-uniform distribution of fibril links in
a Type B row structure with many links in the portions
of lamellae with surfaces normal to the flow direction
and few links in those portions in which the surfaces
are parallel to the flow direction. Therefore, under
an applied stress, those lamellae parallel to the flow
direction can slip whereas those lamellae normal to the
flow direction deform between fibril links as in Figure
3.   There is a distinct difference in the nature of a
Type B row structure and a radial section of a banded
spherulite [10] which it resembles superficially. The
orientation of the lamellae is similar but the distribution of tie molecules is different according to the
present hypothesis.

SOURCE OF RETRACTIVE FORCE

The preceeding sections offer the hypothesis that
these elastic fibers and films are composed of row
structures of lamellae interconnected by short fibril
links such that small deformations of single lamella
units are additive in the aggregate to produce substantial deformation of the macroscopic specimen. The
deformation of the lamella unit is proposed to be
reversible; the degree of strain at any lattice site
is below the yield strain of the polymer crystal. Atten-
tion is now directed at the source of the recovery
force. Various sources have been considered by the
writer including crystallization energy, surface free
energy, mutual repulsion between chain folds and van
der Waal's forces. Although the experimental data do
not seem to exclude any of the above proposals, two
alternatives, based on van der Waal's forces will be
offered as the most acceptable. These two proposals
have the attraction that the lamella surface need not
consist of uniform, tight chain-folds.

Bending of Lamellae

The simplest mechanism consists of lamellae bending
between fibril links in a manner analogous to a leaf-
spring. This mechanism is also proposed by Quynn and
Brody [6]. On bending, the intermolecular separation
is increased from the minimum energy condition on one
side of the lamella and decreased from the minimum
level on the opposite side as illustrated in Figure 9.
Van der Waal's forces will return the crystal to its
original conformation on removal of stress. With increasing temperature, the restoring force should decrease
in agreement with experimental observations.

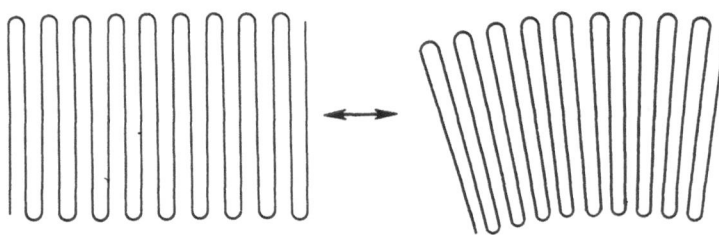

FIG.9   BENDING OF LAMELLAE BETWEEN FIBRIL LINKS

SHEARING OF LAMELLAE

A more complicated mechanism involving van der
Waal's forces and one preferred by the writer involves
shearing of lamellae between interconnecting fibrils
as illustrated in Figure 10. Each chain stem of the
folded-chain lamella slides past its neighboring chain
stem by a distance not exceeding the yield strain of
the crystal. The incremental amount of displacement
will be small but, by virtue of the creation of voids
at the crystal boundries, the small strains are additive
in the macroscopic specimen. In an attempt to place
this mechanism on a more mathematical basis, a calculation is made of the amount of recoverable deformation
which might be expected. Figure 10 represents the
Type A row structure unextended and extended to the
yield point (beyond which the lamellae will begin to
"unravel"). The Figure is idealized; the thickness of
the lamellae should be approximately 25 times the intermolecular separation and thus about 2 1/2 times as thick
as drawn.

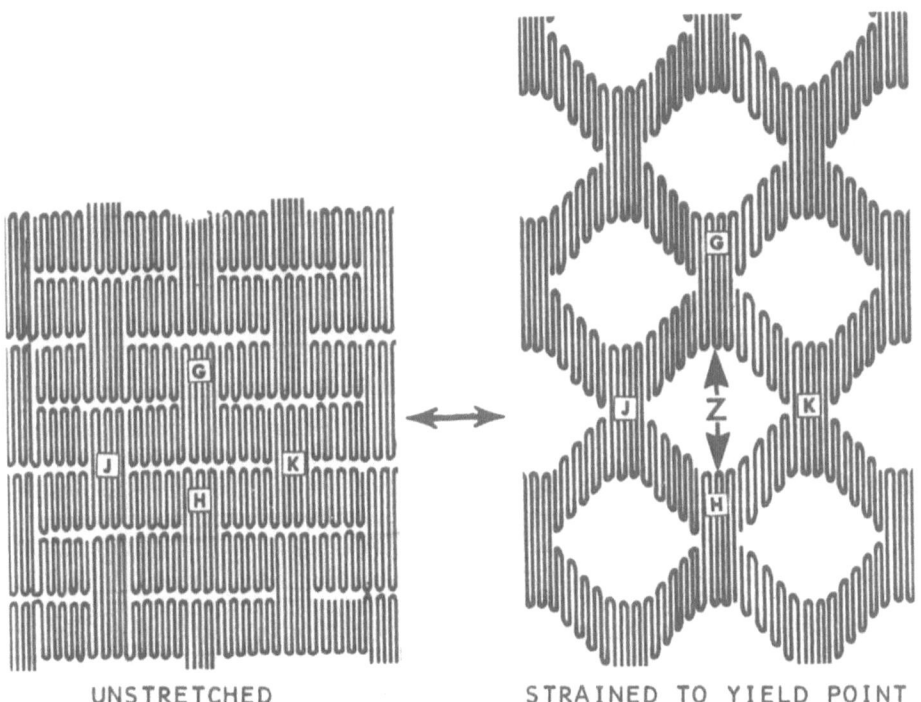

UNSTRETCHED   STRAINED TO YIELD POINT

FIG.10   SHEARING OF LAMELLAE BETWEEN FIBRIL LINKS

Calculation of Elastic Limits

The degree to which a Type A row structure can be elongated without exceeding the yield strain of the crystal can be calculated using Figure 10.

R = repeat distance of chain molecule
N = number of repeat units in molecule stem between chain folds
S = reference distance G-H (unstretched)
T = reference distance G-H (extended to yield point
M = number of chain molecule stems between fibril links (in distance J-K)
Y = yield strain of crystal along chain axis
Z = size of void in extension direction
E = elongation at yield

$$S = 2NR; \quad Z = MYR; \quad T = S + Z$$

$$E = \frac{T - S}{S} = \frac{MY}{2N}$$

Brown [11] reports that the yield strain for linear polymers is in the range of 3-10%. Let us assume a value of 6% for polypropylene (Y = 0.06) and the structure parameters: R = 6.5, N = 20 (130A thick lamella) and M = 200. We calculate that elastic polypropylene should show full recovery to an elongation of 30%, a value in agreement with observations [1b]. Extensions beyond this limit should cause a portion of the lamellae to "unravel" irreversibly.

THE MORPHOLOGICAL VALVE

This concept of row structure elasticity, developed originally from a study of fibers, suggested to the writer that films of this composition should have remarkable permeability behavior. If microvoids in the form of tunnels normal to the surface can open and close in the stretch and recovery process, the permeability of the film to gas should change correspondingly. This concept was reduced to practice with the startling results shown in Figure 11. The permeability of helium at 23°C through elastic films of polypropylene or polyoxymethylene changes by a factor of over one million centibarrers, reversibly, on 40% elongation. Similar effects were found with polybutene-1 films. These films

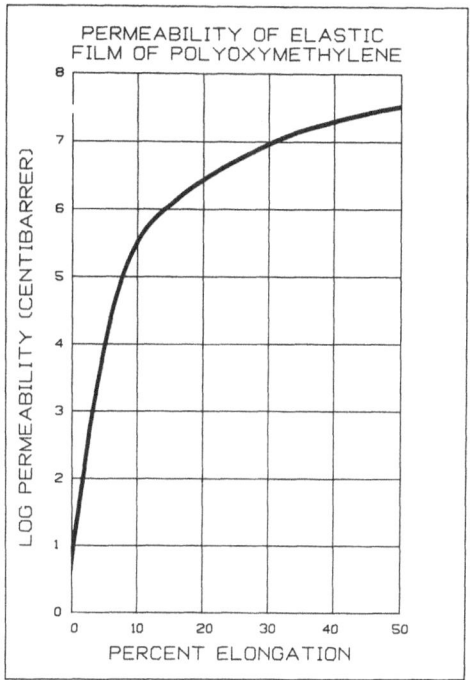

FIG.11   PERMEABILITY OF HELIUM THROUGH ELASTIC
         FILM OF POLYOXYMETHYLENE

tend to split readily when deformed laterally. Additional information on the properties of a "morphological
valve" may be found in a patent by Isaacson and Bierenbaum [12].

CONCLUSIONS

An hypothesis for elasticity, based on an oriented
assemblage of lamellae and short fibril links, is offered. It is proposed that the energy-based recovery
force results from van der Waal's interactions within
the folded-chain lamellae. While the present paper
deals with highly oriented systems which exaggerate this
elastic property, a logical extension is application of
the hypothesis to small deformations and recovery,
"toughness", of less highly oriented cyrstalline systems
such as injection molded plastics.

REFERENCES

1. [a] Belgian Patent 650,890 (Jan. 23, 1965, elastic polyoxymethylene fibers); [b] U.S. Patent 3,256,258 (June 14, 1966, elastic polypropylene fibers); [c] U.S. Patent 3,299,171 (Jan. 17, 1967, elastic polypivalolactone fibers); [d] U.S. Patent 3,382,306 (May 7, 1968, elastic polypropylene film

2. E. S. Clark and C. A. Garber, Intern. J. Polymeric Mater. 1, 31 (1971)

3. A. Keller and M. J. Machin, Polymer Systems: Deformation and Flow, (p. 100), London: Macmillan

4. H. D. Keith, F. J. Padden, Jr., and R. G. Vadimsky, J. Appl. Phys., 37, 4027 (1966)

5. E. S. Clark, SPE Journal 23, 46 (1967)

6. R. G. Quynn and H. Brody, J. Macromol. Sci.-Phys., B5 (4), 721 (1971)

7. A. Peterlin, J. Polymer Sci.,A2, 7, 1151 (1969)

8. C. A. Garber and E. S. Clark, J. Macromol. Sci.-Phys. B4 (3), 499 (1970)

9. J. M. Schultz, W. H. Robinson and G. M. Pound, J. Polymer Sci., A2, 5, 511 (1967)

10. E. W. Fischer, Z. Naturforsch., 12a, 753 (1957)

11. N. Brown, Bulletin of the Amer. Phys. Soc., II, 16, 428 (March 1971)

12. R. B. Isaacson and H. S. Bierenbaum, U.S. Patent 3,558,764 (Jan. 26, 1971, Microporous Film)

MORPHOLOGICAL CHANGES IN NYLON 6 AND EFFECT ON MECHANICAL PROPERTIES

IV. STRESS-STRAIN CYCLES

Koji Hoashi[*], Nobuhiro Kawasaki and Rodney D. Andrews

Department of Chemistry and Chemical Engineering

Stevens Institute of Technology, Hoboken, N.J. 07030

INTRODUCTION

In the present series of papers we have investigated the effects of morphological changes (produced by treating the film with different swelling agents and also by heat treatments) on the mechanical properties of nylon 6 film. Parts I, II and III (1,2,3) have discussed the types of morphological change produced, dynamic mechanical properties, and stress-strain properties, respectively. In the present paper we have examined in more detail the effects produced on the stress-strain behavior, by use of stress-strain cycling. In particular, we have examined the nature of the recoverable and total work of extension at different points of the stress-strain curve (both before and after the yield point) and at different temperatures (both below and above what is generally regarded as the glass transition temperature of the polymer).

The mechanical behavior has been examined in three different morphological states: (1) The original state of the film, in which the crystalline morphology is of an essentially unoriented smectic hexagonal type. This is the type of crystal morphology produced by rapid quenching from the melt. (2) The α-crystalline form produced by dry heat treatment at 212°C for 30 min. This is an α-crystalline form without preferential crystal orientation, in contrast to the α-crystalline form produced by treating the original smectic hexagonal film with phenol solution; the latter gives a preferentially oriented α-form, as discussed in Part I. (3) A plasticized α-crystalline form produced by a preliminary heat treatment of the

---

[*] Present address: Central Research Laboratory, Kuraray Co., Ltd., Kurashiki, Okayama, Japan.

same sort as in (2), followed by swelling in phenol solutions in
carbon tetrachloride. The phenol absorbed was allowed to remain
in during subsequent testing. These samples had an unoriented α-
crystalline form, just as in the case of the heat treatment alone,
and the absorbed phenol is believed to be acting simply as a plasticizer.

EXPERIMENTAL

The original film sample used in these studies was a 5-mil
commercial nylon 6 film obtained from Allied Chemical Corp., designated by the trade name Capran 77C.

The dry heat treatment at 212°C was carried out by sandwich-
ing the film sample between thin glass plates and immersing in a
silicone oil bath at this temperature for 30 min. Phenol plasticization was carried out by immersing samples, already given the
above heat treatment, in phenol solutions of various concentrations in carbon tetrachloride, until equilibrium absorption was
reached. Amounts of phenol absorbed under these conditions have
been tabulated in Part II. The phenol was not extracted (as it
was in the studies of Part I, where the phenol treatment was used
as a method of producing the α-form) but was allowed to remain "in"
during testing.

Stress-strain tests and cycling tests were carried out on an
Instron testing machine with temperature control. Stress-strain
curves were measured and cycling tests were carried out over a range of temperatures. Samples containing phenol were tested only at
20°C, however, in order to avoid evaporation of phenol produced by
heating. The samples tested were 1/4 inch in width and 1/2 inch in
length (between sample clamps). An extension rate of 100%/min. was
used in all cases, and the same rate was used in the recovery part
of the strain cycle, which was carried back to the point of zero
stress. In the measurements of total and recoverable work of straining, only a single strain cycle was measured for any particular
sample; the strain was reversed after reaching the desired point
along the stress-strain curve. A different sample was used for each
maximum strain value, so that there were no complications resulting
from sample history. Total work, W_T, and recoverable work, W_R,
were calculated on the basis of sample weight, as ergs/gram, in
order to normalize the values. These values of total and recoverable work represent the areas under the two parts (the extension
and retraction regions) of the stress-strain cycle.

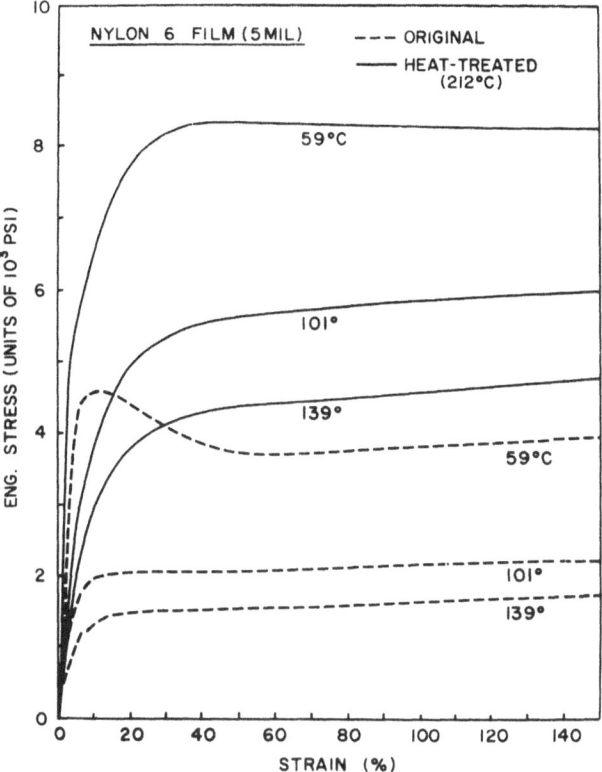

Fig. 1   Stress-strain curves of original and heat-treated films at
three different temperatures. Strain rate: 100%/min.

RESULTS AND DISCUSSION

Stress-Strain Curves

Stress-strain curves for the original and heat-treated films
are shown for three different temperatures in Fig. 1. The temperatures shown here (59°, 101° and 139°C) cover the range of the α-
transition in nylon 6, which is located at about 80°C, and is generally regarded as being the glass transition of the amorphous
phase. We have shown in Part II that the dynamic mechanical loss
peak corresponding to this transition is actually made up of two
components: a larger peak which is shifted to progressively lower
temperatures with increasing amounts of absorbed phenol, and a
smaller peak which shows no shift from a position around 100°C.

The curves in Fig. 1 show a significant decrease in stress

level with increasing temperature in both polymers. Also a significantly higher stress level is observed in the curves for the heat-
treated material in comparison to the original film. There is a
bigger difference in stress level between 59° and 101°, than between 101° and 139°, even though the temperature difference is
nearly equal. This last result can be explained by the presence
of the α-transition between the lower two temperatures.

In terms of the "lattice model" (3,4) which we have used to
discuss the effects of morphology on the mechanical properties of
semicrystalline polymers (see, for example, the discussion in Part
III), the higher stress levels in the heat-treated film can be explained as resulting from the development of a more dense lattice,
containing more junction points, as a result of the transformation

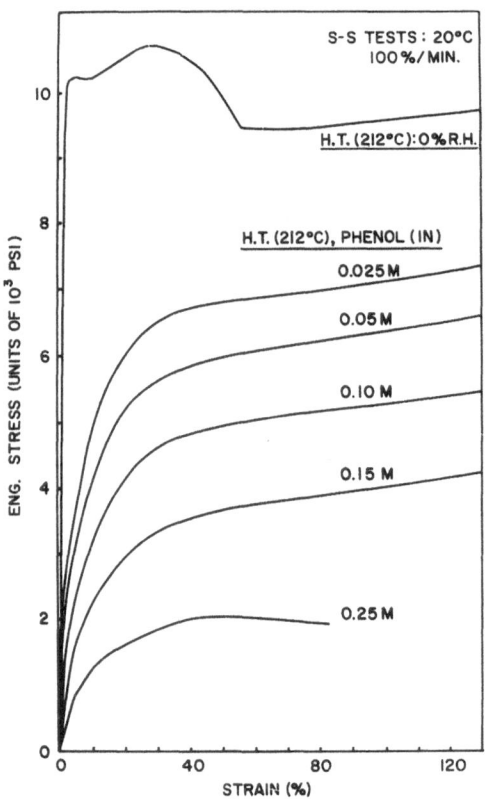

Fig. 2 Stress-strain curves for phenol-plasticized films in the
α-crystalline form. Curve for dry α-crystalline film containing no phenol is shown for comparison.

of the crystalline phase from the smectic hexagonal to the α-crystalline form. Only in the case of the original film at 59°C is a clear "yield peak" seen at the start of the stress-strain curve. However, the other curves all show a well-defined "knee" at the beginning of the curve at about the same strain value; this can be regarded as the yield point in these other curves.

The curves for the plasticized films containing various amounts of absorbed phenol are shown in Fig. 2. The curve for a heat-treated sample at 0% relative humidity which contains no phenol is shown for comparison. These curves were all measured at 20°C, at the standard strain rate of 100%/min. The reference curve has an extremely interesting shape and actually shows a double yield peak. The first sharp peak at a few percent elongation is typically seen in nylon samples which are almost completely dry. The second stress peak is very broad, and the sudden break point (change of slope) in the curve at about 55% elongation represents the onset of necking in the specimen. This unusual curve was accurately reproducible in duplicate experiments.

The presence of absorbed phenol produced a marked lowering of stress level, depending on the amount of phenol absorbed. The effect was very pronounced even for a solution of only 0.01 M concentration. An increase in the amount of absorbed phenol had an effect very similar to that of an increase in temperature, as seen in Fig. 1. This effect is characteristic of normal plasticizer action. In terms of the lattice model, the phenol can be regarded as acting to plasticize the amorphous matrix in which the interconnected crystalline lattice is imbedded.

The sharp stress peak at the beginning of the reference curve quickly disappears when a small amount of either phenol or water is absorbed. The amount of absorbed water corresponding to equilibration with normal atmospheric humidity will cause most of this peak to disappear. This behavior is undoubtedly related to the fact that the first few percent of water absorbed by nylon seems to be absorbed in a different way, and has different effects on properties such as dielectric behavior (5), for example, than additional amounts of absorbed water, which give a more conventional plasticizer effect. The water or other plasticizer first absorbed must be absorbed in special regions of the morphological structure which are quickly saturated. These might be regions such as interlamellar layers. With water present, interlamellar slippage could then take place easily, whereas in a totally dry sample, this slippage would have to be unlocked by mechanical force, giving rise to the high, sharp yield peak. Additional plasticizer absorbed would then go into the bulk amorphous regions, producing the usual plasticizer effects. The formation of a neck at the end of the second stress peak illustrates that plastic yield and neck formation are related in an indirect and complex way, and not in the simple way sometimes assumed.

Stress-Strain Cycles

Stress-strain curves interrupted by a cycle back to zero stress when the elongation had reached 20% are shown at the same three temperatures as in Fig. 1 (59°, 101° and 139°C) for the original and heat-treated specimens in Figs. 3 and 4. Similar curves for phenol-treated specimens at three different phenol concentrations are shown in Fig. 5, where the stress-strain curves were all measured at 20°C. The origins of the successive curves are displaced by 20% along the elongation axis in these graphs to avoid overlapping.

A hysteresis loop is obtained during the strain cycle in all cases, showing the presence of viscoelastic and plastic effects in the stress-strain behavior. In these particular experiments the samples were strained again after being cycled back to zero stress, and an interesting effect is seen in most of these curves: when the sample is re-elongated, a small yield peak or stress "overshoot" is seen before the curve settles back again to its original path. This effect is most pronounced for the original specimen and for lower temperatures. The effect is seen even when the stress-strain curve does not originally have a yield peak or stress maximum.

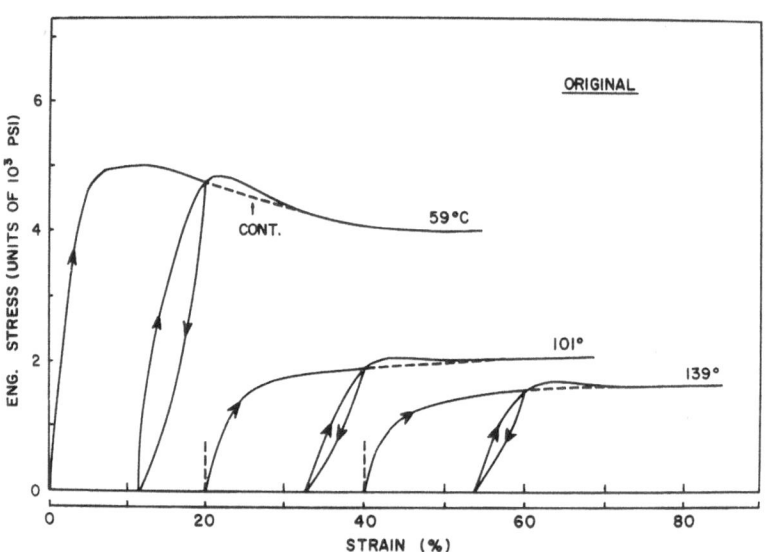

Fig. 3　Stress-strain cycles to 20% elongation with re-extension for original nylon 6 film at three temperatures. Origin points are displaced laterally to avoid overlap.

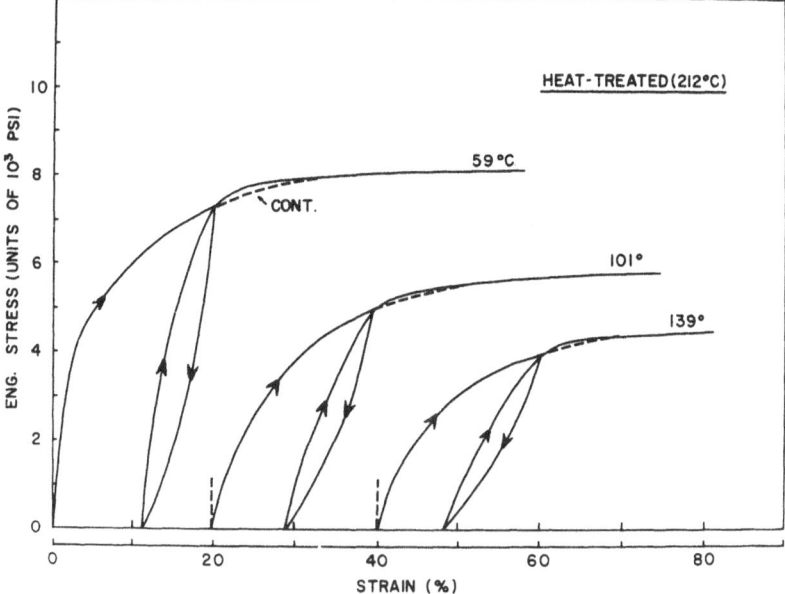

Fig. 4 Stress-strain cycles to 20% elongation with re-extension
for heat-treated film at three temperatures.

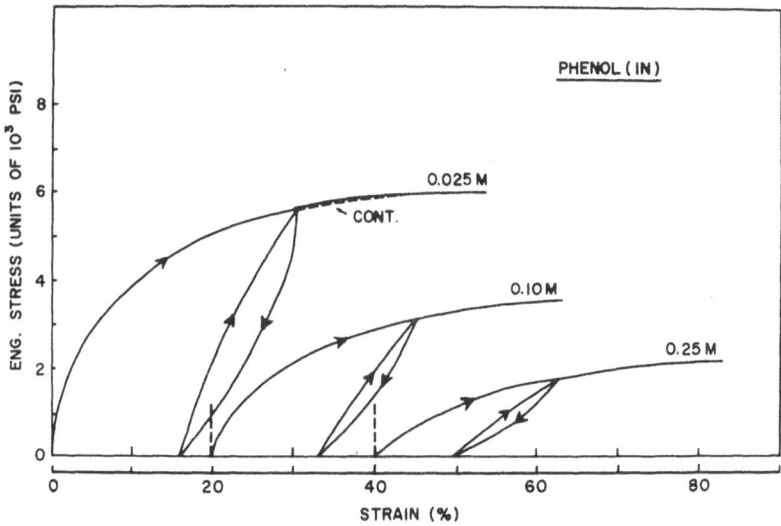

Fig. 5 Stress-strain cycles to 20% elongation with re-extension
for film samples containing different amounts of absorbed
phenol, measured at 20°C. Samples were equilibrated with
carbon tetrachloride solutions of phenol of molarities indicated on the curves.

The effect is almost absent in the phenol-treated specimens except for the sample at the lowest concentration, where a very slight overshoot is seen.

A further point of interest in these curves is the fact that the elongation value at which zero stress is reached on the return part of the cycle increases with increasing temperature for the original film, indicating a decrease in elasticity (or viscoelasticity) and an increase in plasticity. The opposite effect (a decrease in this elongation value) is seen in the heat-treated material when temperature is increased and when phenol concentration is increased (it will be remembered that the sample containing phenol was previously heat-treated), indicating an increase of (visco)elasticity and decrease of plasticity. The latter effect is the one which would ordinarily be expected, on the basis of an increasing mobility of the amorphous phase. The abnormal effect seen in the original film is probably explained by a breakdown of the smectic hexagonal crystallinity and the production of new α-form crystallinity during the straining process; this tends to produce a "permanent set" in the material which would appear to be plastic strain.

Work of Deformation and Recovery

Most of the cycling experiments were single cycle (extension-

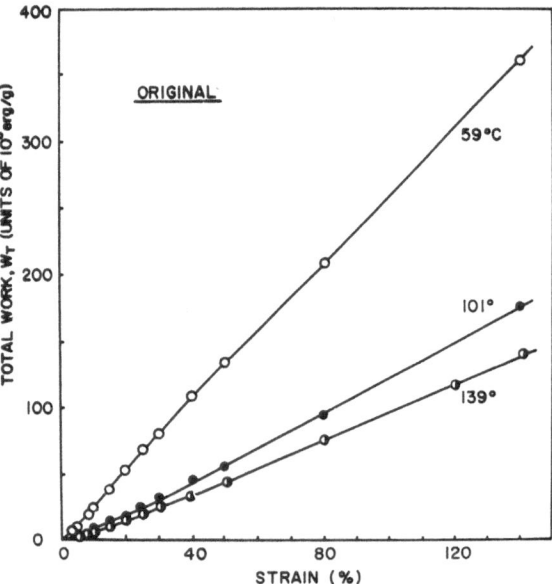

Fig. 6 Curves of total work vs. elongation calculated from stress-strain curves of original film at three temperatures.

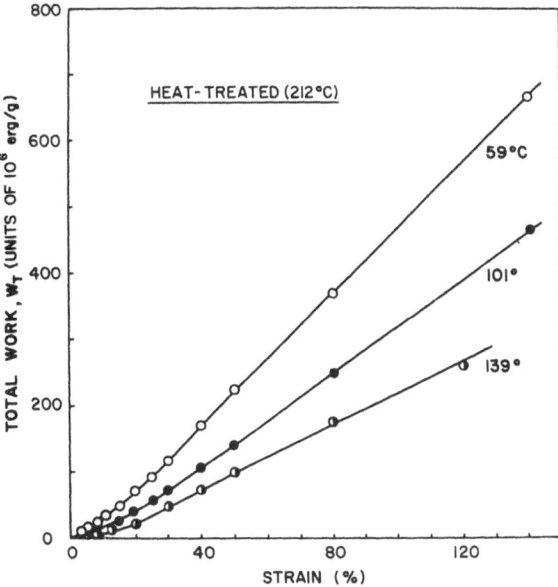

Fig. 7   Curves of total work vs. elongation calculated from stress-
strain curves of heat-treated film at three different temperatures.

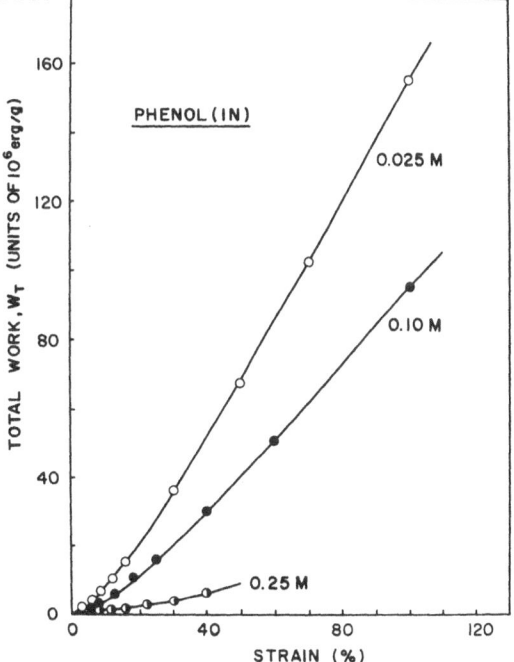

Fig. 8   Curves of total work vs. elongation for α-crystalline film
plasticized with different concentrations of phenol.

retraction) experiments carried out to different points on the
stress-strain curve. The calculation of total work, W_T, and recoverable work, W_R, provided an approximate method for dividing
the strain energy into elastic and plastic components (neglecting
the rate effects associated with the viscoelastic or delayed elastic component of the behavior).

Curves of total work vs. elongation for the three types of
sample, tested under the same conditions as in Figs. 1 and 2, do
not show any striking features, but merely differ in magnitude in
the way that would be expected from the stress magnitudes in Figs.
1 and 2. This can be seen from Figs. 6,7 and 8. Corresponding
curves of work recovered vs. elongation are shown in Figs. 9,10
and 11; these curves seem to have more interesting features. In
particular they show considerable similarity to the overall stress-
strain curve. Note, for example, that the curve for the original
sample at 59°C is the only one which shows a peak (Fig. 9); however, this peak is broader and is observed at a slightly higher
elongation than in the original stress-strain curve (Fig. 1). The
W_R curves also show a continuing increase at higher elongations,
whereas the overall stress-strain curves show more flattening in
this region.

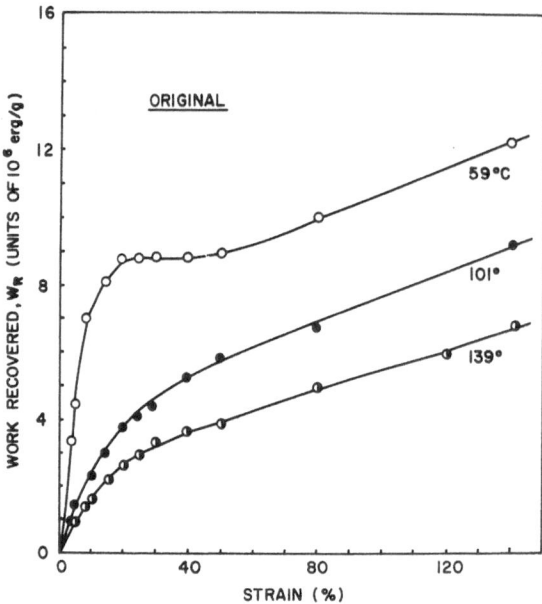

Fig. 9 Curves of work recovered vs. elongation at three different
        temperatures calculated from retraction curves of original
        film.

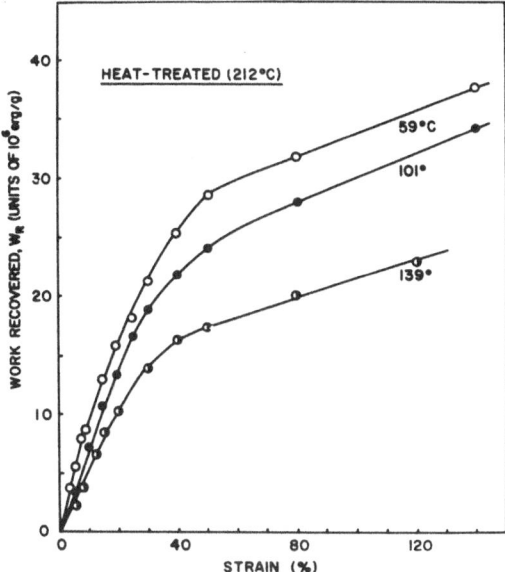

Fig. 10   Curves of work recovered vs. elongation at three different
temperatures calculated from retraction curves of heat-
treated film.

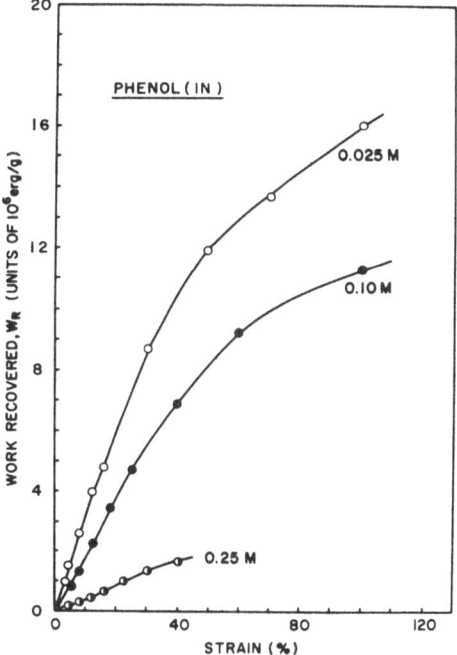

Fig. 11   Curves of work recovered vs. elongation calculated from
retraction curves of α-crystalline film plasticized with
different concentrations of phenol.

These data can also be plotted in the form of the ratio of recovered to total work, W_R/W_T, vs. elongation, as shown in Figs. 12,13 and 14. These curves are plotted on a logarithmic elongation scale, in order to be able to see clearly both the behavior over the whole elongation range and the behavior at low elongations. This ratio shows a progressive decrease with increasing elongation, as would be expected from an increasing proportion of plastic strain as elongation increases. Increasing temperature or increasing phenol content in the previously heat-treated material (Figs. 13 and 14) results in an increased proportion of recovered work, which is consistent with a more elastic character to the strain, as would be anticipated. There does not seem to be any significant difference in behavior above and below the glass transition temperature, except for a change of slope at an earlier strain value in the 59° curve below the glass transition for the work ratio of the heat-treated film (shown in Fig. 13), as compared to the curves above the glass transition temperature in the same plot.

It is interesting that the curves for the original film (Fig. 12) coincide at different temperatures, despite significant differences in the stress level in the stress-strain curves (Fig. 1). This could be explained by assuming that while increased temperature produces a softening of the structure due to enhanced mobility in the amorphous or less ordered regions, this increased mobility also facilitates the recrystallization which takes place on straining, in such a way that the two effects essentially cancel each other.

It is also interesting, and somewhat surprising that the curves of W_R/W_T show no abrupt or even noticeable change in the region

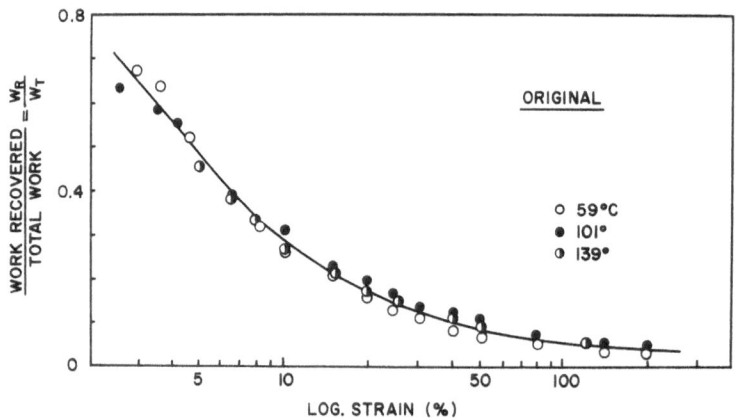

Fig. 12   Ratio of recoverable to total work as a function of
          elongation for original film at three temperatures.

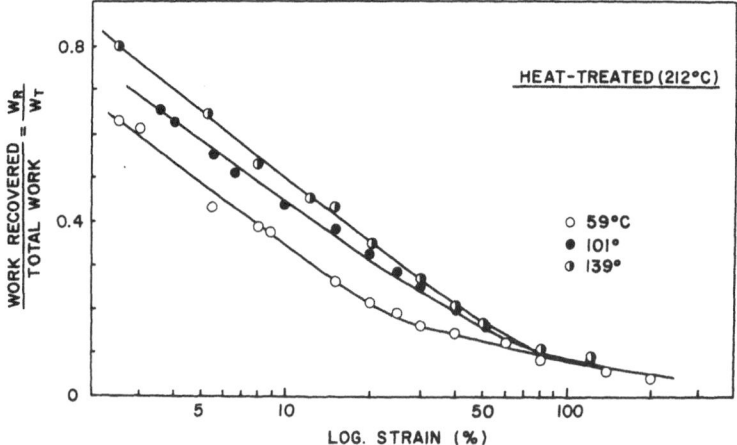

Fig. 13 Ratio of recoverable to total work as a function of
         elongation for heat-treated film at three temperatures.

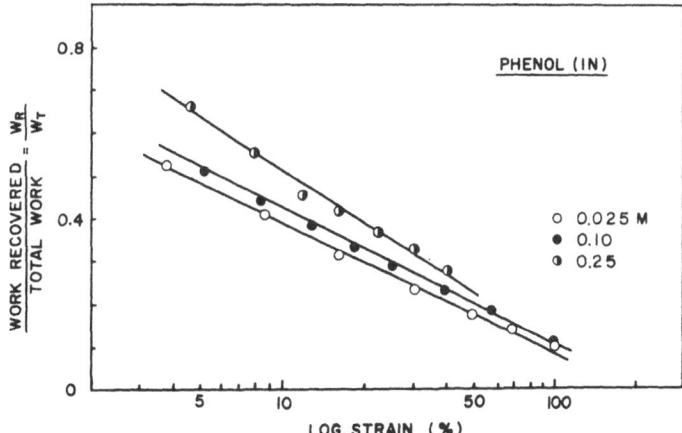

Fig. 14 Ratio of recoverable to total work as a function of
         elongation for α-crystalline film containing different
         amounts of phenol plasticizer.

of the yield point. The yield point may therefore not represent a sudden onset of plastic, rather than viscoelastic, deformation. Some preliminary attempts to calculate the total and recoverable work by use of a viscoelastic theory which does not include plastic yield indicate that such a theory can be usefully applied only at very low strains (up to perhaps 5-6% elongation, which is below the yield point).

ACKNOWLEDGEMENT

This work was supported in part by the Office of Naval Research.

References

(1) (Part I) Paper presented before Division of Cellulose, Wood and Fiber Chemistry at Meeting of American Chemical Society in Chicago, Ill. on September 14, 1970.

(2) (Part II) Paper presented at IUPAC Microsymposium on "Morphology of Polymers" in Prague, Czechoslovakia on September 2, 1971; J. Polymer Sci., Part C (in press).

(3) (Part III) Paper presented before Division of Cellulose, Wood and Fiber Chemistry at Meeting of American Chemical Society in Chicago, Ill. on September 14, 1970.

(4) J. Rubin and R.D. Andrews, Polymer Eng. and Sci. $\underline{8}$, 302 (1968).

(5) L.S. Buchoff, Ph.D. Thesis, Dept. of Chemistry and Chemical Engineering, Stevens Institute of Technology (1971).

TRANSPORT PROCESSES IN POLYMERS

C. E. Rogers, J. R. Semancik* and S. Kapur

Division of Macromolecular Science
Case Western Reserve University
Cleveland, Ohio 44106

ABSTRACT

The dependence of diffusion and relaxation processes on polymer structure is related to the polymer chain segmental mobility and the associated free volume content and distribution. The combination of diffusion and relaxation data for analysis of this mutual dependence has been obtained using a series of isoprene-methyl methacrylate copolymers and several penetrants of increasing molecular size. The effects of specific polymer-penetrant interactions have been studied by concurrent diffusion-relaxation and dynamic mechanical techniques using the Nylon 6-water system. The results of these studies emphasize the marked dependence of transport properties on subtle variations in polymer composition and structure and show the utility of combined transport measurements for elucidation of the complex nature of such systems.

INTRODUCTION

The processes of mass and momentum transport in polymers have a mutual dependence on the presence of void and defect structures and on the nature of polymer chain segmental motions. Variations in chemical composition and physical structure or morphology of a system lead to corresponding changes in the rates of diffusion of low molecular weight penetrants and the rates associated with rheological processes such as stress relaxation or creep. The temperature dependence and magnitude of these transport properties

* Present address: Diamond Shamrock Corporation, Painesville, Ohio

often are jointly affected by transitions which occur in polymer
solutions and their mixtures with sorbed penetrant species.

Studies of these transport processes using otherwise well-
characterized polymer samples can yield information regarding the
different modes by which these processes utilize the available
spectrum of defects and segmental motions in order to accomplish
the unit transport steps. Concurrent diffusion-relaxation studies,
for example, superimpose a known or predictable time-spatial distribution of plasticization due to diffusion-controlled sorption of
a mobile penetrant upon the simultaneous relaxation or dynamic-
mechanical properties of the polymeric material. It is then
possible to gain a measure of the manner or efficiency by which the
different transport processes use, say, the mutual free volume
during the course of diffusion and relaxation. The free volume, in
turn, can be related to the nature of chain segmental motions and
its variation with temperature, concentration, and molecular
interactions.

In order to interpret the results of such studies it is
essential to understand the complications that can arise during
sorption and diffusion processes. The interdependence of those
processes with the polymer structure can lead to wide deviations
from the ideal transport process which is usually considered to
occur during penetrant permeation. The effects of temperature,
polymer composition, molecular size and shape of the penetrant,
and the presence of fillers or crystalline regions can be anticipated and predicted to a fair degree of accuracy (1-3). The
corresponding effects due to polymer-penetrant interactions, leading to a dependence of diffusion coefficients on concentration,
time, or polymer dimensions, cannot be predicted or explained as
easily (1-4). These effects are complicated by our lack of
knowledge of the precise conformations of concentrated solutions
and by the relatively slow approach to solution and mechanical
equilibrium due to coupled diffusion-relaxation processes.

In the absence of suitable specific interactions, the course
of diffusion-relaxation can be predicted by fairly straightforward
time-concentration superposition calculations. When specific
interactions, such as hydrogen-bonding, are present, the predictions and calculations are more difficult. However, such studies
allow certain conclusions to be made concerning the contributions
of various modes of segmental motion to the diffusion and relaxation processes.

A realistic approach to this problem must consider the mutual
dependence of the transport processes on the nature of the
polymeric structure as it changes with experimental time. Approximate solutions, useful for initial studies to determine the major
experimental variables, can be based on modifications of known

rheological relationships such as the WLF equation and the time-temperature-concentration superposition principle for stress relaxation or creep modulus data. The systematic deviation of time-temperature master curves from time-concentration master curves can be interpreted in light of the known or measureable nature of the solution or solvation of polymer groups by a diffusing penetrant. Extension of such studies to a series of polymer-penetrant systems with well-characterized interactions should permit further resolution of the relaxation spectrum in terms of prescribed segmental and conformational motions.

STRESS RELAXATION AND PENETRANT DIFFUSION IN POLYMERS

Polymer chain segmental mobility is a primary factor controlling transport processes, such as diffusion and viscous flow, within a polymeric material. The temperature dependence of viscosity often can be represented by an Arrhenius-type equation:

$$\eta = \eta_o \exp(E_\eta /RT) \tag{1}$$

where E_η is the apparent activation energy for viscous flow. This representation is useful only if the temperature dependence of the pre-exponential factor η_o is small and E_η is at least approximately constant. This is seldom valid for amorphous polymers in the vicinity of the glass temperature where E_η rapidly increases as the temperature decreases.

A better representation of viscosity data is obtained by considering $\ln \eta$ to be a general function of the specific free volume fraction as described by the Doolittle equation:

$$\ln \eta = \ln A_\eta + B_\eta /f \tag{2}$$

where A_η and B_η are constants and f is a fractional free volume defined as the unoccupied volume divided by the specific volume. This equation is a basis for the Williams, Landel and Ferry (WLF) equation (5, 6):

$$\ln \left[\eta (T)/\eta (T_g) \right] = -C_1(T-T_g)/(C_2 + T-T_g) \tag{3}$$

where C_1 and C_2 are characteristic constants. This well-known equation successfully represents viscous flow and viscoelastic moduli data over a range of temperature from T_g to $T_g + 100^{\circ}C$. Modifications of the equation can be used over even wider temperature ranges.

The similarities which have been observed between diffusion and viscous flow processes suggest that the diffusion coefficient also may be formulated in terms of free volume fraction. Fujita, et. al. (7-10) have considered that the rate of diffusion of a given small molecule through a polymer depends primarily on the ease with which polymer chain segments exchange positions with the penetrating species. The mobilities of both the polymer segments

and the diffusant molecules in a polymer-diluent mixture depend
primarily on the free volume of the system. Diffusion is proportional to penetrant mobility within the polymer and is, therefore,
sensitive to the increase in free volume with temperature above T_g
in a fashion similar to the increase in polymer fluidity (inverse
viscosity). Consequently, the mobility can be formulated using a
Doolittle-type equation as a first approximation:

$$M_D = A_D \exp(-B_D/f) \tag{4}$$

where M_D denotes the penetrant mobility, B_D is a measure of the
minimum hole size required to permit considerable displacement of
a diluent molecule, A_D is a proportionality factor dependent primarily on the size and shape of the diluent molecule and independent of temperature and diluent concentration, and f is the
fractional free volume. Equation 4 can be combined with the
definition of molar mobility (11):

$$M_D = D_T/RT \tag{5}$$

to give:

$$\ln\left[D(o)/RT\right] = \ln A_D - B_D/f \tag{6}$$

where D_T is the thermodynamic diffusion coefficient (1) and $D(o)$
is D_T at zero penetrant concentration. This equation and its
various modifications have been used in several studies (4, 7, 9,
12, 13) concerned with elucidation of the concentration dependence
of diffusion coefficients. The theories based on free volume
concepts have proved to be very satisfactory for the interpretation and prediction of the phenomena relating to concentration
dependence of diffusion and permeation processes in polymer
systems (1, 4).

A useful linear relationship between diffusion and relaxation
properties for diluent-polymer systems can be obtained (14) by a
combination of equations 2 (or 3) and 6:

$$\log\left[D(T)/RT\right] = -K \log a_T + \log\left[D(T_g)/RT_g\right] \tag{7}$$

In this case, the reference temperature has been chosen as the
glass temperature, T_g, so that $D(T)$ and $D(T_g)$ are the diffusion
coefficients in the limit of zero penetrant concentration at
temperature T and T_g, respectively, K is the ratio of B_D/B_η, and
a_T (in the form of $\eta(T)/\eta(T_g)$) is the viscoelastic shift factor
from the time-temperature superposition treatment of suitable
stress relaxation (or creep) moduli data.

This equation, and its related forms for other definitions
of reference temperature or for other types of transport processes,
can be used for analysis of the factors affecting transport in
polymeric media. If B_η is assumed to be equal to unity, a common
assumption in practice, then a value of B_D can be recovered from
the slope of a linear plot of equation 7. However, Frisch and
Rogers (15) considered that the slope, $K=B_D/B_\eta$, of such a plot
could better be interpreted as a measure of the efficiency or

inefficiency of utilization of free volume by a mass transport process compared to its utilization by a momentum transfer process in the same polymer. The linear relationship would be obtained if the mechanisms of momentum and mass transport were affected similarly by changes in free volume with changing temperature. Comparable relationships and interpretations apply for other transport processes such as ion flux or thermal conductivity (15).

The magnitude of the slope parameter K is significant, therefore, in relating the nature of polymer segmental motion required for a diffusion process to that required for a viscous process in the same polymer sample, under experimental conditions where variations in free volume are most important. A dependence of K-values on penetrant molecule size and shape would be expected since movement of more bulky penetrants should require more extensive fluctuations in polymer structure. Such motions may involve movements of larger segments of the chain or may entail larger zones of cooperative motions of chain segments. In any case, in order to obtain meaningful relationships, the values of K must be determined from diffusion and viscoelastic property measurements made on the same polymer sample to avoid random variations due to differences in sample preparation, molecular weight, morphology, impurity content, etc.

This concept of related diffusion and relaxation behavior has been studied (14) using a series of well-characterized isoprene-methyl methacrylate copolymers of essentially random sequence distribution. The detailed description of polymer preparation, characterization, and the results and interpretation of mass transport studies, stress relaxation studies, and the application of the diffusion-relaxation relationships have been published elsewhere (14). The general results and interpretations will be illustrated here by selected examples.

Figure 1 illustrates the temperature dependence of diffusion coefficients of methane in samples of different copolymer composition. It is evident that these data do not follow a simple linear Arrhenius relationship except over very limited temperature ranges. This type of variation in the apparent activation energy for diffusion has been interpreted in terms of several related theoretical concepts (1-4, 7) including the Eyring theory of rate processes and the activated zone concept. The change in apparent activation energy due to an increase in the zone of activation for diffusion with decreasing temperature can be visualized in terms of a corresponding decrease in polymer free volume. As the free volume decreases, the extent of motion of long segments decreases and contributions of larger numbers of smaller segments are required to provide sufficient space for a penetrant molecule to make a successful diffusive jump.

This suggests that the dependence of diffusion on temperature may be related in another manner. If we consider a cooperative exchange type of mechanism to be predominant for diffusion of larger size penetrants, their mobility would be a measure of the polymer segmental mobility. Then as the temperature approaches T_g, large changes in diffusion may be expected similar to the observed behavior of mechanical properties, i.e., the change in relaxation moduli in the region of T_g.

The relationship between log D for methane and $T - T_g$ for a range of samples, whose measured values of T_g vary from -70°C for polyisoprene to 107°C for poly(methyl methacrylate) is shown in Figure 2. The variation of T_g with copolymer composition followed the predicted curves for random copolymer materials. The data for all of the polymers, with the significant exception of methyl methacrylate homopolymer, are seen to reduce to a single curve. The relationship is definitely nonlinear in contrast to the linear relationship found for inert gas diffusion through series of rubbers at temperatures in a range well above their T_g values. The increasing nonlinear behavior as T_g is approached, and as the size of the penetrant increases (14), indicates an appreciable

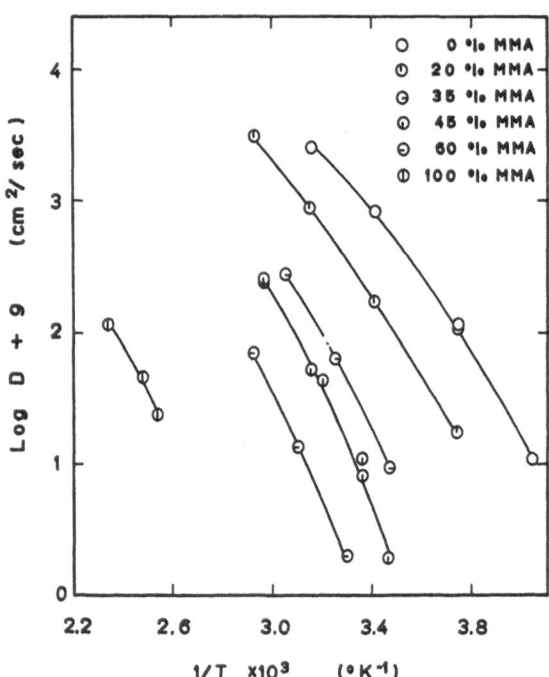

Figure 1. Temperature dependence of diffusion coefficients of methane in isoprene-methyl methacrylate copolymers (mole percent methyl methacrylate indicated).

dependence of diffusion on free volume.

The deviation of the data for poly(methyl methacrylate) from these reduced state curves indicates that free volume, per se, cannot be the single, ubiquitous, factor which correlates and thereby governs the diffusion process through all polymers. A structural property, perhaps related to inherent chain stiffness as it affects the local segmental mobility, relative to the kinetic domains required for unit diffusion processes, must also be considered as a separate factor affecting the overall diffusion process. The ramifications and consequences of this type of complication in mass transport behavior have been discussed (14).

The applications of equation 7 to illustrate the relation-ships between log (D/T) and a_T data for individual polymers and penetrants are shown in Figures 3 and 4. The increase in slope (K-value) with increasing penetrant size is evident in Figure 3 for data for a series of penetrants in a single copolymer sample. Values of K approach unity for the larger size penetrant molecules indicating that the two transport processes are then utilizing similar domains of free volume, or segmental motion size.

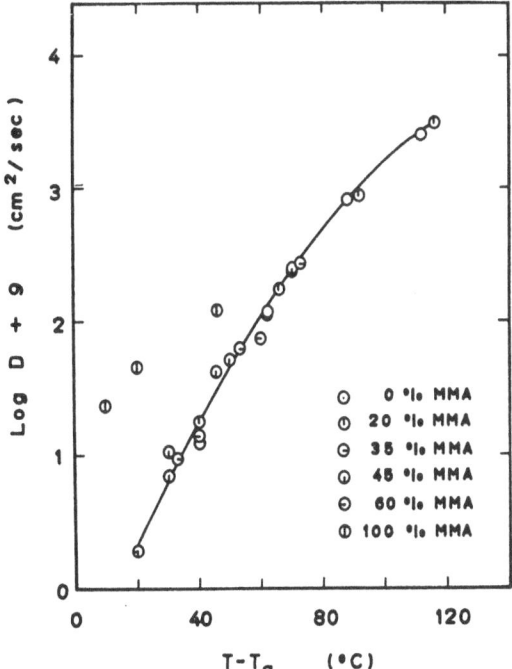

Figure 2.   Dependence of diffusion coefficients of methane on T - T_g for isoprene–methyl methacrylate copolymers (mole percent of methyl methacrylate indicated.)

The nature of the relationship between log D and T - T_g for all copolymers and polyisoprene suggests that a similar corresponding state relation may be obtained between log (D/T) and log a_T. Linear relationships for the larger penetrants and at higher temperatures for all penetrants have been found (14). Curvature becomes apparent for methane data at about 40°C above T_g of the polymers as shown in Figure 4. The temperature region studied using methane extended from 20°C to 112°C above T_g. The curvature became more pronounced as the penetrant molecular size decreased and as the experimental temperature approached the T_g of the samples.

The deviations from linearity may be considered as manifestations of an effect due to the ratio of penetrant volume to volume available in the system for transport. If the available volume for transport greatly exceeds the volume required for diffusion jumps then a small change in available volume, due to temperature change, should affect diffusion only slightly. A confirming mechanistic interpretation of the dependence of diffusion processes on the relative penetrant size and the distribution of local free volume in a polymer has been presented by Frisch (16).

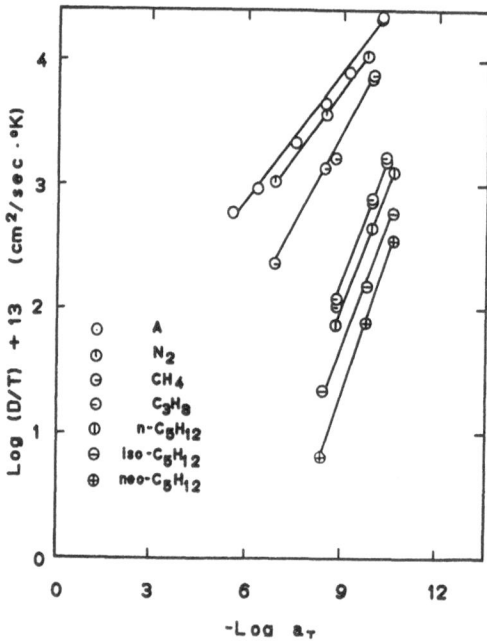

Figure 3. Relationships between stress relaxation shift factor and diffusion coefficients for various penetrants in an isoprene-methyl methacrylate copolymer (45 mole percent MMA).

However, if the available volume approximates that required for viscous flow, a small temperature change should produce a substantial change in log a_T. Consequently, the differing effects of temperature change on the volumes required for diffusion and for viscous transport processes would be more apparent as the penetrant size decreased.

Another interpretation of the nonlinearity can be derived from activation energy considerations. The derivative of equation 7 with respect to 1/T, and substitution of the respective activation energies for the derivative terms leads to the expression (14):

$$E_D - RT = KE_\eta \qquad (8)$$

The RT term is usually small compared to E_D so that equation 8 can be rewritten as:

$$E_D/E_\eta \approx K \qquad (9)$$

The results of stress relaxation measurements show that E_η increases rapidly as T_g is approached. The corresponding decrease in K indicates that E_D remains constant or changes relatively

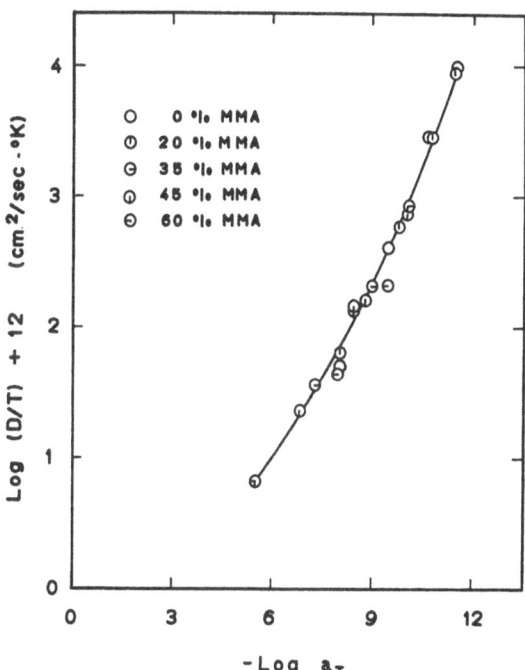

Figure 4. Relationship between stress relaxation shift factor and methane diffusion coefficient for isoprene-methyl methacrylate copolymers (mole percent methyl methacrylate indicated).

little in that same temperature range. In terms of the zone theory interpretation, a progressively larger zone of activation is required for viscous flow relaxation processes as compared with that required for diffusion as the temperature decreases and as the penetrant size decreases.

Both interpretations indicate the importance of the relative sizes of penetrant molecules and polymer kinetic segments to available transport volume (free volume). The opposition to motion of a penetrant molecule or a kinetic segment of polymer within the polymer matrix also is evidenced by the values of friction coefficients derived from the stress relaxation and penetrant diffusion measurements (14).

CONCURRENT DIFFUSION AND RELAXATION PROCESSES

A second approach for the study of the dependence of mass and momentum transport processes on the nature of polymer segmental mobility, or related concepts such as free volume, is to examine the phenomena of diffusion and relaxation processes occurring concurrently in a polymer. This can be regarded as a coupled diffusion-relaxation process (10, 17, 18).

Analysis of such coupled processes is generally both mathematically complex and experimentally difficult. There are two limiting situations which are most amenable to mathematical treatment. These are the cases of relaxation-controlled diffusion and diffusion-controlled relaxation. The first case is often encountered when an organic penetrant is sorbed by a glassy or semicrystalline polymer (1-4). The rates of sorption are controlled not merely by diffusion, as in the case of a rubbery polymer, but also by slow polymeric relaxation processes inherent in these systems.

A study of diffusion-controlled relaxation is of particular value either as a method for the prediction of one property from another or for obtaining additional information which is not available from individual studies of mass or momentum transport. The information obtained should be of relevance for an understanding of relaxation-controlled diffusion phenomena. It is well known that many small penetrant molecules will plasticize a polymer thereby having a profound effect on its mechanical properties. Such an approach is therefore of significance in that not only should it help to characterize the nature of complex polymer-penetrant interactions, but also that a successful description of concurrent behavior should enable us to predict the time dependence of the mechanical properties of a polymer when it is exposed to a given environment.

The general problem can be stated as follows: If we are given individual data for momentum (creep) and mass (sorption) transfer occurring separately in a system - can we predict the behavior due to the two occurring concurrently? As a first step, consider the case of an "ideal" system when the polymer is completely amorphous and is above its glass temperature. Polymeric relaxation processes are then, in general, sufficiently rapid so as not to interfere with the diffusion process. Further, it is known (10) that for many such systems a superposition relationship exists between the time of measurement and the sorbed concentration of penetrant which is analogous to the more widely-known time-temperature superposition. This means that provided it is known how the concentration shift factor a_c and the polymer-penetrant interdiffusion coefficient D vary with concentration, and provided that diffusion occurs in the thickness direction only, then the shift factor profile in a thin polymer film can be computed readily as a function of time. However, to convert this information to a time-dependent compliance or modulus, even for a linear viscoelastic solid, would require an enormous amount of computing time. The alternative then is to establish techniques which may be used to estimate the time-dependent compliance to a reasonable desired degree of accuracy.

One promising approach is to consider the transport of a moving boundary into the polymer film such that the concentration of penetrant in front of the boundary is essentially zero while behind the boundary both the concentration and shift factor profiles can be averaged in some realistic way to yield mean values $\bar{c}$ and $\bar{a}$ respectively. At small times, when the concentration of penetrant at the center of the sheet is truly zero, such a moving boundary is close to reality and can be defined as the plane at which the penetrant concentration has fallen to a given value (e.g. 1% of the surface concentration c_o), Figure 5. Previously, Kishimoto and Fujita (10) have proposed that a hypothetical boundary might be used to advantage. This hypothetical boundary moves at a slower rate than the boundary discussed above and both the concentration and shift factor are assumed to have uniform values of $\bar{c}$ and a_c behind the boundary. Kishimoto and Fujita considered $\bar{c}$ as an integral mean across the film with the reciprocal shift factor as a weighting parameter.

If the boundary has moved a distance x into a sample of thickness 2ℓ at the time t, then from diffusion theory for sorption at small times it follows that:

$$\frac{M_t}{M_\infty} = \frac{\bar{c} \, x}{c_o \ell} = \frac{2}{\ell}\left(\frac{\bar{D}t}{\pi}\right)^{\frac{1}{2}} \tag{10}$$

where M_t and M_∞ are the amounts of penetrant sorbed at time t and at equilibrium, and $\bar{D}$ is an average diffusion coefficient corresponding to the concentration range 0 - c_0. Following Kishimoto and Fujita, we can average the tensile modulus or compliance across the film thickness to obtain

$$H(t) = H_0(t) (1 - x/\ell) + H_0(t/a_{\tau}) \cdot x/\ell \tag{11}$$

where $H(t)$ is modulus or compliance, $H_0(t)$ is the corresponding modulus or compliance of the dry polymer and t refers to length of time after the polymer is first surrounded by vapor. Combination of equations 10 and 11 yields:

$$H(t) = H_0(t) + \frac{2}{\ell}\left(\frac{\overline{D}t}{\pi}\right)^{\frac{1}{2}} \cdot \frac{c_0}{\tau}\left[H_0(t/a_{\tau}) - H_0(t)\right] \tag{12}$$

Equation 12 reveals that $H(t)$ can readily be estimated if $H_0(t)$ and $\overline{D}$ are known and if τ and a_{τ} can be calculated.

In the case of the more realistic boundary at small times, we have for $c/c_0 = 0.01$ at the boundary (19)

$$x \cong 3.6 \ (\overline{D}t)^{\frac{1}{2}} \tag{13}$$

Combining equation 13 with an equation analogous to 11, we now get:

$$H(t) = H_0(t) + \frac{3.6}{\ell} \ (\overline{D}t)^{\frac{1}{2}}\left[H_0(t/a) - H_0(t)\right] \tag{14}$$

Although equation 14 does not invoke the somewhat arbitrary

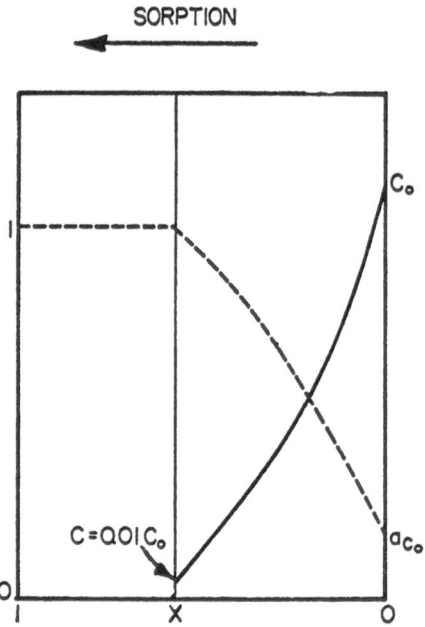

Figure 5. "Realistic" moving boundary used to describe the concentration and shift factor profiles for concurrent diffusion and stress relaxation processes.

parameters τ and a_c, its applicability is restricted to shorter times than is equation 12.

These and more refined related relationships have been used for several studies of the effects of concurrent sorption or desorption of vapor and liquid penetrants on the tensile creep or stress relaxation behavior of polymers (17, 18, 20). The modified moving boundary theory has been generalized to include the effects of dimensional changes through swelling and the larger effects of plasticization associated with the sorption of liquids. The methods of analysis also have been extended to treat longer time relaxation effects in the presence of sorbed plasticizing penetrants. An apparatus for conducting creep or stress-relaxation experiments in the presence of controlled vapor environments has been developed (21).

The results (17, 18) for the concurrent processes of diffusion and creep or stress relaxation in the systems poly(n-butyl methacrylate) with ethanol, methyl ethyl ketone, or benzene vapors at 23°C agree well with the relationships described above. However, at long times the master curves, resulting from the time-concentration superposition procedure, show deviations from the corresponding time-temperature superposition master curve. These deviations, particularly evident for the case of ethanol, are interpreted in terms of the partially specific nature of polymer-penetrant interactions. It is for such non-ideal sorption systems that the study of concurrent diffusion-relaxation processes is most valuable.

The nylon-water system has been the subject of numerous investigations to establish the dependence of transition, mechanical property, and diffusion behavior on polymer structure and the nature of the interactions between the components of the system. It is evident that the hydrogen bonds formed between polymer-polymer and polymer-water moities have a major effect on the system properties. The magnitude and distribution of those hydrogen bonds cause consequent changes in the polymer structure, local chain segmental mobilities, and free volume content and its distribution. It is of interest, therefore, to determine the relationships between the rate and magnitude of water uptake by a nylon sample and the course of the relaxation behavior of the sample. It would be anticipated that the mechanical response of the polymer would be very sensitive to the mode of sorption of water in the polymer as it changes with time, position, temperature and external vapor pressure of water.

The sorption isotherm for water in both oriented (cold rolled) and unoriented Nylon 6 at 25°C is shown in Figure 6. The sigmoid (BET-type) isotherm is a reflection of the conclusion of several different studies (20, 22-24) that water in nylon has three

distinguishable sorption modes. Approximately the first one-third
of the total equilibrium regain of water in nylon undergoes
specific interaction with accessible amide groups to form
"tightly bound" water by hydrogen bond formation. Roughly the
next third is "loosely bound" or in solution within the polymer
while the remaining water is considered to undergo cluster formation (1-4, 24, 25) within the polymeric matrix. Analysis (20) of
the diffusion behavior of water in nylon, considering the different
mobilities of water related to the three apparent sorption modes,
predicts the dependence of the diffusion coefficient on water concentration and accounts for the differences between diffusion
coefficients measured by permeability and various sorption
techniques (20, 26, 27).

The master stress relaxation curve for unoriented Nylon 6 at
25°C (reduced to dry polymer curve) for water uptakes corresponding
to various ambient relative humidities is shown in Figure 7. The
concentration shift factor is a well-defined function of relative
humidity (relative vapor pressure, p/p_0), Figure 8. These relaxation data conform well to a generalized representation in terms of
a WLF-type of equation. The overall relaxation behavior is well
behaved in that sense.

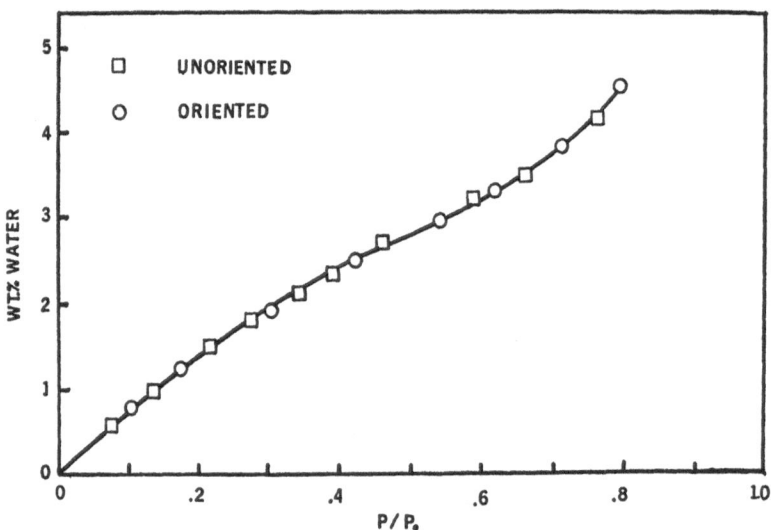

Figure 6. Sorption isotherm for water in unoriented and oriented
(cold rolled) Nylon 6 at 25°C. Weight percent water uptake versus
relative vapor pressure of water (relative humidity).

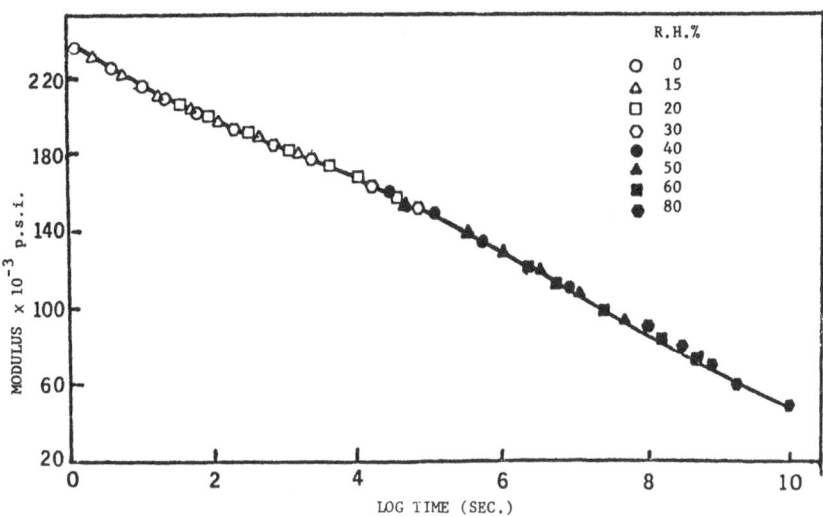

Figure 7. Master stress relaxation curve at 25°C for unoriented
Nylon 6 with varying water content at relative humidities
indicated (reduced to dry polymer curve).

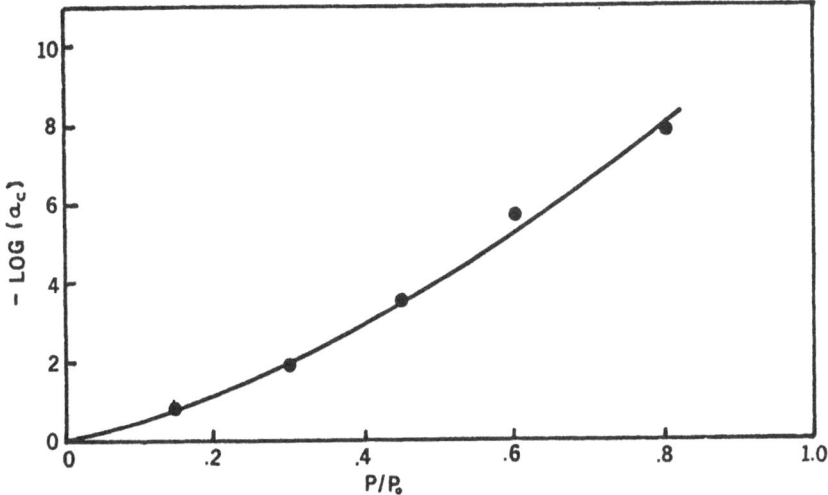

Figure 8. Concentration shift factor versus relative vapor
pressure of water (relative humidity) for unoriented Nylon 6 at
25°C.

Figure 9 shows both the experimentally observed concurrent relaxation curves at various humidities and the curves predicted by equations 10 through 14 using the averaged shift factors from Figure 8 and the known value of the diffusion coefficient for the system. Whereas good agreement is obtained between experimental and calculated curves at 15% and 30% relative humidities, at higher values the experimental curves show much slower relaxation rates than are predicted by the simple theory. This discrepancy is the result of the different effects on relaxation rates by water present in different sorption modes within the polymer.

The differing response of polymer segmental motion to the presence of water sorbed in these various ways can be illustrated by the results obtained in a study of the dynamic mechanical properties of this same nylon-water system. Nylon 6 shows three dynamic mechanical loss maxima at low frequencies (torsion pendulum) ca. 350°, 230°, and 146°K, (28, 29). These have been termed the α, β, and γ transitions, respectively. Several investigations (30-32) have shown that these relaxations occur in the amorphous regions of the polymer. In addition, the presence of three new relaxations in Nylon 6 have been detected recently at temperatures below that of liquid nitrogen (30).

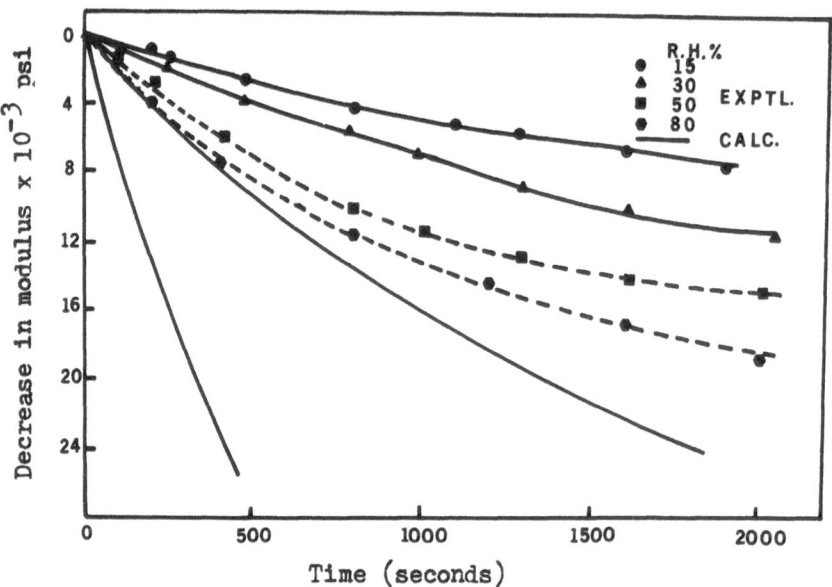

Figure 9. Decrease in relaxation modulus as a function of time after the start of water sorption (at indicated relative humidities) for unoriented Nylon 6 at 25°C. Solid curves calculated as prediction from equation 12.

The effect (30) of varying water content on the α, β, and
γ relaxation spectra is shown in Figure 10 as the change in the
damping characteristics. The γ loss peak, located originally at
146°K for the completely dry specimen, gradually shift toward
125°K. At the same time, its peak height progressively diminishes
by a factor of two. The peak temperature of the β loss shifts
from 230°K toward 190°K. Its height increases below a moisture
content of 1.4% but then decreases slightly as the 8% water content
is attained. The α loss peak also shifts progressively toward
lower temperatures as the amount of sorbed water increases.

The change in the location of these loss peaks with water
content is summarized in Figure 11 which includes the data of
Prevorsek, et. al. (33) for the peak temperature of the α
process as a function of moisture content. For all three transitions, the major change in peak temperature occurs at moisture
contents of less than about 3% by weight. Water sorbed above this
level has a negligible additional effect on the location of the
loss peak temperatures. Similar results are obtained for the
corresponding changes in the relaxation strength (peak heights) as
a function of sorbed water content.

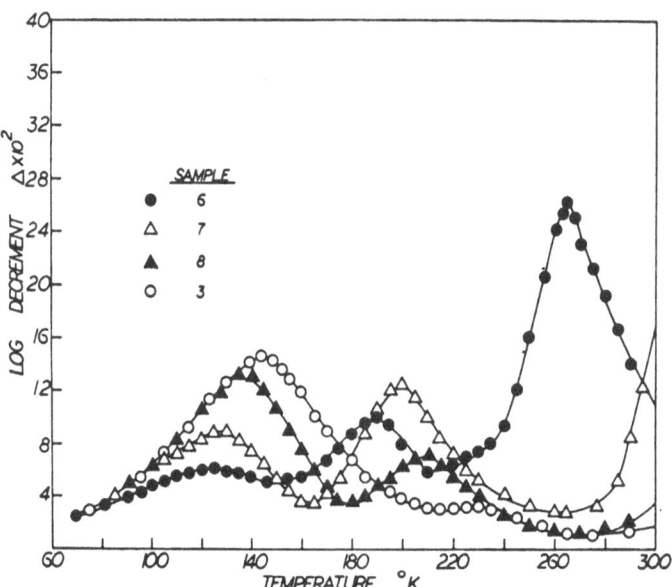

Figure 10. Changes in the α, β, γ relaxations of oriented
(transverse direction) Nylon 6 due to different sorbed water contents (weight percent): sample 3, 0%; sample 8, 0.5%; sample 7,
1.4%; and sample 6, 8%.

A detailed analysis and interpretation of these and other effects of water on the dynamic mechanical properties of Nylon polymers has been presented (20, 30). For our present purposes, these effects can be ascribed as due to the weakening of inter-chain hydrogen bonding and the reduction of steric requirements for hydrogen bonding. Both of these factors are important at low humidities where almost all of the water is in the tightly bound sorption mode which affects all relaxations.

Thus, the tightly bound water, the dominant sorption mode up to about 2.5% by weight representing the equilibrium concentration at 45% relative humidity, is believed to break intermolecular hydrogen bonds and replace them by water bridges or related hydrogen bonded polymer-water configurations.

The mechanical response of the molecular structure to this type of chemical scission of intermolecular hydrogen bonds would be expected to be rapid. This is borne out by the good agreement obtained between experimental results and the theoretical predictions in **Figure 9** for relaxation at the lower relative humidities.

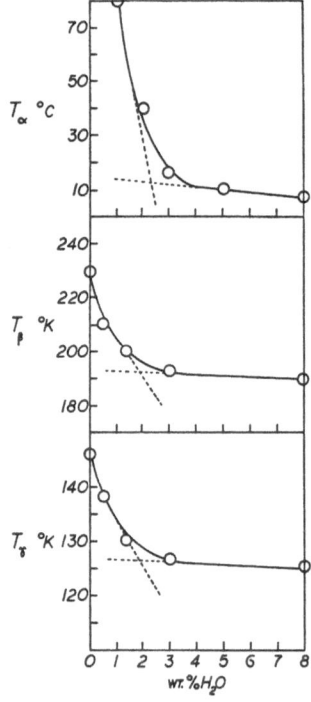

Figure 11. Changes in the temperature of the α, β, and γ loss peak maxima of Nylon 6 due to different sorbed water contents. Data for the α transition are from Prevorsek, et. al. (33).

The equations used assume and require that the time constant associated with the response of the molecular conformation to a change in concentration be small.

The negligible effect on relaxation behavior of water sorbed in excess of that involved in the tightly bound sorption mode is a direct indication that the relaxation time associated with the response of molecular conformations to a change in "loosely bound" water is much greater than that for tightly bound water. The less specific interactions lead to the expectation that any plasticizing action would be due to an increase in interchain spacing. This type of plasticization requires a greater extent of cooperative action among adjacent chain segments and is therefore operative over longer periods of time. This is equivalent to saying that the presence of loosely bound water does not effectively plasticize the nylon structure over short experimental times.

For purposes of prediction of concurrent relaxation behavior, this difference in the plasticizing efficiency of tightly bound and loosely bound water can be incorporated in the concentration shift factor as:

$$a_c = a_c \text{ for } 0 < C \leq 2.5\% \text{ by wt water}$$

$$a_c = a_{2.5} \text{ for } C > 2.5\% \text{ by wt water}$$

where $a_{2.5}$ is the shift factor for the 2.5% by weight water content system. The change in relaxation rates for relative humidities greater than about 40% (corresponding to 2.5% by wt water content) is then due to the different shift factor profiles related to the different advancing boundaries as a function of time and distance. The concentration profiles will be higher at a given time or distance as the overall sorbed amount increases with increasing ambient relative humidity. The concentration, and corresponding shift factor profiles, for these conditions are shown in Figure 12.

This model predicts that the major acceleration of relaxation rates will be observed during sorption involving primarily the tightly bound sorption mode. An additional, but much less efficient, acceleration of relaxation will be observed when sorption involves also the loosely bound sorption mode since the only significant additional effect is a somewhat more rapid attainment of maximum tightly bound water content at any given time or position within the polymeric matrix. The experimental and predicted curves for the concurrent diffusion-relaxation process, using the above assumption as to plasticizing efficiency related to sorption mode, are shown in Figure 13. The agreement is excellent.

In summary, the processes of diffusion and relaxation in
polymer-penetrant systems are very dependent on the precise nature
of the segmental motions which are operative within the time
scale of the transport processes. Studies of these separate
transport processes in the same polymer samples serve to elucidate
their mutual dependence on the structural composition of the
polymer, especially in terms of free volume concepts. Concurrent
diffusion-relaxation studies provide additional information
relating to the effects of polymer-penetrant interactions, or
plasticization, on the transport processes and other structure
sensitive polymer properties. The theories, models, and
techniques which have been developed for these types of investigations are useful initial procedures which should lead to a more
general description of mass and momentum transfer processes in
polymer-penetrant systems.

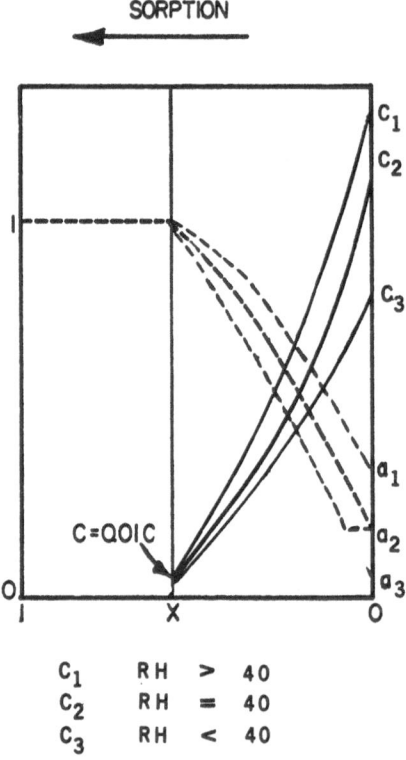

C_1	R H	>	40
C_2	R H	=	40
C_3	R H	<	40

Figure 12. Modification of the moving boundary to account for the
different effects of tightly and loosely bound sorption modes of
water on relaxation behavior of Nylon 6.

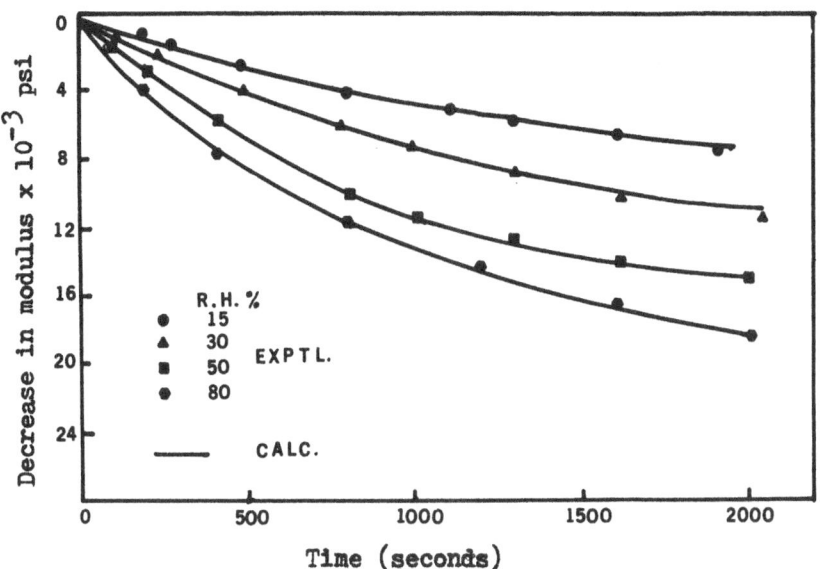

Figure 13. Decrease in relaxation modulus as a function of time
after the start of water sorption (at indicated relative
humidities) for unoriented Nylon 6 at 25°C. Solid curves calculated from equation 12 using modified moving boundary conditions for
tightly and loosely bound water sorption modes.

     Support of this investigation by the National Science
Foundation is gratefully acknowledged.

REFERENCES

1. J. Crank and G. S. Park, Eds., <u>Diffusion in Polymers</u>,
    Academic Press, London, 1968.

2. C. E. Rogers, in <u>Physics and Chemistry of the Organic Solid
    State</u>, Vol. II, D. Fox, M. M. Labes, and A. Weissberger, Eds.,
    Interscience, New York, 1965, Chap. 6.

3. C. E. Rogers, in <u>Engineering Design for Plastics</u>, E. Baer, Ed.,
    Reinhold, New York, 1964, Chap. 9.

4. D. Machin and C. E. Rogers, "The Concentration Dependence of
    Diffusion Coefficients in Polymer-Penetrant Systems", CRC
    Critical Reviews, In press.

5. M. L. Williams, R. F. Landel, and J. D. Ferry, J. Am. Chem. Soc., 77, 3701 (1955).

6. J. D. Ferry, Viscoelastic Properties of Polymers, 2nd Ed., Wiley, New York, 1970.

7. H. Fujita, Fortschr. Hochpolym. Forsch., 3, 1 (1961).

8. H. Fujita and A. Kishimoto, J. Chem. Phys., 34, 393 (1961).

9. H. Fujita, A. Kishimoto, and K. Matsumoto, Trans. Faraday Soc., 56, 424 (1960).

10. H. Fujita and A. Kishimoto, J. Polym. Sci., 28, 547, 569 (1958).

11. R. M. Barrer and R. R. Fergusson, Trans. Faraday Soc., 54, 989 (1958).

12. A. Kishimoto and Y. Enda, J. Polym. Sci., A1, 1799 (1963).

13. A. Kishimoto, J. Polym. Sci., A2, 1421 (1964).

14. J. Semancik and C. E. Rogers, J. Macromol. Sci. Phys., in press.

15. H. L. Frisch and C. E. Rogers, J. Polym. Sci., C12, 297 (1966).

16. H. L. Frisch, J. Polym. Sci., B3, 13 (1965).

17. D. Machin and C. E. Rogers, J. Polym. Sci., A2, 10, 887 (1972).

18. D. Machin and C. E. Rogers, J. Polym. Sci., A2, submitted.

19. J. Crank, The Mathematics of Diffusion, Oxford, London, 1956.

20. S. Kapur and C. E. Rogers, to be published.

21. D. Machin and C. E. Rogers, J. Appl. Polym. Sci., 14, 2833 (1970).

22. R. Puffr and J. Sebenda, J. Polym. Sci., C16, 79 (1967).

23. R. Puffr and J. Sebenda, Coll. Czech. Chem. Comm., 29, 60, 75 (1964).

24. D. Machin and C. E. Rogers, "Sorption", in Encyclopedia of Polymer Sci. and Tech., Wiley, New York, 1970, pp. 679-700.

25. B. H. Zimm and J. L. Lundberg, J. Phys. Chem., 60, 425 (1956).

26. T. Asada and S. Onogi, J. Colloid Sci., 18, 8, 784 (1963).

27. F. H. Muller and E. Hellmuth, Kolloid-Z., 177, 1 (1961).

28. A. E. Woodward and J. A. Sauer, Adv. Polym. Sci., 1, 114 (1958).

29. N. G. McCrum, B. F. Read, and G. Williams, Anelastic and Dielectric Effects in Polymer Solids, Wiley, New York, 1967.

30. Y. S. Papir, S. Kapur, C. E. Rogers, and E. Baer, J. Polym. Sci., A2, in press.

31. A. E. Woodward, J. A. Sauer, C. W. Deely, and D. E. Kline, J. Colloid Sci., 12, 263 (1957).

32. S. Kapur, C. E. Rogers, and E. Baer, J. Polym. Sci., A2, in press.

33. D. C. Pervorsek, R. H. Butler, and H. Reimschuessel, J. Polym. Sci., in press.

GAS, VAPOR AND WATER TRANSPORT IN POLYMER FILMS

V. Stannett, H.B. Hopfenberg and J.L. Williams

Chemical Engineering Department
North Carolina State University, Raleigh, North Carolina
and Camille Dreyfus Laboratory, Research Triangle Park,
North Carolina

Introduction

There has been a huge upsurge of interest in recent years in the transport of gases and vapors through polymer films. More than two hundred papers have been published in this general area in the past five years most of which were concerned, at least in part, with fundamental aspects of the problem.

In addition to the natural growth of interest there has been the stimulation of possible new practical applications. These fall naturally into two categories. 1. The development of polymers with extremely low permeability to oxygen, carbon dioxide and other penetrants. 2. The development of polymers with unusually high permeabilities to gases, water or other penetrants.

The low permeability polymers are in great demand for beverage and other plastic bottles and as possible alternatives to polyvinylidene chloride packaging films. The present approach industrially is to use high acrylonitrile and methacrylonitrile copolymer and blends. Polyacrylonitrile itself has a very low permeability to gases and appears to have a different ratio between the fixed gases such as oxygen and carbon dioxide compared with other plastic films. This leads to a favorably low carbon dioxide permeability of real value for carbonated beverage bottles for example. Little fundamental information concerning gas transport in polyacrylonitrile is available however and such a study is underway in our laboratories.

The high flux materials are of interest for various biomedical applications such as oxygenators and artificial kidney machines and for water treatment in ultrafiltration and reverse osmosis installa-

tions. It is clear that for many of these applications the high
flux for water must be accompanied by adequate selectivity such as
salt rejection.

There are many approaches to changing the transport properties
of polymer films both to decrease and increase penetrant permeability.
Two of these will be briefly discussed in this paper, namely, 1. The
effects of changes in the crystallinity and morphology of semicrystalline polymers on penetrant transport and 2. the effects of graft co-
polymerization on penetrant transport.

1. Effects of Crystallinity and Morphology on Gas and Vapor Transport

The effects of the degree of crystallinity on the permeability
of polyethylene was first discussed in 1958 by the author and his
colleagues[1]. It was suggested that (1) the solubility was directly
proportional to the noncrystalline content (2) the crystallites acted
as impermeable fillers and increased the tortuosity of the diffusion
path. In addition they acted as "crosslinking" agents restricting
the mobilities of the chains. The increase in tortuosities were also
suggested by Klute[2,3,4]. All three factors were introduced independently and developed and discussed in much greater detail by Michaels
and his colleagues[5,6,7]. Considerable subsequent work has shown that
there is a good correlation between the amorphous content estimated
from the density and the solubility of penetrants in most samples of
polyethylene. More recently however it has been shown by Peterlin
et al.[8,9,10] that even this simple relationship breaks down with highly
drawn samples.

The tortuosity or structural factor has been examined theoretically and a number of relationships developed, for different shapes
and size distributions, these have been well presented and summarized
by Barrer[11]. Unfortunately, we still do not yet have an accurate
enough picture of the shape and morphology of polymer films to quantitatively relate the permeability with the degree of crystallinity.
Since the morphology can change drastically without appreciable changes
in density it seems unlikely that any further significant advances will
be made in this area.

The effect of drawing on the transport behavior in semicrystalline polymers has been known for many years to textile chemists. It
was known that drawing could greatly reduce the rate and amount of
dyestuff uptake for example. In the case of polymer films Michaels
et al.[12] first showed the effect with the gas permeability of highly
oriented ethylene-dibutyl maleate copolymer films. The effect with
polyethylene itself was dramatically demonstrated by Peterlin and
Olf[13] using a wide-line NMR. In subsequent papers[8,10] more detailed
information was obtained using direct sorption measurements with
methylene chloride as the penetrant.

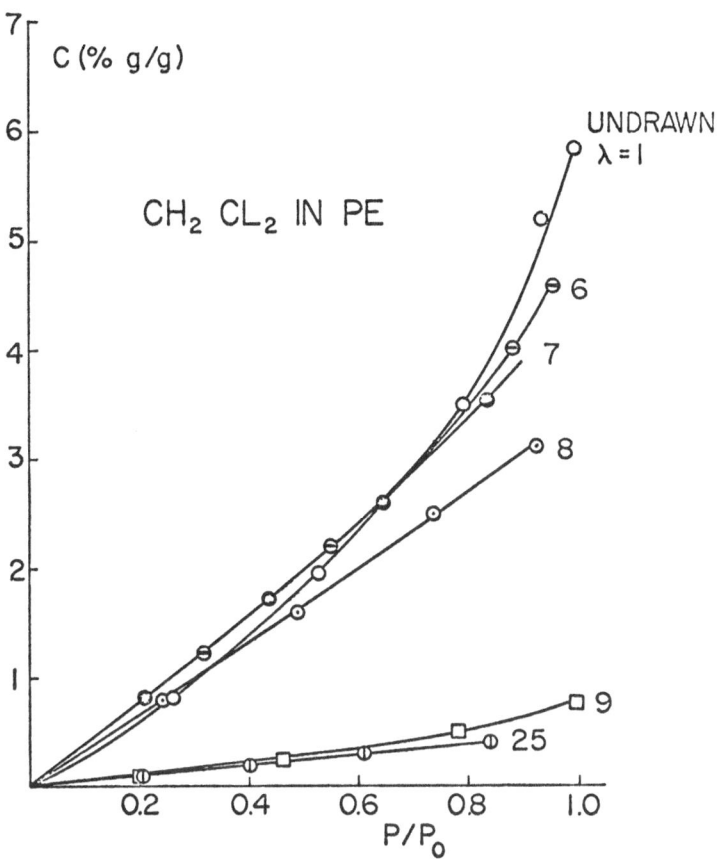

Fig. 1 Methylene chloride sorbed (c in % g/g) at 25°C in undrawn
quenched polyethylene and in films drawn at 60°C to a draw
ratio 6, 7, 8, 9 and 25 as function of vapor activity
p/p_0 (p_0 = 412 torr).

In the first of a series of papers, Peterlin, Williams and Stannett[8] found that the extent of reduction in penetrant solubility and diffusivity is remarkable. For example at 25°C the solubility of methylene chloride at 0.9 activity was found to drop from 45.9% to 6.4% for a draw ratio of 9. The diffusion constants extrapolated to zero concentration dropped from 2.7×10^{-8} to 1×10^{-10} cm^2/sec over the same change in the draw ratio. In terms of permeabilities this could lead to a reduction by a factor of one thousand. In a subsequent paper[9] it was shown that the solubility of water in polyethylene also dropped by a factor of about eight although the diffusivities were not determined. Thus in principle, at least, deep drawing could be an efficient way of greatly improving the barrier properties of polyethylene films.

Later Williams and Peterlin[10], using samples with well-defined draw ratios, found that at draw ratios (λ) between 1 and 8 only samll changes in the rate and degree of sorption occurred as shown in Figs. 1 and 2. However, between 8 and 9 there was a very drastic reduction in both the rate and equilibrium sorption. In fact, the diffusion constant was found to be lowered by a factor of nearly 150. Subsequent drawing to $\lambda = 25$ has very little additional effect on sorption as the experimental points fall nearly on the same curve when $\lambda = 9$. Also, the diffusion constants at draw ratios beyond $\lambda = 9$ are changed only slightly.

These results have been interpreted in terms of Peterlin's well-known model[10,14] for the drawing of crystalline polymers, i.e., a composite material model involving a low permeability fiber structure embedded in a high permeability spherulitic matrix. As the draw ratio is increased the starting spherulitic film is slowly transformed into the fibril structure with the transformation being completed between $\lambda = 8$ and 9. During the subsequent drawing up to $\lambda = 25$ the material arrangement of microfibrils, the basic element of fiber structure, changes by longitudinal sliding. Their transport properties, however, remain nearly constant. The diffusion constant drops slightly as a consequence of the increased fraction of tie molecules which reduces the number of unperturbed sorption sites.

More recently, Peterlin and Williams[15] have found that both the sorption capacity and the rate of diffusion can be partially restored by annealing as can be seen from Figs. 3 and 4. Annealing at 120° and 125°C increases the crystallinity and hence reduces the amount of the amorphous component. In spite of this, it enormously increases the transport properties. It restores the sorption per unit volume of amorphous component to a value which, corresponding to completely relaxed amorphous component, is even higher than that of the starting undrawn but quenched material. The diffusion constant after annealing is a little smaller than in the undeformed original sample and shows the same small dependence on concentration in striking contrast with the drawn material. It can be concluded that the restoration by annealing of the sorption and diffusion coefficients for drawn poly-

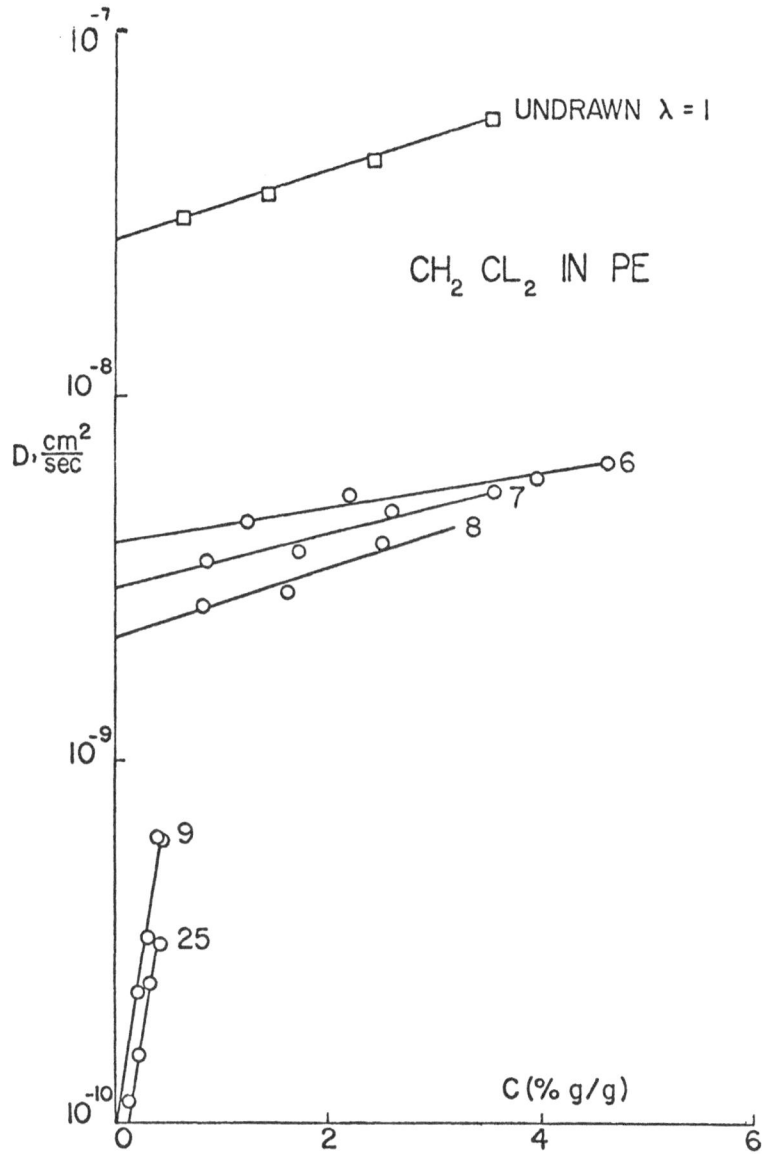

Fig. 2   Diffusion constant of methylene chloride at 25°C into undrawn and drawn polyethylene film as function of concentration of penetrant. The values are derived from the
initial slope of sorption transient curves (Fig. 3). The
extrapolated values D_o are given in Table I.

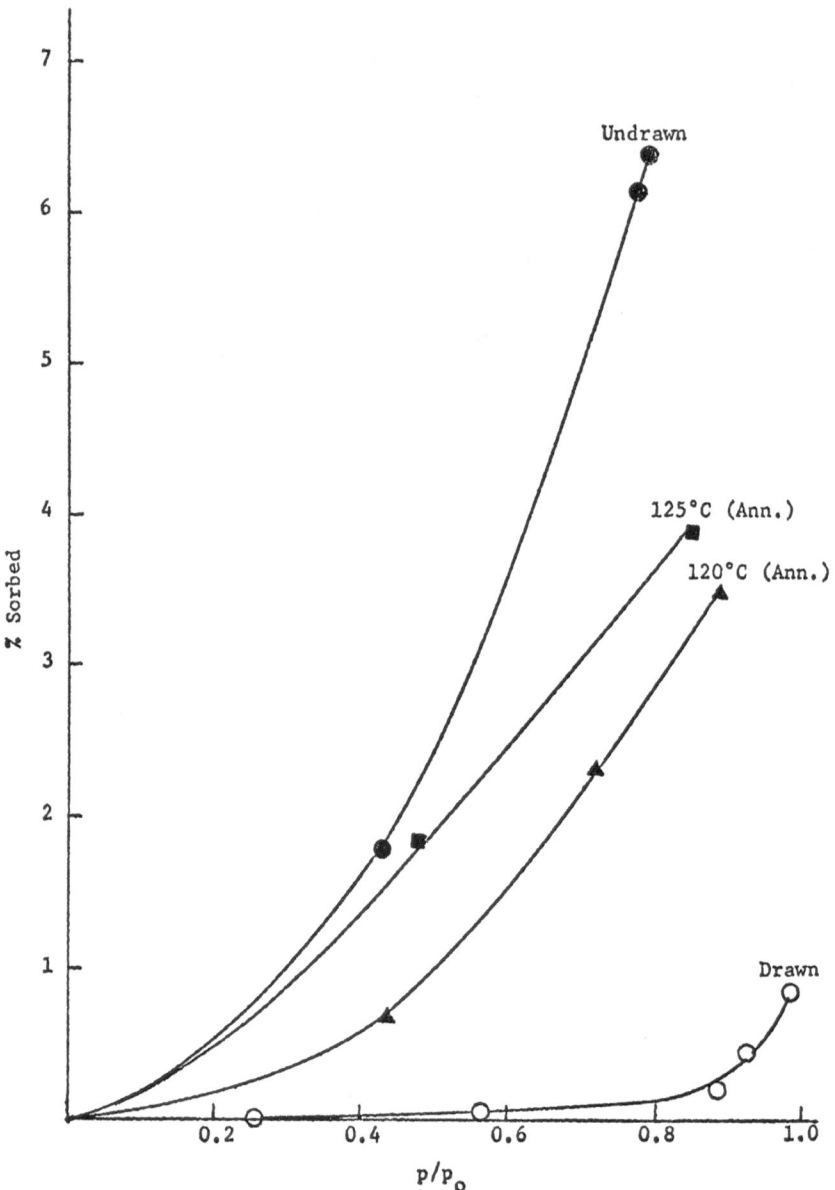

Fig. 3 Sorption isotherms for tetrachloroethylene in 900% drawn poly
ethylene at 25°C.

ethylene to the values of the undrawn material is a consequence of
the complete relaxation of polyethylene chains, in particular of
tie molecules, in the amorphous component. This apparently completely overcompensates the effects of a reduced amorphous fraction
and the increase in crystal perfection. Finally, Warwicker[16] has
reported similar findings in annealed samples of nylon-6 fibers,
although to a much smaller degree, in the case of dye diffusion.

In contrast to these findings Williams and Peterlin[17] have
shown that the deep drawing of polyethylene immersed in, and at
equilibrium with, a solvent such as benzene lead to remarkable increases in sorption and diffusion. Thus with methylene chloride
at 25°C the solubility increased twofold with a draw ratio of only
two and the permeability increased by about one hundred and thirty-
fold. This was clearly shown, however, to be due to the development
of Knudsen type molecular flow, i.e., to the formation of holes or
a porous structure. These results are summarized and contrasted to
data obtained on cold-drawn samples in Table I.

Table 1. Sorption S, diffusion constant D, permeability P, and
the concentration parameter γ of D, at zero sorbent concentration
for methylene chloride into polyethylene, at 25°C

Sample description	S	$D \cdot 10^8$	$P \cdot 10^{10} = SD$	$P_{exp} \cdot 10^{10}$	γ (gm/gm)
Undrawn	0.207	2.66	55.9	–	9.5
Cold drawn (λ=9)	0.0324	0.011	0.0356	–	172.5
Benzene drawn (λ=2)	0.410	178.0	7300.0	580,000	0

Units for $S: cm^3 (STP)/cm^3 (PE) \cdot cm\ Hg; D: cm^2 sec^{-1}; P: cm^3 (STP)/cm \cdot sec \cdot cm\ Hg.$

Again in the context of this paper it is a demonstration of
how permeabilities can be increased or decreased by changes in the
morphology of crystalline polymers, in this case by uniaxial drawing.

Polypropylene behaves in many ways quite differently from polyethylene. Even with undrawn films there is no simple co-relation
between the amorphous fraction, calculated from the density, and
the gas solubilities. The solubilities in quenched films, for example, with below a certain degree of crystallinity tend to become
independent of the amorphous content. This has been attributed by
Vieth and Wuerth[18] to the fact that the true amorphous content remains
essentially constant and that the appearance of a separate crystalline
phase causes the changes in density.

The diffusivities are also much more dependent on the mode of
crystallization of the films compared with polyethylene. With annealed films increasing crystallinity actually leads to enhanced
diffusivities although the activation energies remain the same.

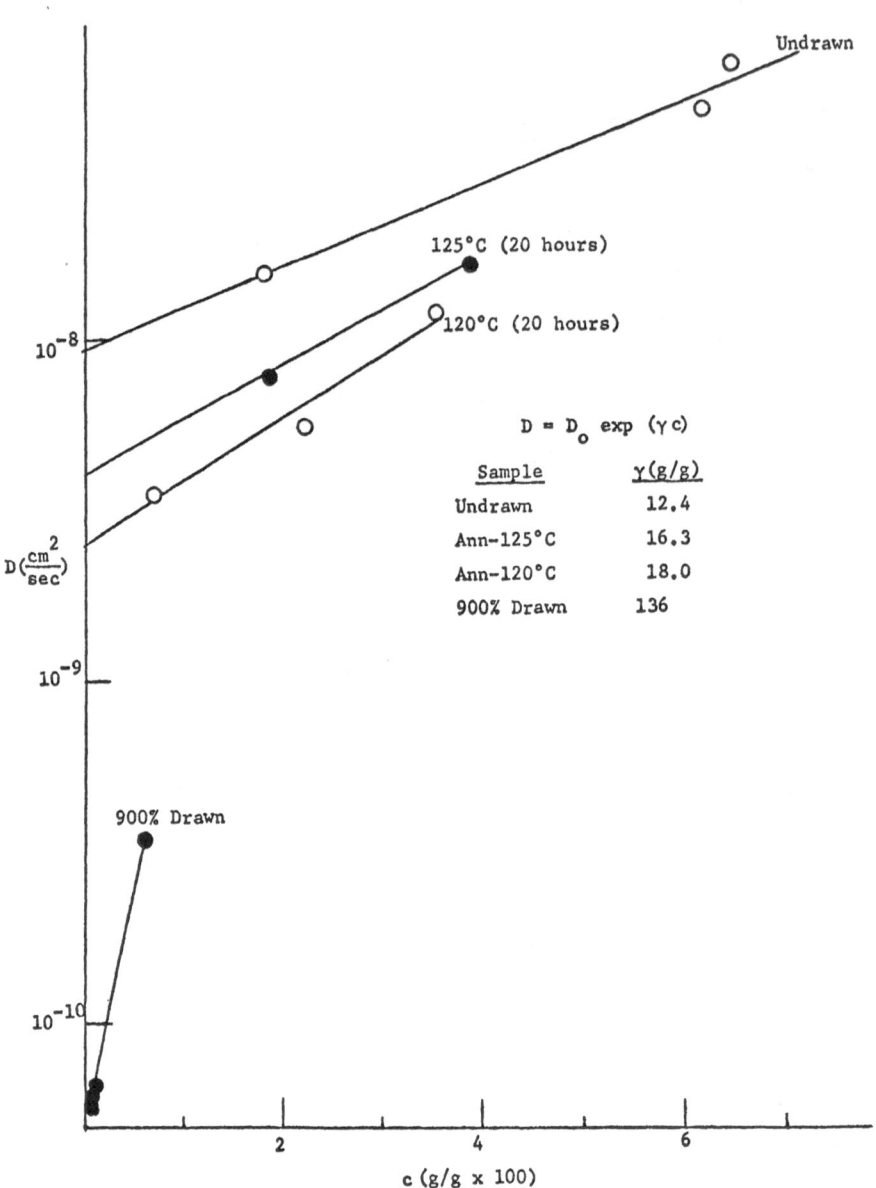

Fig. 4 Concentration dependence for the diffusion of tetrachloro-
        ethylene in 900% drawn polyethylene.

This was attributed to the formation of pore-like defects in the structure by Vieth and Wuerth[18]. Russian workers[19,20] showed clearly that in extreme cases actual pores developed on very slow annealing, for example. It is interesting that this could be avoided by annealing the film on a "nucleating" substrate such as teflon compared with a smooth substrate such as cellophane. This was suggested as a possible means of changing the permeabilities of films, industrially[17].

Polypropylene readily crystallizes to form highly permeable porous films both by very slow cooling and by cooling under uniaxial stretching[21]. It has recently been announced that finely controlled pore size polypropylene films are now in production commercially prepared by "controlled crystallization"[22]. This is another illustration of the use of morphological variations to change the permeability characterization of crystalline polymer films.

2. Effects of Grafting on Gas Vapor and Liquid Transport.

In principle grafting is an attractive way of changing the permeability characteristics of plastic films. The appropriate monomer could be sorbed into the substrate film and rapidly polymerized under an electron beam, for example. Probably homopolymer would not need to be removed as the graft copolymer formed would lead to compatibility. There are many attractive alternative methods of grafting also available.

Early work[23] on grafting styrene and acrylonitrile to polyethylene films showed a progressive decrease in gas permeability with grafting. Subsequent more detailed work by Huang and Kanitz[24,25] with a similar system showed that as the extent of grafting styrene to polyethylene film increased the permeability first decreased and then began to increase again. This was attributed to a decrease in the free volume with lower amounts of grafting and a disruption of the crystallinity at higher degrees of grafting. The densities of the films were also measured and did indeed show an increase slightly above the additive value in the early stages. Interestingly acrylonitrile grafts did not show a minimum but only a progressive decrease in permeability. Since the grafted polyacrylonitrile chains are not even swollen in the grafting solution they are probably inefficient in causing a disruption of the crystallinity. The styrene system on the other hand leads to progressive swelling of the film leading to a disruption of the crystallinity.

A direct demonstration of these effects can be seen in the recent work of Williams and Stannett[26] with polyoxymethylene films. Butadiene and acrylonitrile were radiation grafted onto films each with 18% of dimethyl formamide as a swelling agent in the monomer. The grafting-time curves are shown in Fig. 5. It is clear that butadiene which swells its own polymer leads to a progressive increase in grafting with time. Acrylonitrile, however, grafts as a hole filling type of reaction and the grafting rapidly reaches a maximum and then

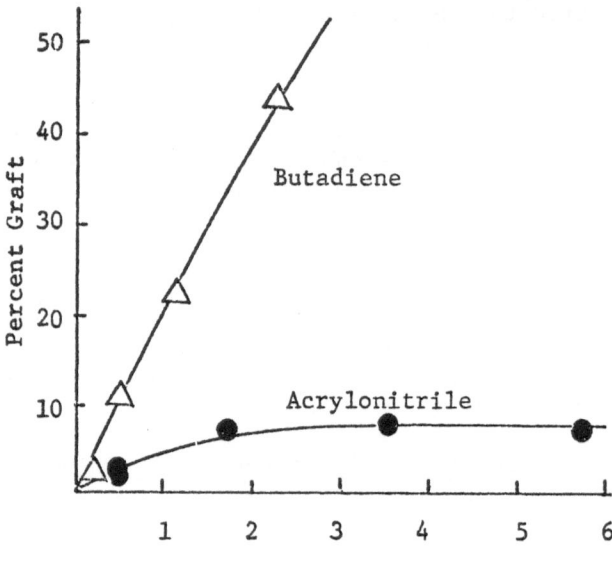

Fig. 5 Percent graft vs. dose curves for butadiene and acrylonitrile
       to polyoxymethylene film. Dose rate 0.1 Mrads per hour, 25°C.
       Both monomers contained 18% dimethyl formamide.

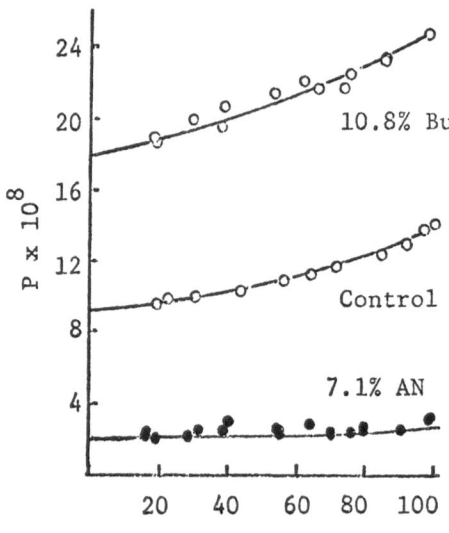

Fig. 6 Water vapor permeability (ccs/cm^2/cm/sec/cm Hg) vs.
       relative humidity for grafted and ungrafted polyoxymethylene
       films.

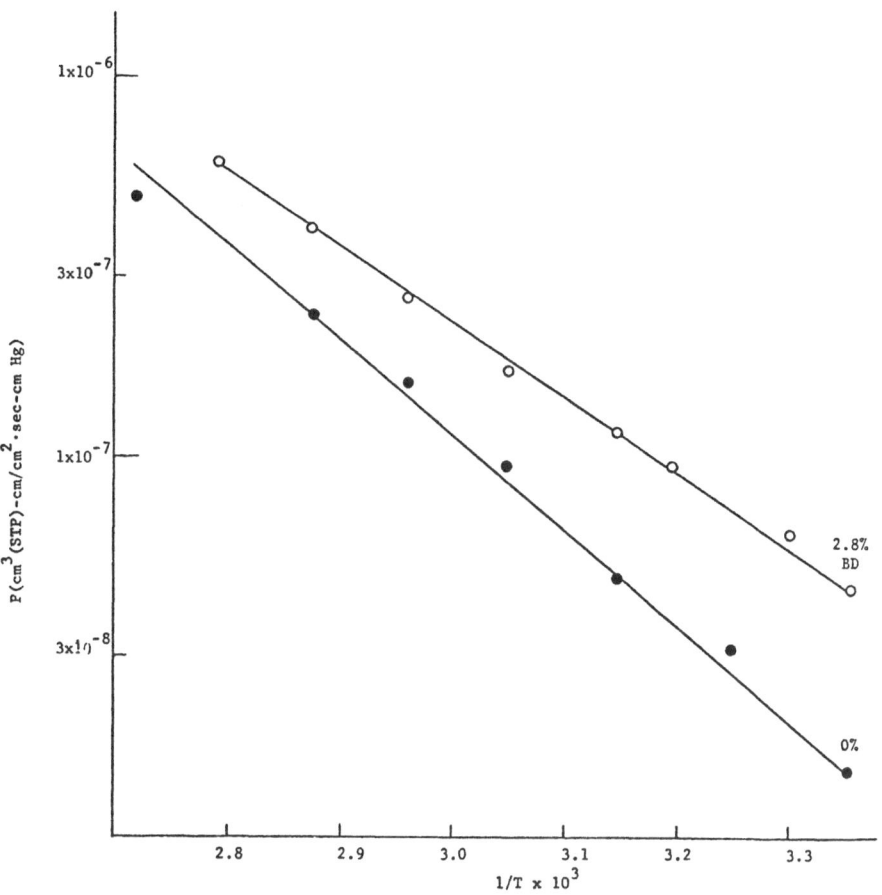

Fig. 7   Temperature dependence for the permeability of water vapor
         in butadiene grafted Celcon.

levels off. This leveling tendency implies that the swelling forces
are not sufficient to overcome the cohesive forces of the polymer in
the case of acrylonitrile, therefore, the polymer structure is not
disrupted during the grafting reaction. The converse is true in the
case of butadiene grafting. This fact is dramatically demonstrated
in terms of the water vapor transport properties presented in Fig. 6.
In the case of the 10% butadiene grafted polyoxymethylene the permeability is actually increased by at least twofold while the acrylonitrile grafting decreased the flux magnitude by sixfold. Further
support to the disruption of the polymer structure in the case of
butadiene grafting is shown in Fig. 7 for the temperature dependence
of 0% and 10% butadiene grafted films. The lower activation energy
would indicate that the barrier to diffusion has been lowered somewhat as a result of the graft.

Somewhat similar changes in the transport behavior of grafted
polymer films was shown in recent work relating to the radiation-
induced grafting of styrene to cellulose acetate[27,28,29]. The ob-
jective of the work was to improve the compaction resistance of membranes for reverse osmosis desalination.

The effects of radiation dose, dose rate, swelling agent type
and concentration, chain transfer agent, and multifunctional monomer
concentration on the flux, rejection, and time dependent properties
of the grafts were studied. Heterogeneous mutual grafting to dense
symmetrical films was the focus of the experimental effort.

The grafts prepared fall rather well into two broad categories.
Grafts prepared in the absence of chain transfer agent (e.g. CCl$_4$)
are characterized by rather long chains of polystyrene (MW = 200,000)
pendant from the cellulose acetate backbone. Grafting recipes including large excesses of chain transfer agent resulted in very short
chains (MW = 4,000).

The kinetics of mutual, radiation induced grafting of styrene
to cellulose acetate were characterized by a leveling of percent
graft versus dose curves if the styrene (monomer) to pyridine (swelling agent) ratio is kept high. This result is presented in Curve 1
of Figure 8 and is quite similar to the grafting kinetics of acrylonitrile to poly(oxymethylene) (Figure 5).

The inability to graft in excess of 40% for the 90:10 styrene:
pyridine recipe was attributed to surface densification consequent
to grafting. The surface layers consequent to grafting offered a
significant resistance to mass transfer of monomer. Consequent to
dissolving the grafted films and recasting, further grafting pro-
ceded linearly with dose. A separate study of the swelling of the
films in the 90% styrene – 10% pyridine grafting solution showed
that the "plateau" grafted films showed almost no swelling and correspondingly little grafting. The recast films on the other hand
swelled more than the original cellulose acetate film and grafted
at a correspondingly faster rate.

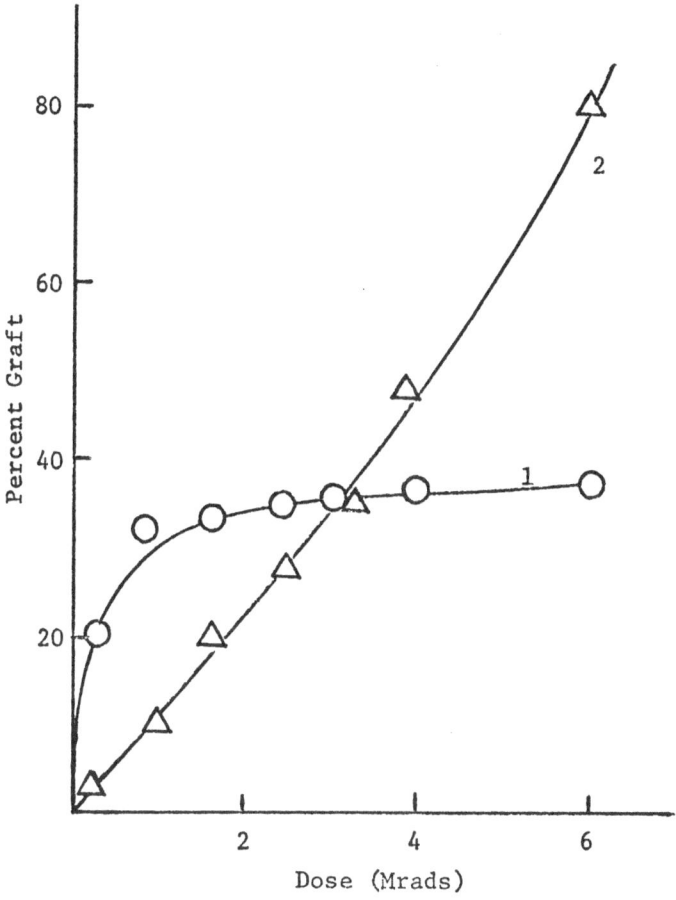

Fig. 8 Percent graft vs. dose curves for styrene to cellulose
acetate film. Dose rate 0.7 Mrads. per hour at 25°C.

Monomer Mixture 1. 90 styrene:10 pyridine
                 2. 20 styrene:60 CCl_4:20 pyridine

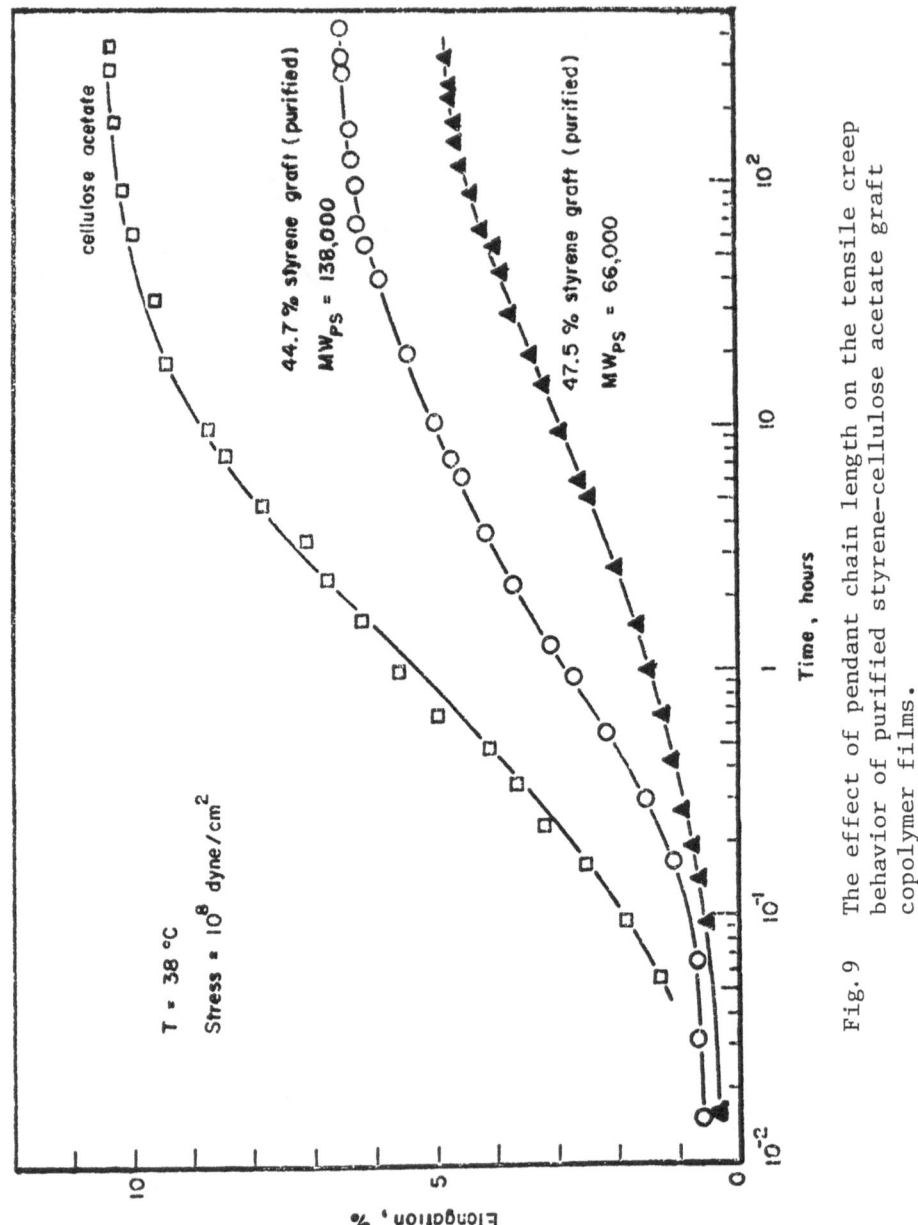

Fig. 9   The effect of pendant chain length on the tensile creep behavior of purified styrene-cellulose acetate graft copolymer films.

Presumably the small but definite degree of crystallinity[30] of the cellulose acetate films exerts a restraining influence on the grafting. The grafting reaction proceeds until the film has become "filled" with polystyrene. Under these conditions it cannot swell or graft. On dissolving and recasting the grafted side chains prevent crystallization and the film is free to continue grafting and expanding indefinitely.

This model of domains of polystyrene filling voids in the cellulose acetate matrix is supported by equilibrium water sorption data and tensile creep behavior of the grafts. The equilibrium water content of the grafts was proportional to the cellulose acetate content suggesting that the polystyrene was confined to non-sorbing, discrete domains. The tensile creep rate of these grafts was markedly reduced as a consequence of grafting as shown in Fig.9.

The form of the grafting kinetics as well as the properties of the resulting grafts were markedly changed as a consequence of introducing large excesses of carbon tetrachloride, which is an active chain transfer agent, to the grafting recipe. For the recipes 20:60:20, styrene:CCl_4:pyridine, percent graft was essentially a linear function of dose and no diffusion controlled limitation on grafting was observed. The molecular weights of the side chains were markedly reduced as a consequence introducing the carbon tetrachloride, a predictable consequence of transferring an active growth center from the pendant chain to the carbon tetrachloride during the grafting process.

The flux and equilibrium water content of these short chain grafts actually increased with percent graft while the salt rejection was virtually unaffected (>99%) up to 40% graft. The rejection dropped markedly at grafting levels in excess of 40% (see Fig. 10).These results suggest that whereas the long chain grafts tend to introduce imobilizing and impermeable domains, the short chain grafting serves to destructure or internally plasticized the resulting graft membranes. The tensile creep properties were, however, improved as a consequence of "short chain grafting" showing that the domain structure probably persists even with the shorter side chains.

The water sorption and transport properties of pure cellulose acetate, long chain grafts, long chain grafts which were dissolved and recast, and short chains grafts are presented in Table II.

The data of Table II contrast the marked changes which accrue from seemingly subtle changes in grafting technique. Styrene grafting in the absence of CCl_4 reduces the flux by a factor of 5. Conversely, styrene grafting in the presence of CCl_4 increases the flux relative to ungrafted cellulose acetate by a factor of 7. The short chain graft, which was characterized by significantly increased flux, exhibited a salt rejection of 98.5%.

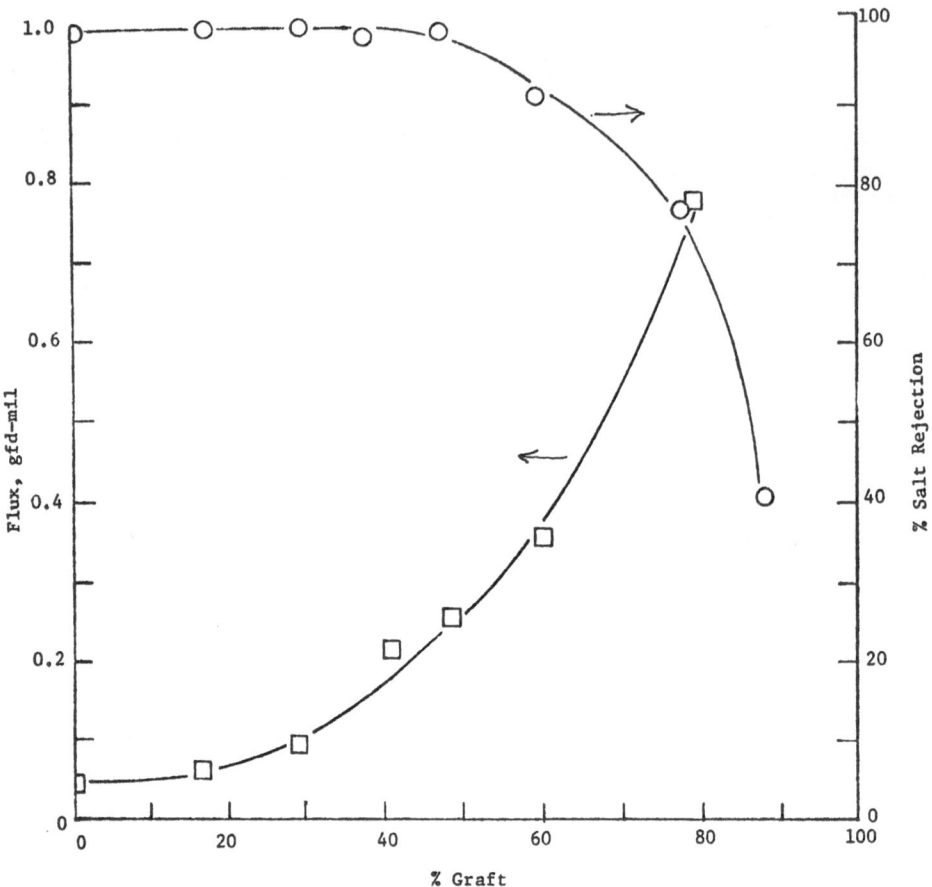

Fig. 10   Product flux and salt rejection as a function of percent
graft for unannealed, dense, short chain graft membranes.

Table II. Water Transport Properties of Styrene-Grafted Cellulose Acetate Films (~35% Graft)

Film	% Water Regain[a]	Diffusivity[a] $cm^2/sec \times 10^8$	Flux[b] gfd-mil	Density
Original C. A. Film	4.5	2.5	0.035	1.285
Long chain Graft	2.3	2.1	0.006	1.223[c]
Long chain Graft	3.5	1.8	0.017	1.202[c]
Short chain Grafted	6.9	3.1	0.23	–

[a] 25°C, 50% RH

[b] 800 psi, 0.5% NaCl

[c] calculated density = 1.217

Moreover, these results viewed in concert with the grafting kinetics and the time dependent mechanical properties suggest rather convincing models for the molecular characteristics of long chain versus short chain grafting.

Hopefully, these few examples will illustrate how grafting technology can also be used to modify the "barrier" properties of films both in the direction of decreased and increased, permeabilities.

References

1. Myers, A. W., Rogers, C. E., Stannett, V. and Szwarc, M., Tappi 41, 716 (1958).

2. Klute, C. H. and Franklin, P. J., J. Poly. Sci., 32, 161 (1958).

3. Klute, C. H., J. Appl. Poly. Sci., 1, 340 (1959).

4. Klute, C. H., J. Poly. Sci., 41, 307 (1959).

5. Michaels, A. S. and Parker, R. B., J. Poly. Sci., 41, 53 (1959).

6. Michaels, A. S. and Bixler, H. J., J. Poly. Sci., 50, 393 (1961).

7. Michaels, A. S. and Bixler, H. J., J. Poly. Sci., 50, 413 (1961).

8. Peterlin, A., Williams, J. L., and Stannett, V., J. Poly. Sci., A-2(5), 957 (1967).

9. Williams, J. L. and Peterlin, A., Macromol. Chem., 120, 215 (1968).

10. Williams, J. L. and Peterlin, A., J. Poly. Sci., A-2(9), 1483
     (1971).

11. Barrer, R. M., Chapter 6. in "Diffusion in Polymers" edited
     by J. Crank and G. S. Park. Academic Press, New York 1968.

12. Michaels, A. S., Vieth, W. R. and Bixler, H. J., J. Appl.
     Poly. Sci., 8, 2735 (1964).

13. Peterlin, A. and Olf, H. G., J. Poly. Sci., A2(4), 587 (1966).

14. Peterlin, A., J. Materials Sci., 6, 490 (1971).

15. Peterlin, A. and J. L. Williams, (to appear in Brit. Polymer
     J.).

16. Warwicker, J., Brit. Polymer J., 3, 68 (1971).

17. Williams, J. L. and Peterlin, A., Makromol. Chem., 135, 41
     (1970).

18. Vieth, W. R. and Wuerth, W. F., J. Appl. Poly. Sci., 13, 685
     (1969).

19. Kosovova, Z. P. and Reitlinger, S. A., Vysokmol. Soed. A9,
     415 (1967).

20. Savin, A. G., Shaposhnikova, T. K., Karpov, V. I., Sogolova,
     T. I. and Kargin, V. A., Vysokomol. Soed. A10, 1584 (1968).

21. Quynn, R. G. et al., J. Macromol. Sci. (Phys1) B4(4), 953 (1970).

22. Chemical Week, Sept. 22, 1971, page 73.

23. Myers, A. W., Rogers, C. E., Stannett, V., Szwarc, M., Patterson,
     G. S., Hoffman, A. S. and Merrill, E. W., J. Appl. Poly. Sci.,
     4, 159 (1960).

24. Huang, R. Y. M. and Kanitz, P. J., J. Appl. Poly. Sci., 13,
     669 (1969).

25. Huang, R. Y. M. and Kanitz, P. J., J. Appl. Poly. Sci., 15,
     67 (1971).

26. Williams, J. L. and Stannett, V., J. Appl. Poly. Sci., 14,
     1949 (1970).

27. Hopfenberg, H. B., Kimura, F., Rigney, P. T. and Stannett, V.,
     J. Appl. Poly. Sci., 28, 243 (1960).

28. Hopfenberg, H. B. Stannett, V., Kimura, F. and Rigney, P. T.,
     Applied Polymer Symposia No. 13, 139 (1970).

29. Kimura-Yeh, F., Bentvelzen, J., Hopfenberg, H. B. and Stannett, V.,
     to be published (O. S. W. Grant 14-01-001-1461. Final Report 1971).

30. Boy, R. E., Schulken, R. M. and Tamblyn, J. W., J. Appl. Poly.
     Sci., 11, 2453 (1967).

INDEX